D0938933

THE VAPOUR PRESSURES OF PURE SUBSTANCES

THE VAPOUR PRESSURES OF PURE SUBSTANCES

Selected Values of the Temperature Dependence of the Vapour Pressures of Some Pure Substances in the Normal and Low Pressure Region

TOMÁŠ BOUBLÍK

Senior Research Scientist, ÚTZCHT
Czechoslovak Academy of Sciences,
Prague, Czechoslovakia

VOJTĚCH FRIED

Professor of Chemistry
Brooklyn College of The City University of New York
Brooklyn, New York

EDUARD HÁLA

Head, Institute of Thermodynamics, ÚTZCHT
Czechoslovak Academy of Sciences,
Prague, Czechoslovak

ELSEVIER SCIENTIFIC PUBLISHING COMPANY

Amsterdam – London – New York 1973

ELSEVIER SCIENTIFIC PUBLISHING COMPANY
335 JAN VAN GALENSTRAAT
P.O. BOX 211, AMSTERDAM, THE NETHERLANDS

AMERICAN ELSEVIER PUBLISHING COMPANY, INC.
52 VANDERBILT AVENUE
NEW YORK, NEW YORK 10017

ISBN: 0-444-41097-X

PRINTED IN THE NETHERLANDS

PREFACE

The saturated vapour pressure is one of the more important physicochemical properties of pure compounds. In this book an extensive set of experimental vapour pressure data is collected; in addition, smoothed values of the vapour pressures, as obtained by fitting the data to the Antoine equation, are presented together with the calculated boiling points and sets of the Antoine constants for each compound.

These constants provide in themselves condensed information on the phase behaviour, very suitable for practical application, while deviations of the smoothed from the experimental pressures and standard deviations (also listed) indicate the accuracy of measurements and the reliability of the equation with the given set of constants.

The choice of systems presented in the following tables was governed by the availability and quality of the experimental data, and partly by the practical importance of the system in question.

We wish to thank Professor V. Bažant for suggesting the preparation of this book and Svatoslava Bernátová for essential help with the manuscript.

THE AUTHORS

CONTENTS

1. VAPOUR PRESSURE TEMPERATURE-DEPENDENCE

The thermodynamic expression of phase equilibrium for a pure substance is given by the Clapeyron equation

$$\frac{dP^\circ}{dT} = \frac{\Delta H}{T\Delta V} \quad , \tag{1}$$

where P° denotes the vapour pressure, T the absolute temperature, ΔH the heat absorbed at constant pressure in transferring one mole from phase ($'$) to phase ($''$) in equilibrium with it, and ΔV the volume change per mole transferred.

The exact equation (1) can be simplified by introducing two approximations:

a) The molar volume of the vapour phase is much greater than that of the liquid (or solid) phase

$$V'' >> V' \ . \tag{2}$$

Hence it is possible to neglect V' and

$$V = V'' - V' \doteq V''. \tag{3}$$

b) At low pressures it may be assumed that the vapour phase behaves ideally (i.e. according to the equation of state of an ideal gas) so that

$$V'' = \frac{RT}{P^\circ} \quad . \tag{4}$$

On substituting the simplifying relations (3) and (4) into equation (1) and rearranging we obtain

$$\frac{1}{P^\circ}\frac{dP^\circ}{dT} = \frac{\Delta H}{RT^2} \quad , \tag{5}$$

or

$$\frac{d \ln P^\circ}{d\,(1/T)} = -\frac{\Delta H}{R} \, . \tag{6}$$

Equations (5) and (6) are differential forms of the Clausius-Clapeyron equation. By integrating (6) under the simplifying assumption that

$$\Delta H = \text{constant} \, , \tag{7}$$

we can write

$$\log P^\circ = A - B/T \, , \tag{8}$$

where A and B ($= -\Delta H/(2.3026R)$) are constants.

Equation (8) will not precisely express the behaviour of substances; the dependence $\log P^\circ = f(1/T)$ is linear only in a narrow range of temperatures. A whole series of semiempirical equations has therefore been proposed, various authors having modified the right-hand side of equation (8) in different ways (for a review see [1]). One suitable equation of this type, published by Antoine[2], has the following form

$$\log P^\circ = A - B/(t + C), \tag{9}$$

where P° is the vapour pressure, t the temperature, and A, B, C are constants characteristic of the substance and the given temperature range.

The Antoine equation (9) represents well the behaviour of most substances over large temperature intervals, as Willingham et al.[3] have shown.

---

1. Miller D.G., Ind. Eng. Chem. 56, (3), 46 (1964)
2. Antoine C., C. R. Acad. Sci., Paris 107, 681, 836, 1143 (1888)
3. Willingham C.B., Taylor W.J., Pignocco J.M., Rossini F.D., J. Res. Natl. Bur. Stand. 35, 219 (1945)

2. GUIDE TO THE TABLES

What follows summarizes information on the vapour pressures of pure substances (up to approximately 3000 mm Hg) in separate tables, each of which displays direct experimental data (temperature in °C and pressure in mm Hg) in the first two columns, smoothed values of the pressure fitted to the Antoine equation in the third column, and deviations (of the smoothed pressures from the experimental data) and percent deviations in the fourth and fifth columns.

The procedure of data fitting to the Antoine equation, which is nonlinear in temperature, consisted of three steps: First, the constant C was assumed to be equal to 230.0 and A and B were evaluated by the least squares method. The set of constants obtained served as a first estimate of the Antoine constants in the second step, where relatively small increments in A, B, and C were determined by the weighted least squares method from the linearized form of the Antoine equation

$$F = (\log P^\circ - A)(C + t) - B.$$

Statistical weight, given by the expression

$$W_i = \left[\left(\frac{B}{C + t_i}\right)^2 s_t^2 + \left(\frac{C + t_i}{2.3026\, P_i}\right)^2 s_P^2 \cdot\right]^{-1},$$

(where s_t and s_P are estimations of the errors in the determination of temperature and pressure) with $s_t = 0.01$ °C and $s_P = 0.1$ mm Hg, was used throughout this work because of the impossibility, in some cases, of specifying the accuracy of the original data and the experimental technique.

Set of constants A, B, C, obtained in the second step served for the preliminary determination of the smoothed values and the standard deviation

$$\text{STD. DEVIATION} = \left[\frac{(\text{PRESSURE DEV.})^2}{\text{NO. EXPTL. POINTS} - 3}\right]^{0.5}$$

Experimental points with deviations exceeding three times the standard deviation were excluded from the set in the further treatment; such points are designated by asterisks before the temperature data.

In the third step the final values of the Antoine constants for the remaining set of points were determined from the linearized form by the repeated weighted least squares method.

Values of constants are given in the heading of each table together with the calculated boiling point at 760.0, 100.0 or 10.0 mm Hg, depending on the measured pressure range.

The tables of compounds are arranged according to the Hill system used in the Chemical Abstracts Formula Index.

TABLES

ARGON

AR

CLARK A.M., DIN F., ROBB J., MICHELS A., WASSERNAAR T., ZWIETERING TH.: PHYSICA, 17, 876(1951).

A = 8.62071 B = 772.351 C = 320.730

B.P.(760) = -186.172

T	P EXPTL.	P CALCD.	DEV.	PERCENT
-182.213	1103.60	1108.82	5.22	0.47
-179.984	1361.69	1358.83	-2.86	-0.21
-176.124	1913.00	1903.90	-9.10	-0.48
-174.813	2132.33	2126.33	-6.00	-0.28
-171.668	2735.92	2749.84	13.92	0.51

STANDARD DEVIATION =13.19

DEUTERODIBORANE B2D6

DITTER J.F.,PERRINE J.C.,SHAPIRO I.:
J.CHEM.ENG.DATA 6,271(1961).

A = 6.48083 B = 545.202 C = 244.727

B.P.(760) = -93.283

T	P EXPTL.	P CALCD.	DEV.	PERCENT
*-154.900	2.50	2.58	0.08	3.14
-152.100	4.20	3.93	-0.27	-6.34
-148.800	6.70	6.27	-0.43	-6.41
-146.100	9.10	8.97	-0.13	-1.40
-144.300	11.50	11.27	-0.23	-1.99
*-143.100	14.30	13.06	-1.24	-8.64
-141.700	15.70	15.45	-0.25	-1.58
-141.100	16.70	16.58	-0.12	-0.71
-139.500	19.90	19.94	0.04	0.18
-137.900	23.70	23.84	0.14	0.57
-136.400	28.10	28.05	-0.05	-0.19
-134.800	33.20	33.20	0.00	0.00
-133.400	38.60	38.33	-0.27	-0.70
-133.300	38.70	38.72	0.02	0.05
-131.800	44.80	44.97	0.17	0.38
-131.100	48.10	48.16	0.06	0.12
-130.400	51.30	51.53	0.23	0.44
-128.300	62.20	62.81	0.61	0.98
-127.900	65.30	65.17	-0.13	-0.19
-126.500	73.70	74.02	0.32	0.43
-126.100	76.60	76.72	0.12	0.15
-125.000	84.50	84.55	0.05	0.06
-123.900	93.20	93.02	-0.18	-0.20
-123.400	96.80	97.09	0.29	0.30
-122.200	107.00	107.44	0.44	0.41
-120.800	120.40	120.63	0.23	0.19
-116.900	163.40	164.31	0.91	0.56
-114.700	194.90	194.01	-0.89	-0.46
-113.400	212.80	213.47	0.67	0.32
-111.700	242.70	241.21	-1.49	-0.61
-110.100	271.10	269.84	-1.26	-0.47
-108.400	305.50	303.11	-2.39	-0.78
*-107.100	334.30	330.66	-3.64	-1.09
-105.300	372.80	371.98	-0.82	-0.22
-103.300	420.10	422.49	2.39	0.57
-102.000	459.30	458.07	-1.23	-0.27
-99.900	520.50	520.38	-0.12	-0.02
-98.000	581.80	582.19	0.39	0.07
-96.000	653.50	653.18	-0.32	-0.05
-94.800	696.90	698.84	1.94	0.28
-94.600	705.50	706.67	1.17	0.17
-93.900	734.90	734.64	-0.26	-0.04

STANDARD DEVIATION = 0.86

DIBORANE B_2H_6

DITTER J.F.,PERRINE J.C.,SHAPIRO I.:
J.CHEM.ENG.DATA 6,271(1961).

A = 6.36638 B = 521.490 C = 241.978

B.P.(760) = -92.364

T	P EXPTL.	P CALCD.	DEV.	PERCENT
-154.900	2.70	2.39	-0.31	-11.64
-152.100	4.10	3.67	-0.43	-10.58
-148.800	6.50	5.88	-0.62	-9.47
-146.100	8.80	8.46	-0.34	-3.88
-144.300	11.00	10.65	-0.35	-3.14
*-143.100	13.80	12.37	-1.43	-10.37
-141.700	15.00	14.65	-0.35	-2.31
-141.100	16.00	15.74	-0.26	-1.65
-139.500	18.80	18.95	0.15	0.80
-137.900	22.70	22.69	-0.01	-0.05
-136.400	26.80	26.73	-0.07	-0.26
-134.800	31.80	31.68	-0.12	-0.39
-133.400	36.60	36.60	0.00	0.00
-133.300	36.90	36.98	0.08	0.20
-131.800	42.80	42.98	0.18	0.41
-131.100	45.80	46.04	0.24	0.52
-130.400	48.90	49.27	0.37	0.77
-128.300	59.60	60.11	0.51	0.86
-127.900	62.10	62.38	0.28	0.45
-126.500	70.80	70.87	0.07	0.10
-126.100	73.50	73.46	-0.04	-0.05
-125.000	80.60	80.98	0.38	0.47
-123.900	89.20	89.11	-0.09	-0.11
-123.400	92.80	93.01	0.21	0.23
-122.200	102.40	102.94	0.54	0.53
-120.800	115.10	115.58	0.48	0.42
-116.900	156.50	157.42	0.92	0.59
-114.700	185.60	185.84	0.24	0.13
-113.400	203.60	204.44	0.84	0.41
-111.700	232.50	230.93	-1.57	-0.67
-110.100	259.30	258.26	-1.04	-0.40
-108.400	292.00	289.99	-2.01	-0.69
-107.100	319.30	316.23	-3.07	-0.96
-105.300	357.30	355.57	-1.73	-0.48
-103.300	402.80	403.60	0.80	0.20
-102.000	438.80	437.40	-1.40	-0.32
-99.900	497.00	496.52	-0.48	-0.10
-98.000	554.80	555.11	0.31	0.06
-96.000	622.30	622.30	0.00	0.00
-94.800	663.70	665.47	1.77	0.27
-94.600	670.20	672.88	2.68	0.40
-93.900	699.70	699.30	-0.40	-0.06
-92.300	760.30	762.62	2.32	0.31

STANDARD DEVIATION = 1.05

B-TRICHLOROBORAZINE B3CL3H3N3

LAUBENGAYER A.W.,SCAIFE W.J.:
J.CHEM.ENG.DATA 11,173(1966).

A = 5.04235 B = 638.951 C = 106.966

B.P.(100) = 123.368

T	P EXPTL.	P CALCD.	DEV.	PERCENT
29.350	0.29	0.31	0.02	6.04
40.250	0.70	0.88	0.18	25.44
50.800	1.70	2.11	0.41	24.22
59.460	3.44	3.99	0.55	16.08
69.540	7.36	7.75	0.39	5.28
80.130	15.20	14.40	-0.80	-5.29
90.490	25.30	24.75	-0.55	-2.19
99.400	37.80	37.75	-0.05	-0.12
109.100	57.20	57.47	0.27	0.48
121.700	93.00	94.07	1.07	1.15
131.400	133.40	132.66	-0.74	-0.56
136.000	154.70	154.66	-0.04	-0.03

STANDARD DEVIATION = 0.61

B-TRIFLUOROBORAZINE B3F3H3N3

LAUBENGAYER A.W.,SCAIFE W.J.:
J.CHEM.ENG.DATA 11,173(1966).

A = 12.55950 B = 4149.672 C = 316.608

B.P.(760) = 112.136

T	P EXPTL.	P CALCD.	DEV.	PERCENT
0.310	0.25	0.29	0.04	16.87
9.820	0.60	0.70	0.10	17.21
20.370	1.62	1.76	0.14	8.54
30.590	3.85	4.05	0.20	5.23
38.840	7.36	7.67	0.31	4.26
50.860	18.70	18.49	-0.21	-1.14
60.740	36.40	36.52	0.12	0.33
70.530	69.90	69.28	-0.62	-0.88
80.230	126.70	126.66	-0.04	-0.03
90.400	230.40	231.17	0.77	0.33
100.230	399.70	402.13	2.43	0.61
110.300	696.40	690.54	-5.86	-0.84
112.200	760.00	762.54	2.54	0.33

STANDARD DEVIATION = 2.19

BROMOCHLORODIFLUOROMETHANE CBRCLF2

GLEN D.N.: .CAN.J.CHEM. 38,208(1960).

A = 6.83998 B = 935.632 C = 240.330

B.P.(760) = -4.010

	T	P EXPTL.	P CALCD.	DEV.	PERCENT
	-95.335	2.45	2.44	-0.01	-0.47
	-78.798	11.19	11.16	-0.03	-0.25
	-45.305	110.29	110.28	-0.01	-0.01
	-21.531	366.06	366.24	0.18	0.05
	-20.002	392.03	392.14	0.11	0.03
	-17.398	439.49	439.58	0.09	0.02
	-15.034	486.39	486.49	0.10	0.02
	-12.447	542.26	542.28	0.02	0.00
	-10.049	598.46	598.38	-0.08	-0.01
	-7.455	664.05	664.10	0.05	0.01
*	-5.074	701.64	729.28	27.64	3.94
	-4.161	755.85	755.57	-0.28	-0.04
	-4.078	758.14	757.99	-0.15	-0.02
	-3.957	761.71	761.54	-0.17	-0.02
	-3.860	764.59	764.39	-0.20	-0.03
	-2.080	818.57	818.23	-0.34	-0.04
	-0.093	882.05	881.77	-0.28	-0.03
	1.914	949.98	949.78	-0.20	-0.02
	3.909	1020.90	1021.34	0.44	0.04
	5.925	1097.60	1097.83	0.23	0.02
	7.907	1177.00	1177.25	0.25	0.02
	9.890	1260.80	1261.07	0.27	0.02

STANDARD DEVIATION = 0.22

CYANOGEN BROMIDE CBRN

LORD G.,WOOLF A.A.:
J.CHEM.SOC. 1954,2546.

A = 8.38753 B = 1499.959 C = 212.709

B.P.(100) = 22.118

T	P EXPTL.	P CALCD.	DEV.	PERCENT
-0.140	22.20	21.44	-0.76	-3.43
11.340	49.30	49.29	-0.01	-0.03
15.040	63.80	63.31	-0.49	-0.76
15.640	66.00	65.89	-0.11	-0.17
15.840	65.80	66.77	0.97	1.47
18.340	77.60	78.63	1.03	1.32
20.140	89.80	88.26	-1.54	-1.72
25.040	119.00	119.82	0.82	0.69
25.140	120.00	120.55	0.55	0.46
29.840	159.80	159.72	-0.08	-0.05
30.340	163.60	164.47	0.87	0.53
35.040	216.00	215.36	-0.64	-0.30
35.440	221.80	220.26	-1.54	-0.70
39.940	281.80	282.22	0.42	0.15

STANDARD DEVIATION = 0.95

DIBROMODIFLUOROMETHANE CBR2F2

MCDONALD R.A.,SHRADER S.A.,STULL D.R.:
J.CHEM.ENG.DATA 4,311(1959).

A = 7.15222 B = 1181.612 C = 253.845

B.P.(760) = 22.789

T	P EXPTL.	P CALCD.	DEV.	PERCENT
-26.150	91.82	91.78	-0.04	-0.04
-11.340	190.30	190.41	0.11	0.06
5.190	389.69	389.60	-0.09	-0.02
21.650	730.14	729.72	-0.42	-0.06
22.390	749.30	749.28	-0.02	0.00
22.790	760.00	760.03	0.03	0.00
23.630	782.57	783.00	0.43	0.06

STANDARD DEVIATION = 0.31

CARBONYL CHLORIDE CCL2O

GIAUQUE W.F.,JONES W.M.:
J.AM.CHEM.SOC. 70,120(1948).

A = 6.97133 B = 998.770 C = 236.619

B.P.(760) = 7.548

T	P EXPTL.	P CALCD.	DEV.	PERCENT
-57.619	24.75	24.64	-0.11	-0.45
-51.651	37.32	37.29	-0.03	-0.07
-45.783	54.67	54.66	-0.01	-0.02
-39.357	80.75	80.94	0.19	0.24
-32.132	122.12	122.20	0.08	0.06
-25.527	173.88	173.74	-0.14	-0.08
-17.778	255.50	255.52	0.02	0.01
-9.087	381.88	381.73	-0.15	-0.04
-0.953	541.28	541.08	-0.20	-0.04
3.946	660.07	660.09	0.02	0.00
8.003	773.12	773.43	0.31	0.04

STANDARD DEVIATION = 0.17

(TRIFLUOROMETHYL)DICHLOROPHOSPHINE CCL2F3P

PETERSON L.K.,BURG A.B.:
J.AM.CHEM.SOC. 86,2587(1964).

A = 7.49029 B = 1381.501 C = 260.513

B.P.(100) = -8.886

T	P EXPTL.	P CALCD.	DEV.	PERCENT
-64.800	2.72	2.70	-0.02	-0.71
-46.000	11.30	11.22	-0.08	-0.68
-19.500	57.20	57.31	0.11	0.19
0.000	154.00	153.91	-0.09	-0.06
10.150	243.40	243.30	-0.10	-0.04
17.300	329.10	329.23	0.13	0.04

STANDARD DEVIATION = 0.13

TRICHLOROFLUOROMETHANE CCL3F

BENNING A.F.,MCHARNESS R.C.:
IND.ENG.CHEM. 32,497(1940).

A = 6.89034 B = 1043.700 C = 236.596

B.P.(760) = 23.709

T	P EXPTL.	P CALCD.	DEV.	PERCENT
-29.650	70.30	70.31	0.01	0.01
0.010	301.49	301.45	-0.04	-0.01
24.020	768.36	768.42	0.06	0.01
61.150	2426.68	2426.64	-0.04	0.00

STANDARD DEVIATION = 0.08

TRICHLOROFLUOROMETHANE CCL3F

OSBORNE D.W.,GARNER C.S.,DOESCHER R.N.,YOST D.M.:
J.AM.CHEM.SOC. 63,3496(1941).

A = 6.88428 B = 1043.004 C = 236.883

B.P.(100) = -23.340

T	P EXPTL.	P CALCD.	DEV.	PERCENT
-36.673	47.35	47.29	-0.06	-0.13
-21.825	108.13	108.25	0.12	0.11
-6.887	223.60	223.57	-0.03	-0.01
2.352	334.61	334.61	0.00	0.00
10.362	463.46	463.22	-0.24	-0.05
15.108	556.15	556.20	0.05	0.01
19.895	664.18	664.35	0.17	0.02

STANDARD DEVIATION = 0.17

CARBON TETRACHLORIDE CCL4

HILDEBRAND D.L.,MCDONALD R.A.:
J.PHYS.CHEM. 63,1521(1959).

A = 6.92180 B = 1235.172 C = 228.937

B.P.(760) = 76.724

T	P EXPTL.	P CALCD.	DEV.	PERCENT
19.880	90.60	90.70	0.10	0.11
36.930	189.10	188.78	-0.32	-0.17
56.160	387.97	388.44	0.47	0.12
75.280	727.30	727.15	-0.15	-0.02
76.160	747.70	747.02	-0.68	-0.09
76.870	763.03	763.37	0.34	0.04
77.710	782.83	783.07	0.24	0.03

STANDARD DEVIATION = 0.50

CARBON TETRACHLORIDE CCL4

BOUBLIK T.:
THESIS,VSCHT PRAHA 1960.

A = 6.87926 B = 1212.021 C = 226.409

B.P.(760) = 76.714

T	P EXPTL.	P CALCD.	DEV.	PERCENT
14.060	69.08	69.03	-0.05	-0.08
16.660	78.20	78.15	-0.05	-0.06
19.480	89.08	89.15	0.07	0.08
21.710	98.70	98.72	0.02	0.02
25.200	115.30	115.39	0.09	0.08
29.130	136.90	136.85	-0.05	-0.03
33.450	164.20	164.10	-0.10	-0.06
36.980	189.50	189.50	0.00	0.00
42.070	231.50	231.66	0.16	0.07
47.210	281.60	281.62	0.02	0.01
52.360	340.20	340.01	-0.19	-0.06
58.610	423.40	423.48	0.08	0.02
64.780	521.20	521.12	-0.08	-0.02
71.430	645.45	645.46	0.01	0.00
76.040	744.50	744.56	0.06	0.01

STANDARD DEVIATION = 0.10

CYANOGEN FLUORIDE CFN

FAWCETT F.S.,LIPSCOMB R.D.: J.AM.CHEM.SOC. 86,2576(1964).

A = 6.77891   B = 697.613   C = 224.954

B.P.(760) = -45.991

T	P EXPTL.	P CALCD.	DEV.	PERCENT
-76.400	120.10	121.03	0.93	0.77
-75.000	136.10	133.88	-2.22	-1.63
-72.500	160.70	159.59	-1.11	-0.69
-71.500	168.70	170.94	2.24	1.33
-66.900	232.00	231.82	-0.18	-0.08
-64.000	278.00	278.40	0.40	0.14
-57.300	414.00	414.84	0.84	0.20
-52.900	528.00	530.02	2.02	0.38
-51.300	578.00	577.63	-0.37	-0.06
-50.000	621.00	618.73	-2.27	-0.37
-49.000	649.00	651.87	2.87	0.44
-47.900	692.00	689.91	-2.09	-0.30
-46.800	731.00	729.66	-1.34	-0.18

STANDARD DEVIATION = 1.91

CARBONYL FLUORIDE CF_2O

PACE E.L.,RENO M.A.: J.CHEM.PHYS. 48,1231(1968).

A = 6.88547   B = 576.699   C = 228.579

B.P.(760) = -84.572

T	P EXPTL.	P CALCD.	DEV.	PERCENT
-109.550	109.60	109.76	0.16	0.15
-104.150	178.60	178.12	-0.48	-0.27
-98.190	289.89	290.11	0.22	0.08
-93.710	406.58	406.89	0.31	0.08
-90.250	520.16	520.51	0.35	0.07
-87.420	631.23	630.98	-0.25	-0.04
-84.950	742.59	741.77	-0.82	-0.11
-83.980	788.73	789.23	0.50	0.06

STANDARD DEVIATION = 0.55

CARBON TETRAFLUORIDE CF4

MENZEL W.,MOHRY F.:
Z.ANORG.CHEM. 210,257(1933).

A = 6.97231 B = 540.502 C = 260.095

B.P.(760) = -127.991

T	P EXPTL.	P CALCD.	DEV.	PERCENT
-180.640	1.50	1.48	-0.02	-1.46
*-178.780	2.20	2.12	-0.09	-3.86
-176.710	3.10	3.09	-0.01	-0.24
-173.670	5.20	5.23	0.03	0.54
-170.890	8.20	8.19	-0.01	-0.14
-168.280	12.20	12.17	-0.03	-0.21
-167.750	13.30	13.16	-0.14	-1.05
-165.720	17.60	17.59	-0.01	-0.08
-163.890	22.50	22.60	0.10	0.44
-163.580	23.50	23.56	0.06	0.25
-163.280	24.50	24.52	0.02	0.07
-138.680	331.70	331.60	-0.10	-0.03
-133.180	517.30	517.06	-0.24	-0.05
-129.680	672.40	672.72	0.32	0.05
-128.280	744.90	744.48	-0.42	-0.06
-127.970	761.40	761.15	-0.25	-0.03
-127.760	772.40	772.62	0.22	0.03
-127.750	773.10	773.17	0.07	0.01
-127.280	798.90	799.33	0.43	0.05

STANDARD DEVIATION = 0.21

TRIFLUOROMETHYL HYPOFLUORITE CF4O

KELLOGG K.B.,CADY G.H.:
J.AM.CHEM.SOC. 70,3986(1948).

A = 6.95060 B = 650.060 C = -18.393

B.P.(760) = 178.121

T	P EXPTL.	P CALCD.	DEV.	PERCENT
144.700	63.00	63.67	0.67	1.07
153.500	139.20	137.77	-1.43	-1.02
158.300	202.20	201.48	-0.72	-0.35
165.900	348.20	349.65	1.45	0.42
178.200	760.90	763.53	2.63	0.35
188.900	1375.00	1374.34	-0.66	-0.05
193.900	1767.00	1764.85	-2.15	-0.12

STANDARD DEVIATION = 2.06

TRIBROMOMETHANE CHBR3

KIREEV V.A.,SIMNIKOV I.P.:
ZH.OBSHCH.KHIM. 14,483(1941).

A = 6.82182 B = 1376.748 C = 200.966

B.P.(100) = 84.559

T	P EXPTL.	P CALCD.	DEV.	PERCENT
30.000	7.40	7.26	-0.14	-1.88
35.000	9.50	9.71	0.21	2.23
40.000	13.10	12.83	-0.27	-2.03
50.100	22.10	21.79	-0.31	-1.41
60.200	35.20	35.50	0.30	0.86
70.300	55.50	55.79	0.29	0.52
80.400	84.80	84.87	0.07	0.08
90.600	126.40	125.87	-0.54	-0.42
100.800	181.50	181.77	0.27	0.15

STANDARD DEVIATION = 0.36

MONOCHLORODIFLUOROMETHANE CHCLF2

KLETSKII A.V.:
INZH.FIZ.ZH.AKAD.NAUK BELORUSSK.SSR 7,40(1964).

A = 6.75769 B = 740.384 C = 231.842

B.P.(760) = -40.868

T	P EXPTL.	P CALCD.	DEV.	PERCENT
-43.555	669.18	669.10	-0.08	-0.01
-41.210	747.84	747.93	0.09	0.01
-41.113	751.30	751.34	0.04	0.01
-36.754	917.48	917.43	-0.05	-0.01

STANDARD DEVIATION = 0.14

TRICHLOROMETHANE CHCL3

BOUBLIK T.,AIM K.:
COLLECTION CZECHOSLOV.CHEM.COMMUN. 36(IN PRESS).

A = 6.95465 B = 1170.966 C = 226.232

B.P.(760) = 61.204

T	P EXPTL.	P CALCD.	DEV.	PERCENT
-10.356	35.93	35.92	-0.02	-0.04
-2.610	52.26	52.27	0.01	0.03
2.882	69.80	69.79	-0.01	-0.01
8.010	90.35	90.30	-0.05	-0.06
11.927	109.13	109.12	-0.01	-0.01
16.417	134.42	134.55	0.13	0.10
20.248	159.87	159.92	0.05	0.03
24.598	193.37	193.32	-0.05	-0.03
28.118	224.44	224.33	-0.11	-0.05
32.289	266.22	266.17	-0.05	-0.02
36.223	311.05	311.21	0.16	0.05
40.913	372.73	372.72	-0.01	0.00
46.583	459.75	459.71	-0.04	-0.01
53.070	578.49	578.32	-0.17	-0.03
60.319	738.15	738.30	0.15	0.02

STANDARD DEVIATION = 0.10

TRICHLOROMETHANE CHCL3

SCATCHARD G.,RAYMOND C.L.:
J.AM.CHEM.SOC. 60,278(1938).

A = 6.49344 B = 929.444 C = 196.030

B.P.(760) = 61.246

T	P EXPTL.	P CALCD.	DEV.	PERCENT
35.000	295.11	295.39	0.28	0.10
40.000	360.20	359.44	-0.76	-0.21
45.000	433.54	433.82	0.28	0.06
50.000	519.18	519.61	0.43	0.08
55.000	617.84	617.90	0.06	0.01
60.000	730.13	729.84	-0.29	-0.04

STANDARD DEVIATION = 0.58

FORMYL FLUORIDE CHFO

FISCHER G.,BUCHANAN A.S.: AUSTRALIAN J.CHEM. 17,481(1964).

A = 5.27001 B = 362.050 C = 174.778

B.P.(100) = -64.059

	T	P EXPTL.	P CALCD.	DEV.	PERCENT
	-94.700	0.79	5.61	4.82	609.82
*	-91.500	10.80	8.37	-2.43	-22.54
	-89.000	13.40	11.20	-2.20	-16.42
	-87.000	17.50	13.98	-3.52	-20.14
	-85.000	22.00	17.27	-4.73	-21.50
	-78.000	38.40	33.80	-4.60	-11.97
	-78.000	39.50	33.80	-5.70	-14.42
	-78.000	39.50	33.80	-5.70	-14.42
	-77.500	36.00	35.33	-0.67	-1.85
	-75.000	43.00	43.80	0.80	1.85
	-74.500	51.50	45.66	-5.84	-11.34
	-74.200	45.10	46.81	1.71	3.78
	-72.400	52.60	54.15	1.55	2.95
	-71.500	63.00	58.13	-4.87	-7.73
	-70.000	61.70	65.25	3.55	5.76
	-67.500	76.00	78.55	2.55	3.35
	-67.000	76.60	81.43	4.83	6.30
	-66.100	83.00	86.82	3.82	4.60
	-59.700	130.00	133.01	3.01	2.31
	-56.500	161.00	161.80	0.80	0.50
	-54.300	188.50	184.03	-4.47	-2.37
	-53.500	186.40	192.62	6.22	3.34
	-51.500	215.00	215.35	0.35	0.16
	-45.500	313.20	294.74	-18.46	-5.89
	-44.800	309.00	305.16	-3.84	-1.24
	-43.500	330.00	325.17	-4.83	-1.46
	-38.000	427.00	419.77	-7.23	-1.69
	-65.900	89.00	88.05	-0.95	-1.07
	-63.200	103.00	105.97	2.97	2.88
	-63.000	104.00	107.40	3.40	3.27
	-62.500	103.00	111.02	8.02	7.79
	-61.600	110.00	117.78	7.78	7.07
	-60.500	113.00	126.43	13.43	11.88

STANDARD DEVIATION = 6.15

TRIFLUOROMETHANE CHF3

VALENTINE R.H.,BRODALE G.E.,GIAUQUE W.R.:
J.PHYS.CHEM. 66,392(1962).

A = 7.08863 B = 705.334 C = 249.782

B.P.(760) = -82.157

T	P EXPTL.	P CALCD.	DEV.	PERCENT
-127.792	20.40	20.26	-0.14	-0.66
-115.060	71.00	71.31	0.31	0.44
-108.639	123.40	123.41	0.01	0.00
-103.958	178.80	178.55	-0.25	-0.14
-100.316	234.50	234.21	-0.29	-0.12
-96.714	302.50	302.45	-0.05	-0.02
-93.791	368.80	368.98	0.18	0.05
-91.140	439.30	439.10	-0.20	-0.05
-89.257	495.10	495.12	0.02	0.00
-87.223	561.20	561.94	0.74	0.13
-85.251	633.40	633.43	0.03	0.00
-83.589	699.20	699.15	-0.05	-0.01
-81.963	768.90	768.58	-0.32	-0.04

STANDARD DEVIATION = 0.32

HYDROGEN CYANIDE CHN

SINOZAKI H.,HARA R.,MITSUKURI S.:
BULL.CHEM.SOC.JAPAN 1, 59(1926).

A = 7.52823 B = 1329.490 C = 260.418

B.P.(760) = 25.653

	T	P EXPTL.	P CALCD.	DEV.	PERCENT
	-16.420	120.05	120.08	0.03	0.02
	-13.240	140.78	141.11	0.33	0.23
	-13.090	142.32	142.17	-0.15	-0.10
	-12.040	149.80	149.81	0.01	0.01
	-11.040	157.58	157.40	-0.18	-0.11
	-10.310	163.16	163.14	-0.02	-0.01
	-7.530	186.54	186.64	0.10	0.05
*	-4.200	219.81	218.44	-1.38	-0.63
	-0.760	256.31	255.90	-0.41	-0.16
	0.020	265.20	265.09	-0.11	-0.04
	4.030	316.80	316.82	0.02	0.01
	6.910	358.60	358.90	0.30	0.08
	9.930	407.80	407.88	0.08	0.02
	13.110	465.40	465.26	-0.14	-0.03
	15.820	519.40	519.26	-0.14	-0.03
	18.810	584.70	584.68	-0.02	0.00
	21.740	654.40	655.19	0.79	0.12
	24.590	730.60	730.26	-0.34	-0.05
	27.620	816.20	817.62	1.42	0.17
	30.590	911.10	911.29	0.19	0.02
	33.920	1028.40	1026.47	-1.93	-0.19
	36.620	1128.20	1128.24	0.04	0.00
	40.090	1272.20	1270.82	-1.38	-0.11
	42.860	1393.20	1394.74	1.54	0.11
*	46.230	1564.10	1558.37	-5.73	-0.37

STANDARD DEVIATION = 0.75

CYANIC ACID CHNO

LINHARD M.: Z.ANORG.ALLGEM.CHEM. 236,200(1938).

A = 7.56859 B = 1251.856 C = 243.787

B.P.(100) = -18.980

	T	P EXPTL.	P CALCD.	DEV.	PERCENT
	-76.270	1.28	1.25	-0.03	-2.64
*	-74.970	1.39	1.42	0.03	2.36
	-73.800	1.57	1.60	0.03	1.92
	-72.120	1.95	1.89	-0.06	-3.13
	-70.250	2.25	2.26	0.01	0.61
	-68.210	2.79	2.75	-0.04	-1.59
	-66.770	3.20	3.14	-0.06	-1.94
	-64.360	3.97	3.91	-0.07	-1.64
	-62.220	4.78	4.72	-0.06	-1.28
	-59.860	5.88	5.79	-0.09	-1.61
	-57.920	6.72	6.81	0.09	1.39
	-54.930	8.80	8.71	-0.09	-1.03
	-52.560	10.55	10.52	-0.03	-0.25
	-50.220	12.71	12.63	-0.08	-0.66
	-48.470	14.47	14.43	-0.04	-0.28
	-45.430	18.02	18.09	0.07	0.39
	-43.460	20.75	20.87	0.12	0.58
	-40.600	25.50	25.56	0.06	0.22
	-39.340	27.71	27.89	0.18	0.65
	-38.520	29.38	29.51	0.13	0.43
	-35.930	35.11	35.15	0.04	0.11
	-34.760	38.10	37.99	-0.11	-0.30
	-32.190	44.97	44.91	-0.06	-0.13
	-30.090	51.33	51.35	0.02	0.03
	-29.740	52.63	52.49	-0.14	-0.27
	-26.800	62.91	63.00	0.09	0.14
	-23.970	74.72	74.75	0.03	0.04
	-21.140	88.30	88.30	0.00	0.01
	-21.120	88.46	88.41	-0.05	-0.06
	-18.000	105.83	105.72	-0.11	-0.10
	-17.990	105.86	105.78	-0.08	-0.07
	-14.970	125.15	125.20	0.05	0.04
	-14.970	125.19	125.20	0.01	0.01
	-14.970	125.19	125.20	0.01	0.01
	-14.950	125.30	125.34	0.04	0.03
	-11.630	150.06	150.07	0.01	0.01
	-11.630	150.06	150.07	0.01	0.01
	-11.630	150.07	150.07	0.00	0.00
	-11.620	150.36	150.15	-0.21	-0.14
*	-8.640	175.38	175.74	0.36	0.21
	-8.630	175.68	175.83	0.15	0.09
	-5.940	201.99	201.98	-0.01	-0.01
	-5.940	201.90	201.98	0.08	0.04

STANDARD DEVIATION = 0.08

BROMOCHLOROMETHANE CH2BRCL

MCDONALD R.A.,SHRADER S.A.,STULL D.R.:
J.CHEM.ENG.DATA 4,311(1959).

A = 6.49606 B = 942.267 C = 192.587

B.P.(760) = 68.050

T	P EXPTL.	P CALCD.	DEV.	PERCENT
15.720	93.34	93.89	0.55	0.59
24.060	141.07	140.20	-0.87	-0.62
44.710	335.52	335.14	-0.38	-0.11
57.180	526.37	529.03	2.66	0.50
66.320	719.65	718.88	-0.77	-0.11
68.000	760.00	758.79	-1.21	-0.16

STANDARD DEVIATION = 1.86

DICHLOROMETHANE CH2CL2

MUELLER C.R.,IGNATOWSKI J.:
J.CHEM.PHYS. 32,1430(1960).

A = 9.72567 B = 2979.516 C = 395.533

B.P.(760) = 39.760

T	P EXPTL.	P CALCD.	DEV.	PERCENT
29.993	529.02	529.31	0.29	0.05
34.993	638.87	638.30	-0.57	-0.09
38.993	739.34	739.15	-0.19	-0.03
39.993	765.94	766.44	0.50	0.07

STANDARD DEVIATION = 0.83

DICHLOROMETHANE CH2CL2

GANEFF J.M.,JUNGERS J.C.:
BULL.SOC.CHIM.BELG. 57,82(1948).

A = 7.40916 B = 1325.938 C = 252.616

B.P.(760) = 40.192

T	P EXPTL.	P CALCD.	DEV.	PERCENT
-40.000	15.00	14.89	-0.11	-0.74
-30.000	28.00	28.38	0.38	1.35
-20.000	51.00	51.17	0.17	0.34
-10.000	88.00	87.90	-0.10	-0.11
-5.000	114.00	113.33	-0.67	-0.59
0.000	145.00	144.65	-0.35	-0.24
5.000	183.00	182.89	-0.11	-0.06
10.000	229.00	229.19	0.19	0.08
15.000	284.00	284.79	0.79	0.28
20.000	350.00	351.07	1.07	0.31
25.000	429.00	429.53	0.53	0.12
30.000	523.00	521.79	-1.21	-0.23
35.000	630.00	629.59	-0.41	-0.07
40.000	755.00	754.80	-0.20	-0.03

STANDARD DEVIATION = 0.64

CHLOROMETHOXYTRICHLOROSILANE CH2CL4OSI

FROST R.E.,ROCHOW E.G.:
J.INORG.NUCL.CHEM. 5,201(1958).

A = 7.31292 B = 1545.708 C = 226.099

B.P.(10) = 18.749

T	P EXPTL.	P CALCD.	DEV.	PERCENT
0.000	3.10	3.00	-0.10	-3.36
10.000	5.82	5.84	0.02	0.26
20.000	10.70	10.77	0.07	0.62
30.000	18.90	18.94	0.04	0.20
40.000	32.00	31.93	-0.07	-0.23
50.000	51.80	51.82	0.02	0.05

STANDARD DEVIATION = 0.09

DIFLUOROMETHANE CH_2F_2

MALBRUNOT P.F.,ET AL.: J.CHEM.ENG.DATA 13, 16(1968).

A = 7.13888 B = 821.676 C = 244.663

B.P.(760) = -51.693

T	P EXPTL.	P CALCD.	DEV.	PERCENT
-81.960	122.66	122.66	0.00	0.00
-68.890	291.08	291.23	0.15	0.05
-58.620	528.05	527.57	-0.48	-0.09
-52.610	726.71	725.25	-1.46	-0.20
-51.360	773.00	772.96	-0.03	0.00
-49.910	829.16	831.39	2.23	0.27
-32.640	1833.12	1834.27	1.15	0.06
-31.820	1900.00	1898.43	-1.57	-0.08

STANDARD DEVIATION = 1.49

FORMALDEHYDE CH_2O

SPENCE R.,WILD W.: J.CHEM.SOC.138,506(1935).

A = 7.19578 B = 970.595 C = 244.124

B.P.(100) = -57.320

T	P EXPTL.	P CALCD.	DEV.	PERCENT
-109.390	0.95	0.98	0.03	3.34
-104.390	1.85	1.78	-0.07	-3.93
-98.290	3.60	3.47	-0.13	-3.62
-95.190	4.85	4.77	-0.08	-1.58
-89.090	8.68	8.62	-0.07	-0.75
-85.590	12.25	11.84	-0.41	-3.32
-78.890	21.02	20.98	-0.04	-0.21
-78.290	22.11	22.03	-0.08	-0.37
-71.890	35.40	36.35	0.95	2.67
-68.490	46.43	46.73	0.30	0.64
-65.290	58.95	58.67	-0.28	-0.47
-64.590	61.65	61.60	-0.05	-0.08
-63.690	65.20	65.55	0.35	0.53
-55.790	111.00	110.20	-0.80	-0.72
-53.990	124.70	123.31	-1.39	-1.12
-49.290	163.10	163.73	0.63	0.39
-40.590	266.60	267.34	0.74	0.28
-39.090	290.60	289.71	-0.89	-0.31
-34.290	368.90	371.74	2.84	0.77
-28.390	496.60	497.44	0.85	0.17
-22.290	664.30	661.39	-2.91	-0.44

STANDARD DEVIATION = 1.13

FORMIC ACID CH2O2

KAHLBAUM G.W.A.:

Z.PHYS.CHEM. 13,14(1894).

A = 4.97536 B = 541.738 C = 137.051

B.P.(10) = -0.777

T	P EXPTL.	P CALCD.	DEV.	PERCENT
0.500	11.00	10.89	-0.11	-1.03
1.000	11.50	11.25	-0.25	-2.17
1.300	11.50	11.47	-0.03	-0.23
2.000	12.00	12.01	0.01	0.05
2.000	12.30	12.01	-0.29	-2.39
3.600	13.30	13.30	0.00	-0.03
5.000	14.60	14.51	-0.09	-0.61
5.100	14.70	14.60	-0.10	-0.68
6.300	15.70	15.71	0.01	0.08
7.200	16.70	16.59	-0.11	-0.66
9.800	19.10	19.33	0.23	1.23
10.900	20.70	20.60	-0.11	-0.51
11.200	21.10	20.95	-0.15	-0.71
12.100	22.20	22.04	-0.16	-0.72
14.100	25.00	24.62	-0.38	-1.52
15.100	26.30	25.99	-0.31	-1.17
15.900	27.40	27.13	-0.27	-0.98
16.500	28.10	28.01	-0.09	-0.32
17.100	29.20	28.91	-0.29	-1.00
17.500	29.50	29.52	0.02	0.07
18.100	30.50	30.46	-0.04	-0.14
18.200	30.80	30.61	-0.19	-0.60
19.000	31.90	31.90	0.00	0.01
20.300	34.00	34.08	0.08	0.24
21.500	36.20	36.19	-0.01	-0.03
23.000	38.80	38.96	0.16	0.40
23.800	40.20	40.50	0.30	0.74
24.600	41.70	42.08	0.38	0.91
25.500	43.30	43.92	0.62	1.43
26.400	45.10	45.81	0.71	1.58
27.100	46.80	47.33	0.53	1.13
28.000	48.80	49.33	0.53	1.09
28.800	50.90	51.16	0.26	0.52
29.800	52.20	53.52	1.32	2.53
30.600	55.10	55.47	0.37	0.66
31.400	57.60	57.46	-0.14	-0.24
33.000	62.10	61.61	-0.49	-0.80
33.800	65.00	63.76	-1.24	-1.91
34.200	66.00	64.86	-1.14	-1.73

STANDARD DEVIATION = 0.46

FORMIC ACID CH2O2

DREISBACH R.R.,SHRADER S.A.:
IND.ENG.CHEM. 41,2879(1949).

A = 7.58178 B = 1699.173 C = 260.714

B.P.(760) = 100.738

T	P EXPTL.	P CALCD.	DEV.	PERCENT
37.350	75.86	76.05	0.19	0.24
48.790	123.76	123.53	-0.23	-0.18
73.550	315.52	315.10	-0.42	-0.13
87.820	507.50	508.83	1.33	0.26
100.700	760.00	759.13	-0.87	-0.11

STANDARD DEVIATION = 1.18

METHYL BROMIDE CH3BR

BEERSMANS J.,JUNGERS J.C.:
BULL.SOC.CHIM.BELG. 56,238(1947).

A = 7.09084 B = 1046.066 C = 244.914

B.P.(760) = 3.556

T	P EXPTL.	P CALCD.	DEV.	PERCENT
-70.000	13.00	12.89	-0.11	-0.82
-60.000	27.20	27.15	-0.05	-0.18
-50.000	52.90	52.97	0.07	0.13
-40.000	96.70	96.81	0.11	0.12
-30.000	167.30	167.29	-0.01	-0.01
-10.000	434.80	434.38	-0.42	-0.10
0.000	660.40	660.21	-0.19	-0.03
2.800	737.70	737.84	0.14	0.02
3.200	749.30	749.50	0.20	0.03
3.600	761.10	761.30	0.20	0.03

STANDARD DEVIATION = 0.22

METHYL CHLORIDE CH3CL

BEERSMANS J.,JUNGERS J.C.:
BULL.SOC.CHIM.BELG. 56,238(1947).

A = 6.98762 B = 899.739 C = 242.921

B.P.(760) = -23.836

T	P EXPTL.	P CALCD.	DEV.	PERCENT
-90.000	12.80	12.70	-0.10	-0.75
-80.000	29.20	29.18	-0.02	-0.07
-70.000	60.80	60.88	0.08	0.13
-60.000	117.10	117.19	0.09	0.08
-50.000	210.80	210.79	-0.01	0.00
-40.000	358.10	357.84	-0.26	-0.07
-30.000	578.30	578.00	-0.30	-0.05
-24.600	735.10	735.28	0.18	0.02
-24.300	744.80	744.91	0.11	0.02
-23.800	761.00	761.20	0.20	0.03

STANDARD DEVIATION = 0.19

METHYL CHLORIDE CH3CL

GANEFF J.M.,JUNGERS J.C.:
BULL.SOC.CHIM.BELG. 57,87(1948).

A = 7.09349 B = 948.582 C = 249.336

B.P.(760) = -24.163

T	P EXPTL.	P CALCD.	DEV.	PERCENT
-75.000	45.00	44.91	-0.09	-0.19
-60.000	121.00	121.18	0.18	0.15
-50.000	216.00	216.16	0.16	0.08
-40.000	365.00	364.84	-0.16	-0.04
-30.000	588.00	587.09	-0.92	-0.16
-20.000	907.00	906.32	-0.68	-0.08
-10.000	1349.00	1349.27	0.27	0.02
-5.000	1622.00	1626.31	4.31	0.27
0.000	1945.00	1945.61	0.61	0.03
5.000	2315.00	2311.25	-3.75	-0.16

STANDARD DEVIATION = 2.22

METHYLTRICHLOROSILANE CH3CL3SI

JENKINS A.C.,CHAMBERS G.F.:
IND.ENG.CHEM. 46,2367(1954).

A = 7.08819 B = 1289.172 C = 239.930

B.P.(760) = 66.477

T	P EXPTL.	P CALCD.	DEV.	PERCENT
13.700	101.00	101.23	0.23	0.23
23.600	157.30	157.13	-0.17	-0.11
36.000	261.60	260.67	-0.93	-0.35
49.800	433.80	435.14	1.34	0.31
57.300	562.30	563.51	1.21	0.22
64.300	710.80	709.09	-1.71	-0.24

STANDARD DEVIATION = 1.54

METHYL FLUORIDE CH3F

MICHELS A.,WASSENAAR T.:
PHYSICA 14,104(1948).

A = 5.96591 B = 403.689 C = 209.836

B.P.(760) = -78.985

T	P EXPTL.	P CALCD.	DEV.	PERCENT
-108.887	98.50	92.68	-5.82	-5.91
-105.887	124.79	120.89	-3.90	-3.12
-101.848	169.56	168.91	-0.65	-0.38
-98.222	220.17	223.41	3.24	1.47
-93.383	306.20	315.78	9.58	3.13
-87.838	437.23	453.89	16.66	3.81
-88.309	509.96	440.68	-69.28	-13.59
-82.743	593.03	616.03	23.00	3.88
-80.359	679.97	704.83	24.86	3.66
-77.321	804.61	830.92	26.31	3.27
-74.120	954.64	980.42	25.78	2.70
-71.699	1083.30	1105.46	22.16	2.05
-69.190	1229.68	1246.45	16.77	1.36
-66.908	1375.45	1385.17	9.72	0.71
-64.820	1519.92	1521.14	1.22	0.08
-62.301	1710.30	1697.06	-13.24	-0.77
-58.988	1987.02	1948.91	-38.11	-1.92
-56.384	2229.69	2163.73	-65.96	-2.96

STANDARD DEVIATION =31.18

METHYL IODIDE CH_3I

BOUBLIK T., AIM K.:
COLLECTION CZECHOSLOV. CHEM. COMMUN. 36 (IN PRESS).

A = 6.98803 B = 1146.340 C = 236.674

B.P.(760) = 42.430

T	P EXPTL.	P CALCD.	DEV.	PERCENT
-13.819	69.80	69.85	0.05	0.07
-8.883	90.35	90.28	-0.07	-0.08
-5.101	109.13	109.09	-0.04	-0.04
-0.788	134.42	134.37	-0.05	-0.04
2.954	159.87	160.03	0.16	0.10
7.146	193.37	193.39	0.02	0.01
10.542	224.44	224.40	-0.04	-0.02
14.567	266.22	266.26	0.04	0.02
18.341	311.05	311.05	0.00	0.00
22.879	372.81	372.76	-0.05	-0.01
28.342	459.75	459.69	-0.06	-0.01
34.600	578.49	578.44	-0.05	-0.01
41.427	734.41	734.50	0.09	0.01

STANDARD DEVIATION = 0.07

METHYL IODIDE CH3I

THOMPSON A.W.,LINNETT J.W.: TRANS.FARAD.SOC.,32,681(1936).

A = 8.00463 B = 1747.541 C = 298.496

B.P.(100) = -7.464

T	P EXPTL.	P CALCD.	DEV.	PERCENT
0.100	142.00	141.94	-0.06	-0.04
4.300	171.00	171.11	0.11	0.07
11.700	235.00	234.94	-0.06	-0.02
34.400	569.00	569.01	0.01	0.00

STANDARD DEVIATION = 0.14

NITROMETHANE CH3NO2

MCCULLOUGH J.P.,SCOTT D.W.,PENNINGTON R.E.,HOSSENLOP I.A.,
WADDINGTON G.: J.AM.CHEM.SOC.76,4791(1954).

A = 7.28166 B = 1446.937 C = 227.600

B.P.(760) = 101.186

T	P EXPTL.	P CALCD.	DEV.	PERCENT
55.711	149.41	149.42	0.01	0.01
61.298	187.57	187.58	0.01	0.00
66.919	233.72	233.76	0.04	0.02
72.557	289.13	289.10	-0.03	-0.01
78.231	355.22	355.21	-0.01	0.00
83.925	433.56	433.47	-0.09	-0.02
89.655	525.86	525.83	-0.03	0.00
95.408	633.99	633.99	0.00	0.00
101.186	760.00	759.98	-0.02	0.00
106.994	906.06	906.14	0.08	0.01
112.826	1074.60	1074.68	0.08	0.01
118.681	1268.00	1268.07	0.07	0.01
124.564	1489.10	1489.19	0.09	0.01
130.473	1740.80	1740.81	0.01	0.00
136.404	2026.00	2025.79	-0.21	-0.01

STANDARD DEVIATION = 0.08

METHANE CH4

STOCK A.,HENNING F.,KUSS E.: BER. 54,1119(1921).

A = 6.34159 B = 342.217 C = 260.221

B.P.(100) = -181.398

T	P EXPTL.	P CALCD.	DEV.	PERCENT
-181.000	105.10	105.16	0.06	0.05
-180.000	118.90	119.04	0.14	0.11
-179.000	134.40	134.34	-0.06	-0.05
-178.000	151.10	151.16	0.06	0.04
-177.000	169.60	169.61	0.01	0.01
-176.000	190.00	189.79	-0.21	-0.11
-175.000	212.10	211.82	-0.28	-0.13
-174.000	236.10	235.80	-0.30	-0.13
-173.000	262.00	261.84	-0.16	-0.06
-172.000	290.00	290.08	0.08	0.03
-171.000	320.40	320.62	0.22	0.07
-170.000	353.20	353.60	0.40	0.11
-169.000	388.80	389.12	0.33	0.08
-168.000	427.20	427.33	0.13	0.03
-167.000	468.20	468.35	0.15	0.03
-166.000	512.20	512.31	0.11	0.02
-165.000	559.30	559.34	0.05	0.01
-164.000	609.60	609.58	-0.02	0.00
-163.000	663.80	663.15	-0.65	-0.10

STANDARD DEVIATION = 0.26

METHANE CH4

HESTERMANS P.,WHITE D.: J.PHYS.CHEM. 65,362(1961).

A = 7.10437 B = 514.120 C = 283.457

B.P.(760) = -161.730

T	P EXPTL.	P CALCD.	DEV.	PERCENT
-163.770	643.72	643.93	0.21	0.03
-158.810	955.62	954.49	-1.13	-0.12
-155.890	1185.30	1186.26	0.96	0.08
-153.170	1439.52	1439.86	0.34	0.02
-146.650	2220.49	2220.15	-0.34	-0.02

STANDARD DEVIATION = 1.12

METHYLDICHLOROSILANE CH4CL2SI

JENKINS A.C.,CHAMBERS G.F.:
IND.ENG.CHEM. 46,2367(1954).

A = 7.02783 B = 1167.794 C = 240.709

B.P.(760) = 40.890

T	P EXPTL.	P CALCD.	DEV.	PERCENT
1.500	160.70	160.84	0.14	0.09
9.100	225.70	225.46	-0.24	-0.11
22.700	393.40	393.03	-0.37	-0.09
28.000	480.00	480.70	0.70	0.15
34.100	600.10	600.26	0.16	0.03
40.700	755.50	755.11	-0.39	-0.05

STANDARD DEVIATION = 0.54

METHANOL CH4O

AMBROSE D.,SPRAKE C.H.S.:
J.CHEM.THERMODYNAMICS 2,631(1970)

A = 8.08097 B = 1582.271 C = 239.726

B.P.(760) = 64.547

T	P EXPTL.	P CALCD.	DEV.	PERCENT
14.899	73.62	73.60	-0.02	-0.03
19.236	93.52	93.52	0.00	0.00
23.323	116.40	116.37	-0.03	-0.02
27.083	141.45	141.46	0.01	0.00
29.911	163.28	163.24	-0.04	-0.03
32.885	189.06	189.16	0.10	0.06
35.858	218.48	218.50	0.02	0.01
40.637	273.72	273.73	0.01	0.00
45.407	340.13	340.20	0.07	0.02
48.876	396.66	396.68	0.02	0.00
53.315	480.31	480.27	-0.04	-0.01
56.428	547.36	547.32	-0.04	-0.01
60.814	655.14	654.96	-0.18	-0.03
63.784	737.54	737.45	-0.09	-0.01
64.717	765.05	765.08	0.03	0.00
68.403	882.93	882.84	-0.09	-0.01
71.770	1003.14	1003.20	0.06	0.01
75.683	1159.90	1159.86	-0.04	0.00
79.626	1337.41	1337.65	0.24	0.02
83.678	1542.53	1543.19	0.66	0.04

STANDARD DEVIATION = 0.09

METHANOL CH4O

DEVER D.F.,FINCH A.,GRUNWALD E.:
J.PHYS.CHEM. 59, 668(1955).

A = 8.53390 B = 1838.010 C = 260.872

B.P.(100) = 20.431

	T	P EXPTL.	P CALCD.	DEV.	PERCENT
	13.880	69.92	69.85	-0.07	-0.09
	14.600	72.81	72.72	-0.09	-0.12
	15.370	75.81	75.91	0.10	0.13
	15.700	77.29	77.31	0.02	0.02
	16.340	80.25	80.09	-0.16	-0.20
	16.730	81.93	81.82	-0.11	-0.13
	17.270	84.36	84.28	-0.08	-0.09
	17.780	86.64	86.66	0.02	0.02
	18.280	89.00	89.05	0.05	0.06
	18.830	91.73	91.75	0.02	0.02
	19.410	94.62	94.66	0.04	0.05
*	19.830	97.65	96.83	-0.82	-0.84
	20.340	99.56	99.51	-0.05	-0.05
	20.910	102.70	102.59	-0.11	-0.11
	21.410	105.30	105.35	0.05	0.05
	21.840	107.70	107.78	0.08	0.08
	22.360	110.80	110.79	-0.01	-0.01
	22.750	112.90	113.09	0.19	0.17
	23.840	119.80	119.73	-0.07	-0.05
	24.140	121.50	121.62	0.12	0.10
	24.840	126.10	126.13	0.03	0.02
	25.130	127.90	128.04	0.14	0.11
	25.540	130.60	130.78	0.18	0.14
	26.110	134.50	134.67	0.17	0.13
	26.340	136.00	136.27	0.27	0.20
	27.090	141.70	141.61	-0.09	-0.07
	27.270	143.00	142.91	-0.09	-0.06
	28.220	150.40	149.98	-0.42	-0.28
	28.520	152.10	152.27	0.17	0.11
	29.070	156.60	156.55	-0.05	-0.03
	29.840	162.80	162.73	-0.07	-0.05
	30.570	169.00	168.77	-0.23	-0.14
	30.770	170.10	170.46	0.36	0.21
	31.320	175.30	175.18	-0.12	-0.07
	31.640	178.00	177.98	-0.02	-0.01
	32.090	182.10	181.98	-0.12	-0.07
	32.610	186.70	186.69	-0.01	0.00
	33.450	194.60	194.54	-0.06	-0.03
	34.110	200.90	200.90	0.00	0.00

STANDARD DEVIATION = 0.15

METHANETHIOL CH4S

RUSSELL H.,OSBORNE D.W.,YOST D.M.:
J.AM.CHEM.SOC. 64,165(1942).

A = 7.06418 B = 1030.117 C = 240.289

B.P.(760) = 5.953

T	P EXPTL.	P CALCD.	DEV.	PERCENT
-51.280	41.13	41.12	-0.01	-0.02
-23.872	201.46	201.51	0.05	0.02
-9.474	399.30	399.22	-0.08	-0.02
0.029	599.40	599.37	-0.03	-0.01
5.977	760.64	760.72	0.08	0.01

STANDARD DEVIATION = 0.09

METHYLAMINE CH5N

ASTON J.G.,SILLER C.W.,MESSERLY G.H.:
J.AM.CHEM.SOC. 59,1743(1937).

A = 7.33690 B = 1011.532 C = 233.286

B.P.(760) = -6.285

T	P EXPTL.	P CALCD.	DEV.	PERCENT
-83.092	4.06	4.00	-0.06	-1.48
-74.196	9.59	9.52	-0.07	-0.73
-66.168	19.19	19.23	0.04	0.23
-58.835	34.75	34.55	-0.20	-0.56
-51.124	60.67	60.81	0.14	0.23
-44.616	94.29	94.51	0.22	0.24
-36.640	155.86	155.94	0.08	0.05
-28.698	247.20	246.97	-0.23	-0.09
-23.565	326.65	326.33	-0.32	-0.10
-17.502	446.01	445.84	-0.17	-0.04
-12.498	569.33	569.40	0.07	0.01
-8.309	693.03	692.99	-0.04	-0.01
-6.226	761.55	762.03	0.48	0.06

STANDARD DEVIATION = 0.23

N-METHYLHYDROXYLAMINE CH5NO

BISSOT T.C.,PARRY R.W.,CAMPBELL D.H.:
J.AM.CHEM.SOC. 79,796(1957).

A = 7.04563 B = 1223.285 C = 172.078

B.P.(10) = 30.264

T	P EXPTL.	P CALCD.	DEV.	PERCENT
40.000	19.10	18.95	-0.15	-0.80
45.000	25.80	25.73	-0.07	-0.28
50.200	34.60	34.85	0.25	0.73
55.000	45.40	45.56	0.16	0.35
60.000	59.80	59.51	-0.29	-0.48
65.000	76.80	76.88	0.08	0.10

STANDARD DEVIATION = 0.26

O-METHYLHYDROXYLAMINE CH5NO

BISSOT T.C.,PARRY R.W.,CAMPBELL D.H.:
J.AM.CHEM.SOC. 79,796(1957).

A = 7.36393 B = 1225.268 C = 225.195

B.P.(760) = 48.112

T	P EXPTL.	P CALCD.	DEV.	PERCENT
-63.500	0.75	0.61	-0.14	-18.49
* -45.300	3.55	3.57	0.02	0.62
-22.800	20.50	20.42	-0.08	-0.38
-20.400	24.20	24.05	-0.15	-0.64
-17.600	29.00	28.96	-0.04	-0.15
-13.300	38.20	38.15	-0.05	-0.13
-10.400	45.60	45.66	0.06	0.14
-6.800	56.40	56.70	0.30	0.54
0.000	83.90	83.75	-0.15	-0.17
9.400	138.80	138.36	-0.44	-0.31
10.100	143.50	143.40	-0.10	-0.07
12.600	162.90	162.67	-0.23	-0.14
14.400	177.30	177.83	0.53	0.30
16.000	192.10	192.28	0.18	0.10
17.700	208.40	208.69	0.29	0.14
19.100	222.80	223.05	0.25	0.11
20.100	233.60	233.80	0.20	0.09
21.100	243.90	244.98	1.08	0.44
24.800	291.10	290.24	-0.86	-0.29
29.800	362.90	362.13	-0.77	-0.21
34.900	451.00	449.86	-1.14	-0.25
40.300	561.00	560.92	-0.08	-0.02
44.800	670.00	669.60	-0.40	-0.06
48.200	761.00	762.52	1.52	0.20

STANDARD DEVIATION = 0.60

METHYLHYDRAZINE CH6N2

ASTON J.G.,FINK H.L.,JANZ G.J.,RUSSELL K.E.:
J.AM.CHEM.SOC. 73,1939(1951).

A = 6.57624 B = 1007.495 C = 181.413

B.P.(10) = -0.736

T	P EXPTL.	P CALCD.	DEV.	PERCENT
1.957	12.11	12.08	-0.03	-0.29
11.739	22.85	22.92	0.07	0.29
17.132	31.76	31.76	0.00	-0.01
20.532	38.72	38.66	-0.06	-0.15
25.166	50.00	50.03	0.03	0.05

STANDARD DEVIATION = 0.07

SILYL METHYL ETHER CH6OSI

STERNBACH B.,MACDIARMID A.G.:
J.AM.CHEM.SOC. 83,3384(1961).

A = 6.45124 B = 722.556 C = 219.775

B.P.(10) = -87.226

T	P EXPTL.	P CALCD.	DEV.	PERCENT
-89.300	8.30	8.19	-0.11	-1.31
-87.100	10.30	10.12	-0.18	-1.75
-85.500	11.70	11.75	0.05	0.43
-77.700	23.10	23.20	0.10	0.43
-72.200	35.70	35.90	0.20	0.55
-67.800	50.30	49.75	-0.55	-1.09
-65.500	58.00	58.57	0.57	0.98
-59.000	90.90	90.58	-0.32	-0.35
-57.600	98.90	99.04	0.14	0.14

STANDARD DEVIATION = 0.38

METHYLPHOSPHINE BORINE CH8BP

BURG A.B.,WAGNER R.I.:
J.AM.CHEM.SOC. 75,3872(1953).

A = 7.52824 B = 1783.835 C = 231.623

B.P.(10) = 41.626

T	P EXPTL.	P CALCD.	DEV.	PERCENT
23.600	3.50	3.46	-0.04	-1.18
35.700	7.10	7.17	0.07	0.93
46.100	12.70	12.74	0.04	0.32
51.000	16.50	16.46	-0.04	-0.22
56.200	21.50	21.41	-0.09	-0.43
66.300	34.80	34.73	-0.07	-0.21
71.300	43.40	43.60	0.20	0.47
76.400	54.60	54.58	-0.02	-0.04
81.300	67.30	67.25	-0.05	-0.08

STANDARD DEVIATION = 0.11

TETRANITROMETHANE CN408

EDWARDS G.:

TRANS.FARADAY SOC. 48, 513(1952).

A = 6.76488 B = 1210.603 C = 185.926

B.P.(100) = 68.142

T	P EXPTL.	P CALCD.	DEV.	PERCENT
40.000	26.50	25.50	-1.00	-3.79
50.000	43.30	43.01	-0.29	-0.66
60.000	68.00	69.54	1.54	2.27
70.000	108.00	108.29	0.29	0.27
80.000	164.00	163.11	-0.89	-0.54
90.000	239.00	238.48	-0.52	-0.22
100.000	339.00	339.55	0.55	0.16

STANDARD DEVIATION = 1.11

TETRANITROMETHANE CN408

NICHOLSON J.C.:

J.CHEM.SOC. 1949,1553.

A = 5.55250 B = 783.184 C = 149.081

B.P.(10) = 22.952

T	P EXPTL.	P CALCD.	DEV.	PERCENT
0.000	1.90	1.99	0.09	4.80
13.800	5.70	5.55	-0.15	-2.65
20.000	8.40	8.33	-0.07	-0.86
30.000	14.90	15.11	0.21	1.39
40.000	25.80	25.73	-0.07	-0.27

STANDARD DEVIATION = 0.20

CARBONYL SULFIDE COS

KEMP J.D.,GIAUQUE W.F.:

J.AM.CHEM.SOC. 59,79(1937).

A = 6.90723 B = 804.480 C = 250.000

B.P.(760) = -50.200

T	P EXPTL.	P CALCD.	DEV.	PERCENT
-111.303	12.82	12.79	-0.03	-0.21
-111.291	12.83	12.81	-0.02	-0.17
-107.003	19.13	19.12	-0.01	-0.07
-102.540	28.26	28.29	0.03	0.11
-97.696	42.19	42.19	0.00	-0.01
-93.027	60.55	60.57	0.02	0.04
-88.000	87.33	87.36	0.03	0.04
-82.895	123.88	123.89	0.01	0.00
-77.386	176.52	176.47	-0.05	-0.03
-71.699	248.51	248.49	-0.02	-0.01
-61.268	441.34	441.25	-0.09	-0.02
-56.472	562.66	562.75	0.09	0.02
-52.578	679.69	679.69	0.00	0.00
-49.309	791.89	791.93	0.04	0.00

STANDARD DEVIATION = 0.05

CARBON DIOXIDE CO2

GIAUQUE W.F.,EGAN C.J.:

J.CHEM.PHYS. 5, 45(1937).

A = 9.66983 B = 1295.524 C = 269.243

B.P.(760) = -78.416

T	P EXPTL.	P CALCD.	DEV.	PERCENT
-118.894	11.32	11.30	-0.02	-0.18
-114.687	19.40	19.39	-0.01	-0.05
-110.493	32.21	32.29	0.08	0.24
-106.386	51.86	51.86	0.00	0.00
-102.399	80.39	80.34	-0.05	-0.06
-98.402	122.20	122.07	-0.13	-0.11
-94.521	179.94	179.91	-0.03	-0.02
-90.757	257.75	257.88	0.13	0.05
-85.734	407.18	407.47	0.29	0.07
-80.427	643.36	643.46	0.10	0.02
-78.108	779.62	779.41	-0.21	-0.03
-77.259	835.30	835.11	-0.19	-0.02

STANDARD DEVIATION = 0.16

CARBON SELENOSULFIDE CSSE

STOCK A.,WILLFROTH E.:
BER. 47,144(1914).

A = 6.69959 B = 1161.967 C = 219.590

B.P.(760) = 84.687

T	P EXPTL.	P CALCD.	DEV.	PERCENT
0.000	26.00	25.59	-0.41	-1.58
5.000	34.00	33.56	-0.44	-1.29
10.000	44.00	43.50	-0.50	-1.13
15.000	55.00	55.77	0.77	1.40
20.000	70.00	70.76	0.76	1.08
25.000	88.00	88.90	0.90	1.03
30.000	112.00	110.68	-1.32	-1.18
40.000	167.00	167.27	0.27	0.16
50.000	246.00	245.16	-0.84	-0.34
60.000	350.00	349.64	-0.36	-0.10
70.000	485.00	486.55	1.55	0.32
80.000	660.00	662.31	2.31	0.35
84.000	748.00	745.01	-2.99	-0.40

STANDARD DEVIATION = 1.48

CARBON DISULFIDE CS2

WADDINGTON G.,SMITH J.C.,WILLIAMSON K.D.,SCOTT D.W.:
J.PHYS.CHEM. 66,1074(1962).

A = 6.94279 B = 1169.110 C = 241.593

B.P.(760) = 46.225

T	P EXPTL.	P CALCD.	DEV.	PERCENT
3.588	149.41	149.43	0.02	0.01
8.772	187.57	187.57	0.00	0.00
13.999	233.72	233.71	-0.01	-0.01
19.269	289.13	289.12	-0.01	0.00
24.582	355.22	355.25	0.03	0.01
29.927	433.56	433.51	-0.05	-0.01
35.318	525.86	525.80	-0.06	-0.01
40.751	633.99	633.96	-0.03	0.00
46.225	760.00	760.00	0.00	0.00
51.744	906.06	906.23	0.17	0.02
57.295	1074.60	1074.63	0.03	0.00
62.885	1268.00	1267.86	-0.14	-0.01
68.531	1489.10	1489.28	0.18	0.01
74.210	1740.80	1740.88	0.08	0.00
79.927	2026.00	2025.79	-0.21	-0.01

STANDARD DEVIATION = 0.11

CARBON DISELENIDE CSE2

IVES D.J.G.,PITTMAN R.W.,WARDLAW W.:
J.CHEM.SOC. 1947,1080.

A = 6.77673 B = 1353.198 C = 219.951

B.P.(10) = 14.299

T	P EXPTL.	P CALCD.	DEV.	PERCENT
0.000	4.70	4.21	-0.49	-10.39
5.070	5.70	5.80	0.10	1.67
10.070	7.70	7.83	0.13	1.69
15.070	10.40	10.45	0.05	0.44
20.070	13.60	13.77	0.17	1.24
25.060	17.90	17.94	0.04	0.20
30.040	23.30	23.11	-0.19	-0.83
35.040	29.60	29.50	-0.10	-0.33
40.020	37.00	37.28	0.28	0.77
50.010	58.20	58.10	-0.10	-0.18

STANDARD DEVIATION = 0.25

TRIFLUOROCHLOROETHYLENE C_2ClF_3

OLIVER G.D.,GRISARD J.W.,CUNNINGHAM C.W.:
J.AM.CHEM.SOC. 73,5719(1951).

A = 6.89616 B = 848.328 C = 239.644

B.P.(760) = -28.372

T	P EXPTL.	P CALCD.	DEV.	PERCENT
-66.820	97.24	97.17	-0.07	-0.07
-52.460	231.17	231.26	0.09	0.04
-44.570	352.35	352.70	0.35	0.10
-40.740	427.87	427.70	-0.17	-0.04
-36.970	513.42	513.43	0.01	0.00
-33.120	614.50	614.48	-0.02	0.00
-30.290	698.40	698.29	-0.11	-0.02
-28.360	760.62	760.41	-0.21	-0.03
-26.470	825.21	825.37	0.16	0.02
-23.050	953.71	953.86	0.15	0.02
-19.000	1126.40	1125.58	-0.82	-0.07
-15.610	1287.30	1286.93	-0.37	-0.03
-10.870	1540.70	1541.73	1.03	0.07

STANDARD DEVIATION = 0.46

TRIFLUOROCHLOROETHYLENE C2CLF3

BOOTH H.S.,BURCHFIELD P.E.,BIXLY E.M.,MAC KELVEY J.B.:
J.AM.CHEM.SOC. 55,2231(1933).

A = 7.20636 B = 990.389 C = 256.963

B.P.(760) = -28.001

T	P EXPTL.	P CALCD.	DEV.	PERCENT
-78.560	45.20	45.18	-0.02	-0.04
-33.560	593.40	593.16	-0.24	-0.04
-28.460	742.10	744.93	2.83	0.38
-27.760	767.30	767.98	0.68	0.09
-26.460	815.00	812.31	-2.69	-0.33
-25.360	853.70	851.39	-2.31	-0.27
-23.460	920.70	922.41	1.71	0.19

STANDARD DEVIATION = 2.45

PENTAFLUOROMONOCHLOROETHANE C2CLF5

ASTON J.G.,WILLS P.E.,ZOLKI T.P.:
J.AM.CHEM.SOC.77,3939(1955).

A = 6.83334 B = 802.969 C = 242.269

B.P.(760) = -39.115

T	P EXPTL.	P CALCD.	DEV.	PERCENT
-95.312	23.45	23.41	-0.04	-0.18
-90.320	35.41	35.39	-0.02	-0.06
-82.422	64.47	64.56	0.09	0.14
-65.084	200.26	200.22	-0.04	-0.02
-56.019	332.74	332.72	-0.02	-0.01
-47.808	506.08	505.97	-0.11	-0.02
-39.847	735.32	735.40	0.08	0.01
-39.225	756.20	756.26	0.06	0.01

STANDARD DEVIATION = 0.08

BIS(TRIFLUOROMETHYL)CHLOROPHOSPHINE C2CLF6P

PETERSON L.K.,BURG A.B.:
J.AM.CHEM.SOC. 86,2587(1964).

A = 7.66106 B = 1386.652 C = 267.137

B.P.(100) = -22.192

T	P EXPTL.	P CALCD.	DEV.	PERCENT
-80.800	1.64	1.66	0.02	1.07
-64.800	6.34	6.43	0.09	1.35
-46.000	24.80	24.58	-0.22	-0.90
-31.400	59.80	60.10	0.30	0.50
-22.700	97.50	97.33	-0.17	-0.18
0.000	295.30	295.31	0.01	0.00

STANDARD DEVIATION = 0.24

DICHLORODIFLUOROETHYLENE C2CL2F2

BOOTH H.S.,ET AL.:
J.AM.CHEM.SOC. 55,2231(1933).

A = 6.92020 B = 1010.134 C = 229.402

B.P.(760) = 20.669

T	P EXPTL.	P CALCD.	DEV.	PERCENT
-33.250	59.05	58.95	-0.10	-0.17
-21.950	113.00	112.45	-0.55	-0.48
-14.950	161.80	162.15	0.35	0.21
-9.950	206.60	207.60	1.00	0.48
-5.950	249.80	250.97	1.17	0.47
-0.050	330.20	328.03	-2.17	-0.66
9.950	502.30	501.09	-1.21	-0.24
14.950	609.90	611.33	1.43	0.23
17.950	685.50	686.14	0.64	0.09
19.850	738.30	737.13	-1.17	-0.16
20.350	750.80	751.03	0.23	0.03
20.850	765.20	765.14	-0.06	-0.01
21.850	793.60	793.97	0.37	0.05

STANDARD DEVIATION = 1.14

TRICHLOROTRIFLUOROETHANE C2CL3F3

RIEDEL L.:

Z.GES.KALTE-IND. 45,221(1938).

A = 6.88342 B = 1111.407 C = 230.008

B.P.(760) = 47.663

T	P EXPTL.	P CALCD.	DEV.	PERCENT
-25.480	28.20	28.14	-0.06	-0.20
-19.820	39.40	39.42	0.02	0.05
-13.800	55.30	55.33	0.03	0.05
-7.100	78.90	78.97	0.07	0.09
0.000	112.50	112.56	0.06	0.05
4.570	139.80	139.80	0.00	0.00
14.430	217.10	217.08	-0.02	-0.01
19.750	271.70	271.31	-0.39	-0.14
26.790	359.50	359.31	-0.19	-0.05
31.040	422.60	422.60	0.00	0.00
35.040	489.90	489.98	0.08	0.02
39.310	571.10	571.03	-0.07	-0.01
44.330	679.10	679.48	0.38	0.06
47.830	763.40	764.21	0.81	0.11
52.550	890.80	891.32	0.52	0.06
57.780	1051.00	1050.78	-0.22	-0.02
61.160	1165.00	1165.05	0.05	0.00
63.990	1269.00	1267.90	-1.10	-0.09
69.990	1510.00	1509.01	-0.99	-0.07
73.010	1642.00	1642.92	0.92	0.06
76.320	1800.00	1799.90	-0.10	-0.01
79.250	1948.00	1948.15	0.15	0.01

STANDARD DEVIATION = 0.48

1,1,1-TRICHLORO-2,2,2-TRIFLUOROETHANE C2CL3F3

HIRAOKA H.,HILDEBRAND J.H.: J.PHYS.CHEM. 67,916(1963).

A = 4.43730 B = 204.133 C = 83.918

B.P.(100) = -0.164

T	P EXPTL.	P CALCD.	DEV.	PERCENT
14.050	230.10	225.75	-4.35	-1.89
15.050	240.10	236.97	-3.13	-1.31
17.550	265.10	266.38	1.28	0.48
20.050	295.00	297.77	2.77	0.94
22.550	325.00	331.12	6.12	1.88
25.050	361.10	366.41	5.31	1.47
27.550	405.20	403.63	-1.57	-0.39
30.050	444.80	442.75	-2.05	-0.46
32.050	478.20	475.39	-2.81	-0.59
34.050	511.00	509.20	-1.80	-0.35
36.050	544.00	544.18	0.18	0.03

STANDARD DEVIATION = 3.90

1,1,2-TRICHLORO-1,2,2-TRIFLUOROMETHANE C2CL3F3

BENNING A.F.,MCHARNESS R.C.: IND.ENG.CHEM. 32,497(1940).

A = 6.88029 B = 1099.897 C = 227.479

B.P.(760) = 47.531

T	P EXPTL.	P CALCD.	DEV.	PERCENT
-25.480	27.44	27.24	-0.20	-0.72
-8.400	72.43	72.40	-0.03	-0.04
-7.100	77.52	77.51	-0.01	-0.01
0.000	110.88	110.95	0.07	0.06
19.750	269.80	270.02	0.22	0.08
20.000	272.61	272.82	0.21	0.08
20.190	274.74	274.97	0.23	0.09
39.310	572.43	572.22	-0.21	-0.04
40.000	586.57	586.41	-0.16	-0.03
47.570	761.52	760.98	-0.54	-0.07
61.160	1174.96	1173.96	-1.00	-0.09
79.250	1969.16	1969.68	0.52	0.03
83.300	2192.60	2193.44	0.84	0.04

STANDARD DEVIATION = 0.50

TRICHLOROACETONITRILE C2CL3N

DAVIES M.,JENKIN D.G.:

J.CHEM.SOC. 1954,2374.

A = 7.18350 B = 1368.276 C = 232.510

B.P.(760) = 85.495

	T	P EXPTL.	P CALCD.	DEV.	PERCENT
	16.800	49.00	49.57	0.57	1.17
	17.300	51.00	50.84	-0.16	-0.31
	20.000	58.00	58.18	0.18	0.32
	20.000	58.50	58.18	-0.32	-0.54
	23.400	68.50	68.67	0.17	0.25
	26.550	80.00	79.76	-0.24	-0.30
	31.000	98.50	97.95	-0.55	-0.56
	35.050	118.50	117.38	-1.12	-0.95
	40.400	147.50	147.86	0.36	0.24
	47.000	192.50	194.19	1.69	0.88
	50.150	222.00	220.18	-1.82	-0.82
	56.000	273.50	276.02	2.52	0.92
	61.650	340.50	340.43	-0.07	-0.02
	65.400	388.00	389.56	1.56	0.40
*	66.900	458.50	410.76	-47.74	-10.41
	73.650	521.00	518.02	-2.98	-0.57
	76.650	572.50	572.41	-0.09	-0.02
	79.750	635.50	633.36	-2.14	-0.34
	83.400	709.50	711.66	2.16	0.31

STANDARD DEVIATION = 1.54

TETRACHLOROETHYLENE C2CL4

BOUBLIK T.,AIM K.:
COLLECTION CZECHOSLOV.CHEM.COMMUN. 36(IN PRESS).

A = 6.97683 B = 1386.915 C = 217.526

B.P.(760) = 121.075

T	P EXPTL.	P CALCD.	DEV.	PERCENT
37.088	35.93	35.86	-0.07	-0.20
46.259	52.26	52.32	0.06	0.12
52.671	69.80	69.80	0.00	0.00
58.697	90.35	90.33	-0.02	-0.02
65.270	109.13	109.04	-0.09	-0.08
68.545	134.42	134.49	0.07	0.05
73.064	159.87	159.98	0.11	0.07
78.165	193.37	193.38	0.01	0.00
82.292	224.44	224.37	-0.07	-0.03
87.183	266.22	266.21	-0.01	0.00
91.791	311.05	311.20	0.15	0.05
97.291	372.81	372.71	-0.10	-0.03
103.930	459.75	459.57	-0.18	-0.04
111.548	578.49	578.40	-0.09	-0.02
119.650	730.09	730.30	0.21	0.03

STANDARD DEVIATION = 0.11

1,1,2,2-TETRACHLORO-1,2-DIFLUOROETHANE C2CL4F2

HOVORKA F.,GEIGER F.E.:
J.AM.CHEM.SOC. 55,4759(1933).

A = 10.99541 B = 4437.098 C = 455.237

B.P.(760) = 91.568

T	P EXPTL.	P CALCD.	DEV.	PERCENT
10.000	28.50	28.72	0.22	0.76
20.000	45.80	45.58	-0.22	-0.47
85.000	604.90	605.56	0.66	0.11
91.000	744.30	745.38	1.08	0.14
91.500	760.00	758.24	-1.76	-0.23

STANDARD DEVIATION = 1.55

TRICHLOROACETYL CHLORIDE C2CL4O

MCDONALD R.A.,SHRADER S.A.,STULL D.R.:
J.CHEM.ENG.DATA 4,311(1959).

A = 6.99075 B = 1390.465 C = 220.106

B.P.(760) = 118.212

T	P EXPTL.	P CALCD.	DEV.	PERCENT
32.170	30.10	30.13	0.03	0.11
42.670	50.15	50.04	-0.11	-0.22
56.300	91.10	91.25	0.15	0.16
74.850	189.20	189.06	-0.14	-0.07
95.870	389.10	389.23	0.13	0.03
116.750	729.40	729.41	0.01	0.00
117.390	742.90	742.67	-0.23	-0.03
118.490	766.10	765.92	-0.18	-0.02
119.430	785.90	786.23	0.33	0.04

STANDARD DEVIATION = 0.21

ETHYLENE-D4 OXIDE C2D4O

LEITCH L.C.,MORSE A.T.:

CAN.J.CHEM. 30,924(1952).

A = 5.09430 B = 638.380 C = 187.308

B.P.(10) = -31.389

T	P EXPTL.	P CALCD.	DEV.	PERCENT
-42.700	5.32	4.78	-0.54	-10.09
-37.000	7.13	7.03	-0.10	-1.36
-29.500	11.14	11.19	0.05	0.49
-22.600	16.30	16.54	0.24	1.46
-17.900	21.04	21.18	0.14	0.68
-13.900	25.31	25.88	0.57	2.24
-10.950	29.96	29.82	-0.14	-0.47
-7.900	34.86	34.36	-0.50	-1.44
0.000	48.47	48.54	0.07	0.15

STANDARD DEVIATION = 0.40

BIS(MONOFLUOROCARBONYL)PEROXIDE C2F2O4

VON ARVIA A.J.,AYMONINO P.J.,SCHUMACHER H.J.:
Z.ANORG.ALLG.CHEM. 316,325(1962).

A = 9.60839 B = 2247.636 C = 319.827

B.P.(100) = -24.411

T	P EXPTL.	P CALCD.	DEV.	PERCENT
-47.250	24.90	23.04	-1.86	-7.47
-36.750	46.00	46.60	0.60	1.30
-30.150	69.20	70.68	1.48	2.13
-23.750	105.00	103.99	-1.01	-0.96
-18.050	143.30	144.67	1.37	0.96
-7.150	264.60	263.04	-1.56	-0.59

STANDARD DEVIATION = 1.94

TRIFLUOROACETONITRILE C2F3N

PACE E.L.,BOBKA R.J.:
J.CHEM.PHYS. 35,454(1961).

A = 7.12759 B = 773.815 C = 249.917

B.P.(760) = -67.704

T	P EXPTL.	P CALCD.	DEV.	PERCENT
-131.390	3.01	3.97	0.96	31.95
-121.600	13.13	12.51	-0.63	-4.76
-111.570	34.05	34.22	0.17	0.50
-101.160	84.12	84.28	0.16	0.19
-94.250	143.73	143.42	-0.31	-0.21
-86.190	251.82	251.96	0.14	0.06
-79.880	377.23	377.33	0.10	0.03
-74.990	505.65	505.75	0.10	0.02
-71.730	607.81	609.35	1.54	0.25
-67.590	766.55	764.68	-1.87	-0.24

STANDARD DEVIATION = 1.03

TETRAFLUOROETHYLENE C2F4

FURUKAVA G.T.,MCCOSKEY R.E.,REILLY M.L.:
J.RES.NATL.BUR.STANDARDS 51,69(1953).

A = 6.89659 B = 683.840 C = 245.931

B.P.(760) = -75.643

T	P EXPTL.	P CALCD.	DEV.	PERCENT
-131.160	8.70	8.68	-0.02	-0.28
-125.180	17.10	17.11	0.01	0.09
-120.830	26.90	26.93	0.03	0.12
*-125.460	16.70	16.60	-0.10	-0.58
-120.240	28.60	28.57	-0.03	-0.10
-117.390	37.80	37.72	-0.08	-0.21
-114.270	50.50	50.43	-0.07	-0.15
-113.820	52.50	52.52	0.02	0.04
-109.010	79.70	79.83	0.13	0.17
-104.600	114.30	114.29	-0.01	-0.01
-103.470	124.70	124.85	0.15	0.12
-100.260	159.20	159.29	0.09	0.05
-97.810	190.50	190.47	-0.03	-0.02
-97.340	197.10	196.98	-0.12	-0.06
-95.140	229.80	229.92	0.12	0.05
-94.890	233.90	233.93	0.03	0.01
-93.930	249.90	249.85	-0.05	-0.02
-92.930	267.30	267.35	0.05	0.02
-92.200	280.70	280.74	0.04	0.01
-91.150	301.00	300.94	-0.06	-0.02
-90.200	320.40	320.21	-0.19	-0.06
-88.790	350.80	350.62	-0.18	-0.05
-86.620	402.20	401.89	-0.31	-0.08
-84.890	447.00	446.91	-0.09	-0.02
-83.050	499.20	499.10	-0.10	-0.02
-81.420	549.10	549.27	0.17	0.03
-79.860	601.00	600.95	-0.05	-0.01
-77.100	701.40	701.70	0.30	0.04
-77.100	701.80	701.70	-0.10	-0.01
-74.750	797.50	797.55	0.05	0.01
-72.550	895.80	896.29	0.49	0.05
-70.320	1005.60	1005.84	0.24	0.02
-67.720	1146.10	1146.42	0.32	0.03
-66.500	1217.80	1217.40	-0.40	-0.03
* -66.170	1236.30	1237.17	0.87	0.07
-65.660	1268.60	1268.21	-0.39	-0.03
-64.750	1325.10	1325.09	-0.01	0.00

STANDARD DEVIATION = 0.19

TRIFLUOROACETYL FLUOROSULFATE — C2F4O3S

DELFINO J.J.,SHREEVE J.M.:
INORG.CHEM. 5,308(1966).

A = 7.15129 B = 1112.787 C = 214.287

B.P.(760) = 46.290

T	P EXPTL.	P CALCD.	DEV.	PERCENT
-23.350	24.60	21.05	-3.55	-14.43
-10.650	49.20	48.61	-0.59	-1.20
-4.450	68.30	70.50	2.20	3.22
4.050	116.00	113.41	-2.59	-2.24
7.250	133.40	134.36	0.96	0.72
10.650	160.60	160.02	-0.58	-0.36
11.450	167.40	166.61	-0.79	-0.47
12.950	171.90	179.58	7.68	4.47
13.350	184.00	183.17	-0.83	-0.45
15.450	208.30	203.02	-5.28	-2.53
22.250	280.30	279.76	-0.54	-0.19
28.550	356.90	370.55	13.65	3.82
32.750	439.30	443.35	4.05	0.92
35.650	505.70	500.05	-5.65	-1.12
41.250	632.80	626.02	-6.78	-1.07
43.050	675.60	671.50	-4.10	-0.61
44.850	721.50	719.58	-1.92	-0.27
46.450	770.00	764.59	-5.41	-0.70
47.350	785.80	790.88	5.09	0.65

STANDARD DEVIATION = 5.42

HEXAFLUOROETHANE — C2F6

PACE E.L.,ASTON J.G.:
J.AM.CHEM.SOC.70,566(1948).

A = 6.79335 B = 657.061 C = 246.213

B.P.(760) = -78.275

T	P EXPTL.	P CALCD.	DEV.	PERCENT
-93.190	315.88	315.84	-0.04	-0.01
-92.930	321.21	321.18	-0.03	-0.01
-84.850	526.38	526.50	0.12	0.02
-83.100	582.11	582.22	0.11	0.02
-78.280	759.84	759.81	-0.03	0.00
-78.250	761.07	761.03	-0.04	-0.01
-78.050	769.27	769.23	-0.04	-0.01
-77.940	773.83	773.77	-0.06	-0.01

STANDARD DEVIATION = 0.09

BIS(TRIFLUOROMETHYL)IODOPHOSPHINE C2F6IP

PETERSON L.K.,BURG A.B.:
J.AM.CHEM.SOC. 86,2587(1964).

A = 6.90139 B = 1180.723 C = 222.952

B.P.(100) = 17.944

T	P EXPTL.	P CALCD.	DEV.	PERCENT
0.000	40.30	40.32	0.02	0.05
15.600	89.60	89.50	-0.10	-0.11
25.400	140.30	140.33	0.03	0.02
32.300	188.70	188.65	-0.05	-0.02
32.700	191.50	191.82	0.32	0.17
41.500	273.70	273.27	-0.43	-0.16
47.400	341.80	342.00	0.20	0.06

STANDARD DEVIATION = 0.29

2-BROMO-2-CHLORO-1,1,1-TRIFLUOROETHANE C2HBRCLF3

BOTTOMLEY G.A.,SEIFLOW G.H.F.: J.APPL.CHEM. 13,399(1963).

A = 6.94502 B = 1127.856 C = 227.341

B.P.(760) = 50.168

T	P EXPTL.	P CALCD.	DEV.	PERCENT
-50.700	3.38	3.63	0.25	7.42
-48.700	4.02	4.28	0.26	6.48
-46.400	5.03	5.15	0.12	2.37
-43.800	6.30	6.31	0.01	0.16
-41.400	7.37	7.57	0.20	2.77
-39.650	8.21	8.63	0.42	5.09
-38.650	8.84	9.28	0.44	5.03
-36.550	10.52	10.80	0.28	2.69
-36.350	10.87	10.96	0.09	0.81
-32.200	14.07	14.63	0.56	4.00
-28.000	19.50	19.37	-0.13	-0.67
* -25.050	24.50	23.42	-1.08	-4.40
-21.500	29.00	29.23	0.23	0.78
-18.100	36.40	35.88	-0.52	-1.44
-16.200	40.90	40.11	-0.79	-1.92
* -12.500	51.40	49.58	-1.82	-3.54
-10.300	55.90	56.04	0.14	0.25
-8.000	63.80	63.53	-0.27	-0.42
-4.000	79.30	78.54	-0.76	-0.96
-0.600	93.80	93.50	-0.30	-0.32
1.700	104.70	104.90	0.20	0.19
4.550	120.70	120.58	-0.12	-0.10
6.150	130.20	130.20	0.00	0.00
7.550	138.60	139.13	0.53	0.38
9.550	152.10	152.74	0.64	0.42
10.800	162.00	161.78	-0.22	-0.13
11.000	164.50	163.27	-1.23	-0.75
11.950	170.00	170.49	0.49	0.29
13.250	181.50	180.79	-0.71	-0.39
15.500	199.40	199.80	0.40	0.20
17.600	218.40	218.99	0.59	0.27
18.100	224.40	223.77	-0.63	-0.28
18.650	228.80	229.12	0.32	0.14
21.050	252.80	253.73	0.93	0.37
23.000	274.20	275.26	1.06	0.39
25.450	304.10	304.37	0.27	0.09
27.800	335.00	334.58	-0.42	-0.13
30.050	365.40	365.71	0.31	0.09
32.100	395.30	396.06	0.76	0.19
33.700	421.20	421.12	-0.08	-0.02
34.500	434.20	434.12	-0.08	-0.02
35.650	453.60	453.36	-0.24	-0.05
37.150	479.10	479.48	0.38	0.08
37.850	492.00	492.07	0.07	0.01
38.750	508.00	508.64	0.64	0.13
39.450	520.90	521.83	0.93	0.18
40.050	533.90	533.35	-0.55	-0.10
41.050	554.30	553.01	-1.29	-0.23
41.100	554.30	554.01	-0.29	-0.05
42.000	571.80	572.21	0.41	0.07

(CONTINUED)

	T	P EXpTL.	P CALCD.	DEV.	PERCENT
	42.050	573.30	573.23	-0.07	-0.01
	43.000	592.20	592.98	0.78	0.13
	43.900	612.60	612.19	-0.41	-0.07
	43.900	612.60	612.19	-0.41	-0.07
	44.850	632.10	632.99	0.89	0.14
	44.950	635.10	635.21	0.11	0.02
	45.550	649.50	648.68	-0.82	-0.13
	46.400	668.50	668.13	-0.37	-0.06
	46.650	673.00	673.94	0.94	0.14
	47.300	690.40	689.23	-1.17	-0.17
	48.200	711.80	710.85	-0.95	-0.13
	48.800	725.30	725.55	0.25	0.04
	49.250	737.30	736.74	-0.56	-0.08
	49.700	747.70	748.06	0.36	0.05
	49.850	751.20	751.87	0.67	0.09
	49.850	751.20	751.87	0.67	0.09
*	49.950	756.70	754.41	-2.29	-0.30
	50.300	762.70	763.37	0.67	0.09
	50.300	762.70	763.37	0.67	0.09
	50.400	767.70	765.95	-1.75	-0.23
	50.450	767.70	767.24	-0.46	-0.06
	50.800	776.10	776.32	0.22	0.03
	50.950	780.60	780.23	-0.37	-0.05
	51.300	788.60	789.43	0.83	0.11
	52.000	808.00	808.09	0.09	0.01
	53.000	836.00	835.34	-0.67	-0.08
	54.300	872.30	871.83	-0.47	-0.05
	55.350	903.20	902.21	-0.99	-0.11

STANDARD DEVIATION = 0.62

TRICHLOROETHYLENE C2HCL3

MCDONALD H.J.:

J.PHYS.CHEM. 48,47(1944).

A = 6.51827 B = 1018.603 C = 192.731

B.P.(760) = 87.301

T	P EXPTL.	P CALCD.	DEV.	PERCENT
17.800	46.80	47.87	1.07	2.28
19.850	54.20	53.29	-0.91	-1.67
20.990	56.30	56.52	0.22	0.40
24.400	66.10	67.16	1.06	1.60
25.500	71.60	70.91	-0.69	-0.96
27.480	77.40	78.11	0.71	0.92
30.200	90.00	88.95	-1.05	-1.17
31.200	93.20	93.23	0.03	0.03
34.390	107.20	108.00	0.80	0.75
38.490	129.80	129.70	-0.10	-0.07
39.290	135.20	134.32	-0.88	-0.65
42.410	154.10	153.60	-0.50	-0.32
44.510	168.20	167.78	-0.42	-0.25
46.440	182.40	181.71	-0.69	-0.38
51.050	219.50	218.74	-0.76	-0.35
51.780	225.10	225.11	0.01	0.00
54.820	254.70	253.25	-1.45	-0.57
55.860	264.10	263.49	-0.61	-0.23
57.550	280.60	280.82	0.22	0.08
60.500	314.10	313.22	-0.88	-0.28
63.270	346.50	346.23	-0.27	-0.08
64.940	366.90	367.41	0.51	0.14
65.070	366.70	369.10	2.40	0.66
67.190	394.10	397.54	3.44	0.87
68.050	408.10	409.54	1.44	0.35
71.440	458.40	459.65	1.25	0.27
72.930	482.20	483.11	0.91	0.19
74.270	505.40	505.00	-0.40	-0.08
75.490	524.70	525.59	0.89	0.17
76.930	547.40	550.71	3.31	0.60
77.950	569.90	569.06	-0.84	-0.15
78.260	574.40	574.73	0.33	0.06
79.220	593.10	592.56	-0.54	-0.09
79.280	592.70	593.68	0.98	0.17
80.380	614.80	614.67	-0.13	-0.02
81.270	633.20	632.05	-1.15	-0.18
81.310	631.80	632.84	1.04	0.16
84.220	694.90	692.39	-2.51	-0.36
85.820	727.70	726.90	-0.80	-0.11
86.320	742.00	737.95	-4.05	-0.55
86.470	742.50	741.30	-1.20	-0.16

STANDARD DEVIATION = 1.40

TRICHLOROACETIC ACID C2HCL3O2

MCDONALD R.A.,SHRADER S.A.,STULL D.R.: J.CHEM.ENG.DATA 4,311(1959).

A = 7.27296 B = 1594.314 C = 165.393

B.P.(760) = 197.599

T	P EXPTL.	P CALCD.	DEV.	PERCENT
112.620	34.65	34.54	-0.11	-0.33
118.130	44.62	44.64	0.02	0.05
122.750	54.95	54.94	-0.01	-0.02
129.990	74.99	75.08	0.09	0.12
168.750	317.29	317.39	0.10	0.03
183.920	511.62	511.46	-0.16	-0.03
194.910	706.23	704.74	-1.49	-0.21
196.490	736.37	736.79	0.42	0.06
197.020	746.94	747.81	0.87	0.12
197.480	756.82	757.48	0.66	0.09
197.930	767.45	767.03	-0.42	-0.06

STANDARD DEVIATION = 0.69

PENTACHLOROETHANE C2HCL5

NELSON O.A.: IND.ENG.CHEM.22,971(1930).

A = 6.74011 B = 1378.096 C = 197.367

B.P.(760) = 159.718

T	P EXPTL.	P CALCD.	DEV.	PERCENT
25.100	3.50	3.51	0.01	0.33
* 41.100	13.00	9.14	-3.86	-29.66
50.000	16.00	14.76	-1.24	-7.76
55.000	20.00	19.03	-0.97	-4.85
65.000	33.40	30.73	-2.67	-8.00
69.600	37.40	37.85	0.45	1.20
75.000	49.30	47.91	-1.39	-2.83
80.000	59.80	59.10	-0.70	-1.17
88.200	82.20	82.09	-0.11	-0.14
88.900	88.40	84.35	-4.05	-4.58
95.700	106.60	109.09	2.49	2.33
100.200	125.50	128.50	3.00	2.39
105.000	148.90	152.20	3.30	2.22
110.000	178.30	180.53	2.23	1.25
115.600	213.70	217.16	3.46	1.62
119.900	248.00	249.15	1.15	0.46
124.900	293.80	290.97	-2.83	-0.96
125.600	299.70	297.25	-2.45	-0.82
128.700	329.10	326.35	-2.75	-0.84
132.400	367.90	364.00	-3.90	-1.06
134.600	376.90	387.97	11.07	2.94
139.400	446.70	444.60	-2.10	-0.47
139.600	454.10	447.09	-7.01	-1.54
145.300	530.60	522.91	-7.69	-1.45
151.300	605.00	613.25	8.25	1.36
154.100	653.70	659.36	5.66	0.87
157.300	711.50	715.32	3.82	0.54
158.400	735.40	735.38	-0.02	0.00
159.200	752.30	750.25	-2.05	-0.27
161.800	805.10	800.17	-4.93	-0.61
162.200	811.10	808.07	-3.03	-0.37

STANDARD DEVIATION = 4.29

TRIFLUOROACETIC ACID C2HF3O2

KREGLEWSKI A.:

BULL.ACAD.POL.SCI. 10,629(1962).

A = 6.14776 B = 1228.595 C = 216.089

B.P.(10) = 22.577

T	P EXPTL.	P CALCD.	DEV.	PERCENT
12.000	5.80	5.77	-0.03	-0.49
28.780	13.44	13.50	0.06	0.46
38.680	21.17	21.15	-0.02	-0.08
48.070	31.45	31.39	-0.06	-0.19
55.210	41.64	41.61	-0.03	-0.07
61.100	51.80	51.93	0.13	0.25
65.940	61.92	61.87	-0.05	-0.08
71.650	75.51	75.50	-0.01	-0.02

STANDARD DEVIATION = 0.08

ACETYLENE C2H2

AMBROSE D.:

TRANS.FARADAY SOC. 52,772(1956).

A = 6.57935 B = 536.808 C = 229.819

B.P.(760) = -84.678

T	P EXPTL.	P CALCD.	DEV.	PERCENT
-80.563	961.58	961.13	-0.45	-0.05
-80.515	965.24	963.69	-1.55	-0.16
-80.269	977.20	976.91	-0.29	-0.03
-80.247	978.37	978.10	-0.27	-0.03
-79.947	994.44	994.41	-0.03	0.00
-79.640	1011.20	1011.32	0.12	0.01
-79.369	1027.08	1026.42	-0.66	-0.06
-79.031	1045.26	1045.50	0.24	0.02
-78.672	1065.60	1066.05	0.45	0.04
-78.464	1078.49	1078.10	-0.39	-0.04
-77.822	1114.95	1115.94	0.99	0.09
-77.560	1131.88	1131.66	-0.22	-0.02
-76.654	1187.36	1187.33	-0.03	0.00
-76.614	1189.06	1189.84	0.78	0.07
-75.754	1244.68	1244.65	-0.03	0.00
-75.403	1266.43	1267.55	1.12	0.09
-74.866	1303.08	1303.21	0.13	0.01
-74.206	1346.89	1348.05	1.16	0.09
-73.923	1367.13	1367.63	0.50	0.04
-73.044	1429.37	1429.80	0.43	0.03
-73.040	1429.67	1430.09	0.42	0.03
-73.004	1431.37	1432.68	1.31	0.09
-72.146	1495.10	1495.47	0.37	0.02
-71.000	1582.58	1582.50	-0.08	-0.01
-69.522	1701.74	1700.23	-1.51	-0.09
-68.047	1826.51	1824.07	-2.44	-0.13
* -66.849	1934.44	1929.46	-4.98	-0.26

STANDARD DEVIATION = 0.90

CIS-2-CHLOROVINYLDICHLOROARSINE C2H2ASCL3

WHITING G.H.:

J.CHEM.SOC. 1948,1209.

A = 5.48788 B = 785.094 C = 115.612

B.P.(100) = 109.479

T	P EXPTL.	P CALCD.	DEV.	PERCENT
67.900	16.90	16.21	-0.69	-4.09
72.200	19.90	20.31	0.41	2.05
72.400	20.70	20.52	-0.18	-0.88
76.800	26.20	25.56	-0.64	-2.43
76.800	26.10	25.56	-0.54	-2.06
78.000	27.70	27.10	-0.60	-2.18
80.900	31.30	31.10	-0.20	-0.64
85.300	37.50	38.04	0.54	1.44
85.900	39.10	39.07	-0.03	-0.07
87.100	39.80	41.20	1.40	3.53
91.200	49.10	49.17	0.07	0.15
94.700	56.60	56.87	0.27	0.48
100.100	70.70	70.53	-0.18	-0.25
102.600	76.20	77.63	1.43	1.88
107.100	92.60	91.78	-0.82	-0.89
109.300	100.00	99.36	-0.64	-0.64

STANDARD DEVIATION = 0.75

BETA-CHLOROVINYLDICHLOROARSINE C2H2ASCL3

GOULD C.,HOLZMAN G.,NIEMANN C.:

ANAL.CHEM. 19,204(1947).

A = 2.81045 B = 97.174 C = -27.508

B.P.(10) = 81.182

T	P EXPTL.	P CALCD.	DEV.	PERCENT
66.500	3.00	2.08	-0.92	-30.63
79.200	11.00	8.52	-2.48	-22.52
87.000	14.00	15.03	1.03	7.38
92.200	19.00	20.34	1.34	7.05
98.200	26.00	27.28	1.28	4.92
103.000	32.00	33.36	1.36	4.25
105.300	36.00	36.42	0.42	1.16
110.000	45.00	42.90	-2.10	-4.66

STANDARD DEVIATION = 1.89

TRANS-2-CHLOROVINYLDICHLOROARSINE C2H2ASCL3

MATTHEWS J.B.,SUMNER J.F.,MOELWYN - HUGHES E.A.:
TRANS.FARADAY SOC. 46,797(1950).

A = 6.81403 B = 1465.067 C = 178.526

B.P.(100) = 125.806

T	P EXPTL.	P CALCD.	DEV.	PERCENT
50.000	2.30	2.53	0.23	9.99
55.000	3.40	3.47	0.07	2.07
60.000	4.60	4.70	0.10	2.12
65.000	6.20	6.28	0.08	1.29
70.000	8.10	8.30	0.20	2.45
75.000	10.60	10.85	0.25	2.32
80.000	13.90	14.03	0.13	0.93
85.000	17.90	17.97	0.07	0.39
90.000	22.80	22.81	0.01	0.03
95.000	28.80	28.70	-0.10	-0.36
100.000	36.10	35.81	-0.29	-0.81
105.000	45.10	44.33	-0.77	-1.70
110.000	55.70	54.48	-1.22	-2.18
115.000	67.50	66.49	-1.01	-1.49
120.000	81.00	80.61	-0.39	-0.49
125.000	96.30	97.10	0.80	0.83
130.000	113.70	116.26	2.56	2.25
135.000	136.90	138.41	1.51	1.10
140.000	163.20	163.87	0.67	0.41
145.000	194.10	193.02	-1.08	-0.56
150.000	228.00	226.21	-1.79	-0.78

STANDARD DEVIATION = 1.00

CIS-1,2-DIBROMOETHYLENE C2H2BR2

NOYES R.M.,NOYES W.A.,STEINMETZ H.:
J.AM.CHEM.SOC.72,33(1950).

A = 7.03874 B = 1349.835 C = 209.256

B.P.(100) = 58.635

T	P EXPTL.	P CALCD.	DEV.	PERCENT
26.240	20.80	20.27	-0.53	-2.55
34.640	32.00	31.94	-0.06	-0.20
43.440	50.00	49.77	-0.23	-0.45
52.540	75.30	76.33	1.03	1.37
60.840	110.60	109.93	-0.67	-0.60
64.540	127.30	128.43	1.13	0.89
70.340	164.00	162.53	-1.47	-0.90
71.240	168.00	168.43	0.43	0.26
75.840	202.00	201.41	-0.59	-0.29
77.640	215.00	215.67	0.67	0.31

STANDARD DEVIATION = 0.95

TRANS-1,2-DIBROMOETHYLENE C2H2BR2

NOYES R.M.,NOYES W.A.,STEINMETZ H.:
J.AM.CHEM.SOC.72,33(1950).

A = 4.58111 B = 393.641 C = 103.564

B.P.(100) = 48.944

T	P EXPTL.	P CALCD.	DEV.	PERCENT
3.840	12.20	8.24	-3.96	-32.44
19.740	26.60	24.47	-2.13	-8.00
26.040	31.40	34.98	3.58	11.41
34.340	54.50	53.29	-1.21	-2.22
41.640	74.40	74.16	-0.24	-0.32
49.340	100.00	101.55	1.55	1.55
51.440	105.30	110.04	4.74	4.50
57.740	138.30	138.28	-0.03	-0.02
62.340	160.00	161.59	1.59	0.99
64.940	181.70	175.80	-5.90	-3.25
70.440	208.00	208.38	0.38	0.18

STANDARD DEVIATION = 3.48

)IBROMODICHLOROETHANE C2H2BR2CL2

MULLER K.L.,SCHUMACHER H.-J.: Z.PHYS.CHEM. 42,327(1939).

A = 5.19753 B = 763.442 C = 110.812

B.P.(100) = 127.949

	T	P EXPTL.	P CALCD.	DEV.	PERCENT
	25.000	0.90	0.38	-0.52	-58.13
	33.900	1.60	0.84	-0.76	-47.78
	44.400	2.60	1.90	-0.70	-26.91
	52.500	4.10	3.33	-0.77	-18.72
*	28.500	5.50	0.52	-4.98	-90.51
	64.300	7.60	6.88	-0.72	-9.43
	71.000	10.00	9.96	-0.04	-0.36
	79.600	15.20	15.42	0.22	1.45
	86.800	21.80	21.59	-0.21	-0.98
	90.200	23.90	25.09	1.19	4.99
	91.400	26.60	26.43	-0.17	-0.65
	93.700	29.20	29.14	-0.06	-0.20
	102.000	39.30	40.75	1.45	3.69
	107.800	50.20	50.73	0.53	1.06
	110.400	56.10	55.76	-0.34	-0.60
	112.300	59.80	59.67	-0.13	-0.22
	118.000	72.50	72.61	0.11	0.15
	120.300	80.20	78.38	-1.82	-2.28
	129.700	104.80	105.51	0.71	0.68

STANDARD DEVIATION = 0.83

1,1-DICHLOROETHYLENE C2H2CL2

HILDENBRAND D.L.,MC DONALD R.A.,KRAMER W.R.,STULL D.R.: J.CHEM.PHYS. 30,930(1959).

A = 6.97215 B = 1099.446 C = 237.164

B.P.(760) = 31.561

T	P EXPTL.	P CALCD.	DEV.	PERCENT
-28.360	50.95	50.90	-0.05	-0.10
-17.720	91.54	91.62	0.08	0.09
-2.960	189.61	189.56	-0.05	-0.02
13.750	389.38	389.39	0.01	0.00
30.390	729.37	729.29	-0.08	-0.01
31.560	760.00	759.96	-0.04	0.00
31.730	764.95	764.50	-0.45	-0.06
32.500	784.76	785.33	0.57	0.07

STANDARD DEVIATION = 0.33

CIS 1,2-DICHLOROETHYLENE C2H2CL2

KETELAAR J.A.A.,VAN VELDEN P.F.,ZALM J.S.:
REC.TRAV.CHIM. 66,721(1947).

A = 7.02230 B = 1205.436 C = 230.620

B.P.(760) = 60.443

	T	P EXPTL.	P CALCD.	DEV.	PERCENT
	0.760	64.70	64.94	0.24	0.38
	0.810	64.80	65.11	0.31	0.48
	1.680	68.00	68.10	0.10	0.15
	1.690	67.70	68.14	0.44	0.65
	3.850	76.40	76.07	-0.33	-0.43
	4.800	80.20	79.79	-0.41	-0.51
	6.450	86.80	86.61	-0.19	-0.22
	8.160	94.40	94.19	-0.21	-0.22
	8.170	94.50	94.23	-0.27	-0.28
	10.730	106.30	106.60	0.30	0.28
	13.120	120.40	119.32	-1.08	-0.89
	15.030	130.70	130.37	-0.33	-0.25
	16.530	139.20	139.63	0.43	0.31
	18.510	152.50	152.66	0.16	0.11
	20.190	164.00	164.49	0.49	0.30
	22.900	184.90	185.15	0.25	0.13
	23.740	191.80	191.96	0.16	0.08
	26.630	217.20	217.00	-0.20	-0.09
	26.730	217.90	217.91	0.01	0.00
	30.060	250.40	250.10	-0.30	-0.12
	30.110	250.10	250.61	0.51	0.20
	31.410	264.40	264.20	-0.20	-0.08
	33.590	287.60	288.33	0.73	0.25
	36.350	321.20	321.41	0.21	0.06
	37.530	335.40	336.46	1.06	0.31
	40.020	369.90	370.07	0.17	0.05
	41.510	391.50	391.45	-0.05	-0.01
	43.460	421.10	420.91	-0.19	-0.05
	45.520	455.50	453.94	-1.56	-0.34
	45.670	457.00	456.42	-0.58	-0.13
	45.690	456.80	456.75	-0.05	-0.01
*	47.660	508.20	490.42	-17.78	-3.50
	50.340	539.90	539.37	-0.53	-0.10
	52.040	572.90	572.39	-0.51	-0.09
	52.060	572.90	572.79	-0.11	-0.02
	54.450	622.40	621.93	-0.47	-0.08
	55.280	639.00	639.77	0.77	0.12
	58.790	718.60	719.71	1.11	0.15
	59.370	734.60	733.64	-0.96	-0.13
*	59.970	742.50	748.29	5.79	0.78
	62.850	822.30	821.82	-0.48	-0.06
	63.460	837.10	838.10	1.00	0.12
	66.560	923.70	924.82	1.12	0.12
	71.580	1079.10	1080.03	0.93	0.09
	71.580	1079.90	1080.03	0.13	0.01
	71.840	1088.30	1088.59	0.29	0.03
	76.760	1260.40	1260.83	0.43	0.03
	77.400	1286.40	1284.71	-1.69	-0.13
	79.020	1347.50	1346.73	-0.77	-0.06
*	83.630	1529.60	1535.99	6.39	0.42

TRANS 1,2-DICHLOROETHYLENE C2H2CL2

KETELAAR J.A.A.,VAN VELDEN P.F.,ZALM J.S.:
REC.TRAV.CHIM. 66,721(1947).

A = 6.96513 B = 1141.984 C = 231.930

B.P.(760) = 47.673

	T	P EXPTL.	P CALCD.	DEV.	PERCENT
*	-38.190	10.40	11.77	1.37	13.15
	-30.140	20.40	20.22	-0.18	-0.87
	-28.300	22.50	22.75	0.25	1.11
*	-19.190	41.20	39.55	-1.65	-4.01
	-15.120	50.50	49.88	-0.62	-1.23
	-7.290	75.70	76.12	0.42	0.56
	0.520	113.00	112.80	-0.20	-0.18
	1.100	115.20	116.02	0.82	0.71
	3.100	128.10	127.71	-0.39	-0.30
	4.290	135.20	135.12	-0.08	-0.06
	5.980	145.80	146.24	0.44	0.30
	7.900	159.90	159.77	-0.13	-0.08
	9.510	172.00	171.88	-0.12	-0.07
	11.120	185.40	184.74	-0.66	-0.35
	13.040	200.50	201.09	0.59	0.29
	14.720	216.60	216.34	-0.26	-0.12
	16.910	237.40	237.63	0.23	0.10
	19.000	259.10	259.49	0.39	0.15
	19.060	260.70	260.14	-0.56	-0.22
	22.380	298.00	298.27	0.27	0.09
	25.320	334.90	335.68	0.78	0.23
	26.510	352.30	351.85	-0.45	-0.13
	28.230	377.50	376.34	-1.16	-0.31
	32.150	438.40	437.25	-1.15	-0.26
	35.840	501.70	501.57	-0.13	-0.03
	39.520	571.90	572.99	1.09	0.19
	43.720	662.10	664.12	2.02	0.30
	47.170	745.30	747.23	1.93	0.26
	48.830	792.00	790.04	-1.96	-0.25
	50.820	845.60	843.87	-1.73	-0.20
	54.520	950.60	951.58	0.98	0.10
	56.120	1000.50	1001.35	0.85	0.09
	59.870	1125.30	1126.00	0.70	0.06
	63.580	1258.70	1260.88	2.18	0.17
	67.800	1430.10	1429.16	-0.94	-0.07
	68.190	1447.00	1445.54	-1.46	-0.10
	73.760	1700.00	1695.76	-4.24	-0.25
	77.160	1864.80	1864.05	-0.75	-0.04
	78.330	1925.80	1924.83	-0.97	-0.05
	80.780	2059.70	2056.98	-2.72	-0.13
	82.420	2146.70	2149.22	2.52	0.12
	84.020	2238.00	2242.22	4.22	0.19
*	84.840	2284.40	2291.06	6.66	0.29

STANDARD DEVIATION = 1.50

CHLOROACETYL CHLORIDE C2H2CL2O

MCDONALD R.A.,SHRADER S.A.,STULL D.R.: J.CHEM.ENG.DATA 4,311(1959).

A = 7.14977 B = 1340.785 C = 208.102

B.P.(760) = 105.975

T	P EXPTL.	P CALCD.	DEV.	PERCENT
28.350	30.08	30.12	0.04	0.14
38.130	50.67	50.65	-0.02	-0.04
50.170	90.87	90.87	0.00	0.00
67.080	189.50	189.42	-0.08	-0.04
85.970	389.20	389.40	0.20	0.05
104.660	729.60	729.22	-0.38	-0.05
105.260	742.90	743.13	0.23	0.03
106.220	766.10	765.83	-0.27	-0.04
107.070	786.10	786.39	0.29	0.04

STANDARD DEVIATION = 0.26

1,1,1,2-TETRACHLOROETHANE C2H2CL4

DREISBACH R.R.,SHRADER S.A.: IND.ENG.CHEM. 41,2879(1949).

A = 6.89875 B = 1365.876 C = 209.744

B.P.(760) = 130.201

T	P EXPTL.	P CALCD.	DEV.	PERCENT
59.310	66.39	66.40	0.01	0.01
62.390	75.86	75.79	-0.07	-0.09
74.470	123.76	123.86	0.10	0.08
100.690	315.52	315.39	-0.13	-0.04
115.990	507.50	507.60	0.10	0.02
130.200	760.00	759.98	-0.02	0.00

STANDARD DEVIATION = 0.12

1,1,2,2-TETRACHLOROETHANE C2H2CL4

MATTHEWS J.B.,SUMNER J.F.,MOELWYN-HUGHES E.A.:
TRANS.FARADAY SOC. 46,797(1950).

A = 6.63168 B = 1228.062 C = 179.942

B.P.(100) = 85.202

	T	P EXPTL.	P CALCD.	DEV.	PERCENT
	25.000	4.20	4.36	0.16	3.80
*	30.000	5.00	6.06	1.06	21.11
	35.000	8.10	8.28	0.18	2.27
	40.000	11.20	11.17	-0.03	-0.25
	45.000	15.00	14.87	-0.13	-0.89
	50.000	20.10	19.54	-0.56	-2.78
	55.000	26.10	25.39	-0.71	-2.73
	60.000	33.30	32.62	-0.68	-2.03
	65.000	42.20	41.50	-0.70	-1.67
	70.000	52.00	52.28	0.28	0.53
	75.000	64.00	65.26	1.26	1.97
	80.000	80.00	80.78	0.78	0.98
	85.000	98.00	99.19	1.19	1.22
	90.000	120.60	120.87	0.27	0.23
	95.000	146.50	146.24	-0.26	-0.18
	100.000	177.50	175.73	-1.77	-1.00
	105.000	212.50	209.81	-2.69	-1.27
	110.000	248.50	248.97	0.47	0.19
	115.000	292.00	293.73	1.73	0.59
	120.000	342.50	344.63	2.13	0.62
	125.000	400.00	402.24	2.24	0.56
	130.000	471.00	467.15	-3.85	-0.82

STANDARD DEVIATION = 1.56

CIS-1,2-DIIODOETHYLENE $C_2H_2I_2$

NOYES M.R.,NOYES A.W.,STEINMETZ H.: J.AM.CHEM.SOC. 72,33(1950).

A = 5.52174 B = 797.816 C = 106.369

B.P.(100) = 120.171

T	P EXPTL.	P CALCD.	DEV.	PERCENT
28.950	1.30	0.42	-0.88	-67.49
26.150	1.60	0.32	-1.28	-80.17
38.250	4.20	1.01	-3.19	-75.91
41.050	4.70	1.29	-3.41	-72.60
44.050	6.60	1.65	-4.95	-74.98
61.350	6.90	5.82	-1.08	-15.66
69.550	11.00	9.70	-1.30	-11.85
74.250	16.10	12.72	-3.38	-20.97
85.450	24.40	23.04	-1.36	-5.56
88.550	25.50	26.83	1.33	5.23
94.050	33.00	34.75	1.75	5.31
98.650	44.40	42.69	-1.71	-3.85
105.450	52.30	56.92	4.62	8.83
111.850	70.30	73.40	3.10	4.41
119.750	99.90	98.50	-1.40	-1.40
132.350	147.00	151.24	4.24	2.88
132.850	148.50	153.69	5.19	3.50
136.850	175.60	174.38	-1.22	-0.69
144.850	214.00	221.80	7.80	3.64
151.850	286.00	270.43	-15.57	-5.45

STANDARD DEVIATION = 5.18

TRANS 1,2-DIIODOETHYLENE $C_2H_2I_2$

NOYES R.M.,NOYES W.A.,STEINMETZ H.: J.AM.CHEM.SOC. 72,33(1950).

A = 6.09311 B = 1196.998 C = 172.289

B.P.(100) = 120.154

T	P EXPTL.	P CALCD.	DEV.	PERCENT
77.240	20.80	19.77	-1.03	-4.94
89.640	32.90	33.36	0.46	1.39
102.040	52.70	53.67	0.97	1.84
105.940	62.90	61.79	-1.11	-1.77
112.440	76.90	77.47	0.57	0.74
118.540	94.60	94.90	0.30	0.32
130.040	136.50	136.10	-0.40	-0.30

STANDARD DEVIATION = 1.00

KETENE C_2H_2O

REUBEN B.G.:
J.CHEM.ENG.DATA 14,235(1969).

A = 7.61480 B = 1036.407 C = 268.744

B.P.(760) = -49.815

T	P EXPTL.	P CALCD.	DEV.	PERCENT
-87.800	79.50	77.09	-2.41	-3.03
-72.550	207.50	214.90	7.40	3.56
-68.500	275.30	274.83	-0.47	-0.17
-63.800	358.90	361.21	2.31	0.64
-62.400	393.00	390.91	-2.09	-0.53
-58.050	500.60	496.33	-4.27	-0.85
-55.350	577.00	572.81	-4.19	-0.73
-53.100	646.80	643.71	-3.09	-0.48
-50.150	745.40	747.39	1.99	0.27
-49.350	774.30	777.74	3.44	0.44

STANDARD DEVIATION = 4.34

BROMOETHYLENE C_2H_3BR

GUYER A.,SCHUTZE H.,WEIDENMANN M.:
HELV.CHIM.ACTA 20,936(1937).

A = 6.99738 B = 1099.886 C = 251.631

B.P.(760) = 15.554

T	P EXPTL.	P CALCD.	DEV.	PERCENT
-87.500	2.20	1.98	-0.22	-10.11
-66.000	12.50	11.81	-0.69	-5.52
-53.000	27.00	28.84	1.84	6.83
-39.000	69.00	66.78	-2.22	-3.22
-27.700	124.00	121.81	-2.20	-1.77
-19.600	177.00	180.77	3.77	2.13
-13.600	237.80	238.02	0.22	0.09
-8.600	292.20	296.26	4.07	1.39
-4.500	357.20	352.18	-5.02	-1.41
-0.900	408.80	408.01	-0.79	-0.19
2.300	460.90	463.39	2.49	0.54
3.600	485.00	487.54	2.54	0.52
7.600	568.30	568.20	-0.10	-0.02
10.900	642.80	642.44	-0.36	-0.06
16.000	776.60	772.09	-4.51	-0.58

STANDARD DEVIATION = 2.94

ACETYL BROMIDE C2H3BRO

DEVORE J.A.,O'NEAL H.E.:
J.PHYS.CHEM. 73,2644(1969).

A = 5.19702 B = 545.784 C = 150.396

B.P.(100) = 20.320

T	P EXPTL.	P CALCD.	DEV.	PERCENT
2.400	44.50	42.17	-2.33	-5.23
9.690	62.50	61.34	-1.16	-1.86
16.060	81.30	82.83	1.53	1.88
20.970	102.90	102.83	-0.07	-0.07
24.640	120.10	119.92	-0.18	-0.15
28.820	141.20	141.78	0.58	0.41
32.950	159.50	166.05	6.55	4.10
37.380	198.50	195.19	-3.31	-1.67
40.480	218.20	217.60	-0.60	-0.27
45.380	256.70	256.58	-0.12	-0.05
51.800	314.90	314.59	-0.31	-0.10
55.690	357.30	353.75	-3.55	-0.99
60.500	404.40	406.53	2.13	0.53

STANDARD DEVIATION = 2.84

CHLOROETHYLENE C2H3CL

MCDONALD R.A.,SHRADER S.A.,STULL D.R.:
J.CHEM.ENG.DATA 4,311(1959).

A = 6.89117 B = 905.008 C = 239.475

B.P.(760) = -13.807

T	P EXPTL.	P CALCD.	DEV.	PERCENT
-64.900	50.77	50.94	0.17	0.34
-56.040	91.11	90.67	-0.44	-0.48
-43.370	188.37	188.91	0.54	0.28
-29.090	388.92	388.58	-0.34	-0.09
-14.830	728.55	728.70	0.15	0.02
-13.640	765.40	765.20	-0.20	-0.03
-13.000	785.30	785.41	0.11	0.01

STANDARD DEVIATION = 0.42

ACETYL CHLORIDE C_2H_3ClO

MCDONALD R.A.,SHRADER S.A.,STULL D.R.:
J.CHEM.ENG.DATA 4,311(1959).

A = 6.94887 B = 1115.954 C = 223.554

B.P.(760) = 50.768

T	P EXPTL.	P CALCD.	DEV.	PERCENT
-6.290	64.23	64.93	0.70	1.09
-3.730	74.99	74.52	-0.47	-0.63
6.170	123.89	123.33	-0.56	-0.45
27.410	317.29	317.83	0.54	0.17
39.680	511.62	512.22	0.60	0.12
48.660	706.23	706.83	0.60	0.08
49.830	736.37	735.96	-0.41	-0.06
50.230	746.94	746.14	-0.80	-0.11
50.640	756.82	756.69	-0.13	-0.02
51.050	767.45	767.35	-0.10	-0.01

STANDARD DEVIATION = 0.64

CHLOROACETIC ACID $C_2H_3ClO_2$

MCDONALD R.A.,SHRADER S.A.,STULL D.R.:
J.CHEM.ENG.DATA 4,311(1959).

A = 7.55016 B = 1723.365 C = 179.978

B.P.(760) = 189.103

T	P EXPTL.	P CALCD.	DEV.	PERCENT
104.470	31.06	31.01	-0.05	-0.16
114.600	50.07	50.10	0.03	0.06
128.220	90.78	90.87	0.09	0.09
146.780	189.00	188.81	-0.19	-0.10
167.510	389.40	389.63	0.23	0.06
187.690	729.10	729.22	0.12	0.02
189.350	766.30	765.47	-0.83	-0.11
190.270	785.60	786.19	0.59	0.07

STANDARD DEVIATION = 0.48

1,1,1-TRICHLOROETHANE C2H3CL3

RUBIN T.R.,LEVEDAHL B.H.,YOST D.M.:
J.AM.CHEM.SOC. 66,279(1944).

A = 8.64344 B = 2136.621 C = 302.769

B.P.(10) = -23.233

T	P EXPTL.	P CALCD.	DEV.	PERCENT
-5.364	28.80	28.79	-0.01	-0.04
5.109	50.50	50.54	0.04	0.07
11.304	69.30	69.26	-0.04	-0.05
16.923	91.20	91.22	0.02	0.02

STANDARD DEVIATION = 0.06

1,1,2-TRICHLOROETHANE C2H3CL3

DREISBACH R.R.,SHRADER S.A.:
IND.ENG.CHEM.41,2879(1949).

A = 6.95185 B = 1314.410 C = 209.197

B.P.(760) = 113.671

T	P EXPTL.	P CALCD.	DEV.	PERCENT
49.970	75.86	75.89	0.03	0.04
61.290	123.76	123.72	-0.04	-0.04
85.990	315.52	315.53	0.01	0.00
100.340	507.50	507.54	0.04	0.01
113.670	760.00	759.96	-0.04	-0.01

STANDARD DEVIATION = 0.06

VINYLTRICHLOROSILANE C2H3CL3SI

JENKINS A.C.,CHAMBERS G.F.: IND.ENG.CHEM. 46,2367(1954).

A = 7.14095 B = 1401.922 C = 238.406

B.P.(100) = 34.291

T	P EXPTL.	P CALCD.	DEV.	PERCENT
17.700	46.30	46.45	0.15	0.32
28.600	78.00	77.70	-0.30	-0.38
44.800	154.90	155.16	0.26	0.17
68.500	374.30	374.14	-0.16	-0.04
82.900	599.40	599.45	0.05	0.01

STANDARD DEVIATION = 0.32

1,2-DICHLOROETHYLTRICHLOROSILANE C2H3CL5SI

JENKINS A.C.,CHAMBERS G.F.: IND.ENG.CHEM. 46,2367(1954).

A = 7.82600 B = 2144.973 C = 253.082

B.P.(760) = 180.667

T	P EXPTL.	P CALCD.	DEV.	PERCENT
102.200	61.20	61.46	0.26	0.43
128.000	158.10	157.53	-0.57	-0.36
147.300	295.30	294.23	-1.07	-0.36
148.400	303.40	304.34	0.94	0.31
164.000	480.40	482.16	1.76	0.37
180.500	758.00	756.66	-1.34	-0.18

STANDARD DEVIATION = 1.56

1,1,1-TRIFLUOROETHANE C2H3F3

RUSSELL H.,JR.,GOLDING D.R.W.,YOST D.M.:
J.AM.CHEM.SOC. 66,16(1944).

A = 6.90378 B = 788.204 C = 243.229

B.P.(760) = -47.303

T	P EXPTL.	P CALCD.	DEV.	PERCENT
-99.546	26.17	26.19	0.02	0.06
-87.303	70.62	70.60	-0.02	-0.03
-77.730	138.41	138.41	0.00	0.00
-70.947	213.14	213.15	0.01	0.01
-64.321	314.91	314.87	-0.04	-0.01
-58.945	423.26	423.31	0.05	0.01
-54.315	538.79	538.87	0.08	0.01
-50.618	648.04	647.99	-0.05	-0.01
-47.370	757.62	757.58	-0.04	-0.01

STANDARD DEVIATION = 0.05

2,2,2-TRIFLUOROETHANOL C2H3F3O

MEEKS A.C.,GOLDFARB I.J.:
J.CHEM.ENG.DATA 12,196(1967).

A = 6.78815 B = 978.129 C = 173.056

B.P.(10) = -4.067

T	P EXPTL.	P CALCD.	DEV.	PERCENT
-0.400	13.55	13.27	-0.28	-2.05
0.300	13.80	13.99	0.19	1.38
2.250	16.05	16.17	0.12	0.72
4.100	18.55	18.49	-0.06	-0.34
5.850	21.00	20.93	-0.07	-0.31
7.550	23.55	23.57	0.02	0.07
8.750	25.60	25.59	-0.01	-0.04
9.300	26.50	26.56	0.06	0.24
11.100	29.90	29.97	0.07	0.24
12.400	32.85	32.66	-0.20	-0.59
15.000	38.65	38.63	-0.03	-0.06
16.800	43.15	43.27	0.12	0.28
19.800	52.05	52.04	-0.01	-0.02
20.600	54.40	54.61	0.21	0.39
23.100	63.35	63.34	-0.01	-0.02
25.400	72.50	72.35	-0.15	-0.21

STANDARD DEVIATION = 0.14

ACETYL IODIDE C2H3IO

DEVORE J.A.,O'NEAL H.E.:
J.PHYS.CHEM. 73,2644(1969).

A = 4.18144 B = 355.452 C = 108.160

B.P.(10) = 3.567

T	P EXPTL.	P CALCD.	DEV.	PERCENT
3.480	9.90	9.94	0.04	0.44
6.890	12.70	12.36	-0.34	-2.70
12.880	17.30	17.57	0.27	1.57
16.160	20.90	21.00	0.10	0.49
22.170	28.50	28.45	-0.05	-0.17
28.200	37.60	37.56	-0.04	-0.11

STANDARD DEVIATION = 0.26

METHYL ISOTHIOCYANATE C2H3NS

BAUER H.,BURSCHKIES K.:
BER. 68,1243(1935).

A = 2.89679 B = 103.627 C = 45.370

B.P.(10) = 9.263

T	P EXPTL.	P CALCD.	DEV.	PERCENT
10.000	12.50	10.60	-1.90	-15.21
12.000	13.60	12.32	-1.28	-9.43
14.000	14.20	14.17	-0.03	-0.22
16.000	16.80	16.15	-0.65	-3.86
18.000	18.70	18.26	-0.44	-2.35
20.000	20.70	20.49	-0.21	-1.01
22.000	22.80	22.84	0.04	0.16
24.000	25.00	25.29	0.29	1.16
26.000	27.50	27.85	0.35	1.27
28.000	30.10	30.51	0.41	1.35
30.000	33.10	33.26	0.16	0.47
32.000	36.20	36.09	-0.11	-0.30
34.000	39.20	39.01	-0.19	-0.49
35.000	41.40	40.50	-0.90	-2.18
37.000	42.00	43.52	1.52	3.63
40.000	45.00	48.19	3.19	7.08
42.000	48.60	51.37	2.77	5.70
44.000	52.80	54.61	1.81	3.43
46.000	57.50	57.90	0.40	0.69
48.000	62.80	61.23	-1.57	-2.51
50.000	68.50	64.60	-3.90	-5.70

STANDARD DEVIATION = 1.64

ETHYLENE C_2H_4

EGAN C.J.,KEMP J.D.: J.AM.CHEM.SOC. 59,1265(1937).

A = 6.76809 B = 590.952 C = 255.777

B.P.(760) = -103.754

T	P EXPTL.	P CALCD.	DEV.	PERCENT
-149.690	15.78	15.76	-0.02	-0.11
-144.605	28.38	28.34	-0.04	-0.14
-137.232	60.67	60.68	0.01	0.01
-132.215	96.56	96.70	0.14	0.15
-127.611	143.65	143.63	-0.02	-0.02
-122.148	221.70	221.69	-0.01	-0.01
-116.990	323.74	323.66	-0.08	-0.02
-111.530	469.37	469.10	-0.27	-0.06
-108.036	586.44	586.35	-0.09	-0.02
-105.398	689.00	689.16	0.16	0.02
-102.801	803.37	803.59	0.22	0.03

STANDARD DEVIATION = 0.15

ETHYLENE C_2H_4

MICHELS A.,WASSENNAR T.: PHYSICA 16,221(1950).

A = 6.74819 B = 584.291 C = 254.862

B.P.(760) = -103.780

T	P EXPTL.	P CALCD.	DEV.	PERCENT
-123.783	195.24	195.27	0.03	0.01
-117.037	322.83	322.71	-0.12	-0.04
-113.407	414.51	414.57	0.06	0.02
-109.549	533.41	533.66	0.25	0.05
-106.531	644.26	644.28	0.02	0.00
-104.112	745.25	745.22	-0.03	0.00
-104.113	745.32	745.18	-0.14	-0.02
-104.106	745.39	745.49	0.10	0.01
-104.109	745.39	745.35	-0.04	0.00
-103.437	775.44	775.47	0.03	0.00
-103.441	775.44	775.28	-0.16	-0.02
-103.443	775.47	775.19	-0.28	-0.04
-100.704	907.65	907.76	0.11	0.01
-93.376	1348.82	1348.85	0.03	0.00
-89.182	1664.83	1665.55	0.72	0.04
-84.581	2074.73	2074.18	-0.55	-0.03

STANDARD DEVIATION = 0.28

1,2-DIBROMOETHANE C2H4BR2

DREISBACH R.R.,SHRADER S.A.: IND.ENG.CHEM. 41,2879(1949).

A = 6.72148 B = 1280.820 C = 201.751

B.P.(760) = 131.737

T	P EXPTL.	P CALCD.	DEV.	PERCENT
52.560	57.04	48.42	-8.62	-15.11
60.430	66.39	68.59	2.20	3.31
63.590	75.86	78.42	2.56	3.37
75.640	123.76	127.09	3.33	2.69
101.940	315.52	319.13	3.61	1.15
117.210	507.50	508.02	0.52	0.10
131.410	760.00	753.42	-6.58	-0.87

STANDARD DEVIATION = 6.19

1,1-DICHLOROETHANE C2H4CL2

LI J.C.M.,PITZER K.S.: J.AM.CHEM.SOC. 78,1077(1956).

A = 6.97702 B = 1174.022 C = 229.060

B.P.(100) = 6.829

T	P EXPTL.	P CALCD.	DEV.	PERCENT
-38.770	6.44	6.42	-0.02	-0.35
-29.600	12.37	12.33	-0.04	-0.31
-20.200	22.65	22.69	0.04	0.20
-19.170	24.26	24.18	-0.08	-0.32
-18.850	24.76	24.66	-0.10	-0.40
-17.680	26.38	26.48	0.10	0.38
-16.810	27.87	27.91	0.04	0.13
-12.600	35.79	35.75	-0.04	-0.11
-12.190	36.52	36.60	0.08	0.23
-2.220	63.19	63.31	0.12	0.19
1.170	75.58	75.45	-0.13	-0.17
11.090	122.51	122.55	0.04	0.03
11.330	124.03	123.93	-0.10	-0.08
17.490	164.08	164.14	0.06	0.04
17.610	165.02	165.02	0.00	0.00

STANDARD DEVIATION = 0.08

1,2-DICHLOROETHANE $C_2H_4Cl_2$

PEARCE J.N.,PETERS P.E.:
J.PHYS.CHEM. 33,873(1929).

A = 7.02530 B = 1271.254 C = 222.927

B.P.(760) = 83.807

	T	P EXPTL.	P CALCD.	DEV.	PERCENT
*	-30.820	3.20	2.56	-0.64	-20.07
	-24.920	4.60	4.03	-0.57	-12.44
	-19.320	6.60	6.05	-0.55	-8.36
	-15.020	8.20	8.14	-0.06	-0.70
	-10.420	11.60	11.04	-0.56	-4.79
	-5.120	15.90	15.44	-0.46	-2.88
	0.070	21.00	21.11	0.11	0.54
	5.090	28.20	28.19	-0.01	-0.05
	10.070	36.70	37.09	0.39	1.05
	15.520	49.10	49.42	0.32	0.65
	20.210	62.10	62.63	0.53	0.85
	25.250	79.80	79.97	0.17	0.22
	30.230	100.90	100.86	-0.04	-0.04
	35.240	126.30	126.23	-0.07	-0.06
	39.790	153.20	153.61	0.41	0.27
	45.130	191.60	191.79	0.19	0.10
	50.120	234.80	234.15	-0.65	-0.28
	55.220	285.00	285.01	0.01	0.00
	60.280	344.80	343.97	-0.83	-0.24
	65.270	412.80	411.38	-1.42	-0.34
	70.270	490.90	489.18	-1.72	-0.35
	75.360	579.20	580.03	0.83	0.14
	80.430	682.90	683.41	0.51	0.07
	85.450	794.60	799.65	5.05	0.64
*	91.030	934.60	946.60	12.00	1.28
	96.030	1095.60	1095.57	-0.03	0.00
	99.400	1208.60	1205.90	-2.70	-0.22

STANDARD DEVIATION = 1.38

BETA BETA'-DICHLOROETHYL SULFIDE C2H4CL2S

BALSON E.W.,DENBIGH K.G.,ADAM N.K.:
TRANS.FARADAY SOC. 43,42(1948).

A = 8.58741 B = 2588.225 C = 246.055

B.P.(10) = 95.066

T	P EXPTL.	P CALCD.	DEV.	PERCENT
14.900	0.05	0.05	0.00	1.48
19.700	0.07	0.07	0.00	0.73
25.000	0.11	0.11	0.00	3.13
30.100	0.16	0.16	0.00	0.66
35.200	0.24	0.24	0.00	1.53
41.800	0.39	0.39	0.00	0.63
47.600	0.61	0.59	-0.02	-2.83
68.400	2.26	2.27	0.01	0.57
75.800	3.52	3.51	-0.01	-0.16

STANDARD DEVIATION = 0.01

BISTRICHLOROSILYLETHANE C2H4CL6SI2

JENKINS A.C.,CHAMBERS G.F.:
IND.ENG.CHEM. 46,2367(1954).

A = 7.83511 B = 2241.769 C = 249.837

B.P.(100) = 134.349

T	P EXPTL.	P CALCD.	DEV.	PERCENT
91.200	18.50	18.27	-0.23	-1.25
114.700	48.20	48.47	0.27	0.56
147.100	154.20	153.97	-0.23	-0.15
159.900	231.00	231.14	0.14	0.06

STANDARD DEVIATION = 0.44

ETHYLENE OXYDE C2H4O

GIAUQUE W.F.,GORDON J.: J.AM.CHEM.SOC.71,2176(1949).

A = 7.12843 B = 1054.542 C = 237.762

B.P.(760) = 10.504

T	P EXPTL.	P CALCD.	DEV.	PERCENT
-49.372	33.94	33.95	0.01	0.02
-43.311	50.70	50.73	0.03	0.06
-38.242	69.68	69.67	-0.01	-0.01
-33.273	93.73	93.64	-0.09	-0.09
-28.169	124.93	125.04	0.11	0.09
-21.847	175.58	175.54	-0.04	-0.02
-16.956	225.28	225.20	-0.08	-0.04
-14.825	250.14	250.16	0.02	0.01
-9.828	317.68	317.63	-0.05	-0.02
-4.904	397.77	397.88	0.11	0.03
0.194	497.41	497.47	0.06	0.01
5.644	625.07	625.17	0.10	0.02
9.210	721.95	722.03	0.08	0.01
11.934	804.01	803.78	-0.23	-0.03

STANDARD DEVIATION = 0.10

ETHYLENE OXIDE C2H4O

COLES K.F.,POPPER F.:
IND.ENG.CHEM. 42,1434(1950).

A = 8.69016 B = 2005.779 C = 334.765

B.P.(760) = 10.502

	T	P EXPTL.	P CALCD.	DEV.	PERCENT
	0.300	506.00	505.73	-0.27	-0.05
	0.400	508.00	507.82	-0.18	-0.04
	1.100	522.00	522.61	0.61	0.12
	1.200	526.00	524.76	-1.24	-0.24
	10.500	760.00	759.93	-0.07	-0.01
	10.700	765.00	765.84	0.84	0.11
	11.100	775.00	777.77	2.77	0.36
	12.550	823.00	822.36	-0.64	-0.08
	15.700	928.00	926.76	-1.24	-0.13
	16.050	939.00	939.03	0.03	0.00
	16.600	958.00	958.58	0.58	0.06
	16.750	964.00	963.98	-0.02	0.00
	21.600	1156.00	1152.71	-3.29	-0.28
	24.900	1294.00	1298.27	4.27	0.33
*	29.000	1518.00	1500.45	-17.55	-1.16
	30.350	1574.00	1572.56	-1.44	-0.09
	31.800	1654.00	1653.24	-0.76	-0.05

STANDARD DEVIATION = 1.85

ACETALDEHYDE C2H4O

COLES K.F., POPPER F.:
IND.ENG.CHEM. 42,1434(1950).

A = 8.00552 B = 1600.017 C = 291.809

B.P.(760) = 20.407

T	P EXPTL.	P CALCD.	DEV.	PERCENT
-0.200	332.00	330.11	-1.89	-0.57
2.700	375.00	373.85	-1.15	-0.31
6.700	443.00	442.07	-0.93	-0.21
9.300	494.00	491.78	-2.22	-0.45
11.600	531.00	539.58	8.58	1.62
13.300	577.00	577.35	0.35	0.06
17.600	682.00	682.84	0.84	0.12
20.700	766.00	768.45	2.45	0.32
30.800	1120.00	1111.48	-8.52	-0.76
34.400	1259.00	1260.78	1.78	0.14

STANDARD DEVIATION = 4.88

ACETIC ACID C2H4O2

POTTER A.E., RITTER H.L.:
J.PHYS.CHEM. 58,1040(1954).

A = 7.38782 B = 1533.313 C = 222.309

B.P.(760) = 117.897

T	P EXPTL.	P CALCD.	DEV.	PERCENT
29.800	20.00	20.22	0.22	1.12
40.250	35.20	35.31	0.11	0.32
51.080	60.10	60.16	0.06	0.09
55.540	74.10	74.01	-0.09	-0.12
60.930	94.50	94.26	-0.24	-0.25
65.500	115.00	114.89	-0.11	-0.09
71.040	145.10	144.84	-0.26	-0.18
75.500	173.30	173.45	0.15	0.09
80.410	210.30	210.23	-0.07	-0.04
85.690	256.70	256.75	0.05	0.02
90.590	307.00	307.24	0.24	0.08
95.660	367.40	367.80	0.40	0.11
100.290	431.20	431.34	0.14	0.03
105.450	512.40	512.45	0.05	0.01
110.000	594.00	593.89	-0.11	-0.02
115.120	697.50	697.78	0.28	0.04
118.410	772.40	771.96	-0.44	-0.06
122.440	871.10	871.37	0.27	0.03
123.860	909.50	908.75	-0.75	-0.08
126.450	980.10	980.26	0.16	0.02

STANDARD DEVIATION = 0.29

METHYL FORMATE C2H4O2

NELSON O.A.: IND.ENG.CHEM.20,1382(1928).

A = 3.02742 B = 3.018 C = -11.880

B.P.(760) = 32.466

T	P EXPTL.	P CALCD.	DEV.	PERCENT
21.000	525.80	497.17	-28.63	-5.45
25.000	624.50	627.18	2.68	0.43
26.600	647.20	664.35	17.15	2.65
26.800	654.70	668.56	13.86	2.12
27.100	663.00	674.73	11.73	1.77
28.300	678.10	697.62	19.52	2.88
28.900	711.40	708.11	-3.29	-0.46
30.100	722.90	727.41	4.51	0.62
31.400	760.80	746.12	-14.68	-1.93
31.700	767.70	750.15	-17.55	-2.29

STANDARD DEVIATION =18.46

THIACYCLOPROPANE C2H4S

GUTHRIE G.B.,SCOTT D.W.JR.,WADDINGTON G.:
J.AM.CHEM.SOC. 74,2795(1952).

A = 7.03379 B = 1192.361 C = 232.178

B.P.(760) = 54.932

	T	P EXPTL.	P CALCD.	DEV.	PERCENT
	18.290	187.58	187.61	0.03	0.02
	23.410	233.72	233.68	-0.04	-0.02
	28.580	289.13	289.14	0.01	0.00
	33.780	355.22	355.24	0.02	0.00
	39.010	433.56	433.49	-0.07	-0.02
	44.290	525.86	525.96	0.10	0.02
	49.590	633.99	633.97	-0.02	0.00
	54.930	760.00	759.94	-0.06	-0.01
	60.310	906.06	906.09	0.03	0.00
	65.720	1074.60	1074.50	-0.10	-0.01
*	71.120	1268.00	1266.09	-1.91	-0.15
	76.660	1489.10	1489.32	0.22	0.01
	82.180	1740.80	1740.93	0.13	0.01
	87.730	2026.00	2025.75	-0.25	-0.01

STANDARD DEVIATION = 0.13

ETHYL BROMIDE C2H5BR

ZMACZYNSKI M.A.: J.CHIM.PHYS. 27,503(1930).

A = 6.98860 B = 1121.941 C = 234.739

B.P.(760) = 38.387

T	P EXPTL.	P CALCD.	DEV.	PERCENT
28.145	525.76	525.74	-0.02	0.00
33.247	633.90	633.90	0.00	0.00
43.560	906.10	906.08	-0.02	0.00
48.771	1074.60	1074.65	0.05	0.00
54.018	1268.10	1268.16	0.06	0.00
59.300	1489.20	1489.26	0.06	0.00
64.619	1741.00	1740.88	-0.13	-0.01
69.974	2026.30	2025.96	-0.34	-0.02
75.365	2347.40	2347.69	0.29	0.01

STANDARD DEVIATION = 0.19

ETHYL CHLORIDE C2H5CL

GORDON J.,GIAUQUE W.F.: J.AM.CHEM.SOC. 70,1506(1948).

A = 6.98647 B = 1030.007 C = 238.612

B.P.(760) = 12.263

T	P EXPTL.	P CALCD.	DEV.	PERCENT
-55.937	22.45	22.28	-0.17	-0.74
-43.644	50.45	50.53	0.08	0.15
-35.441	82.45	82.57	0.12	0.14
-28.085	124.13	124.15	0.02	0.02
-19.994	188.41	188.37	-0.04	-0.02
-13.479	257.86	257.85	-0.01	-0.01
-7.636	336.61	336.59	-0.02	-0.01
-1.878	432.19	432.08	-0.11	-0.03
3.448	538.99	538.63	-0.36	-0.07
6.491	608.30	608.31	0.01	0.00
10.490	710.37	710.54	0.17	0.02
12.511	766.83	767.12	0.29	0.04

STANDARD DEVIATION = 0.18

ETHYLTRICHLOROSILANE C2H5CL3SI

JENKINS A.C.,CHAMBERS G.F.: IND.ENG.CHEM. 46,2367(1954).

A = 6.60608 B = 1118.310 C = 201.370

B.P.(760) = 98.826

T	P EXPTL.	P CALCD.	DEV.	PERCENT
28.500	55.30	55.09	-0.21	-0.37
38.700	88.20	88.68	0.48	0.54
53.700	167.10	166.63	-0.47	-0.28
70.200	307.90	307.70	-0.20	-0.07
90.000	583.40	586.09	2.69	0.46
95.600	694.70	692.37	-2.33	-0.34

STANDARD DEVIATION = 2.10

ETHYL IODIDE C2H5I

SMYTH C.P.,ENGEL E.W.: J.AM.CHEM.SOC.51,2646(1929).

A = 6.95939 B = 1231.550 C = 229.388

B.P.(100) = 18.939

T	P EXPTL.	P CALCD.	DEV.	PERCENT
30.000	162.60	162.73	0.13	0.08
40.000	244.80	244.19	-0.61	-0.25
50.000	355.00	355.92	0.92	0.26
60.000	505.90	505.45	-0.45	-0.09

STANDARD DEVIATION = 1.20

N-METHYLFORMAMIDE C2H5NO

HEINRICH J.,ILAVSKY J.,SUROVY J.:
CHEM.ZVESTI 15,414(1961).

A = 7.49739 B = 1849.402 C = 201.089

B.P.(760) = 199.511

T	P EXPTL.	P CALCD.	DEV.	PERCENT
96.400	19.90	19.08	-0.82	-4.10
112.100	39.90	39.11	-0.79	-1.97
122.000	59.80	59.33	-0.47	-0.79
129.300	79.70	79.38	-0.32	-0.40
135.400	99.60	100.28	0.68	0.68
140.500	119.60	121.14	1.54	1.28
145.900	145.50	147.07	1.57	1.08
151.700	180.30	179.95	-0.35	-0.19
155.000	201.20	201.25	0.05	0.03
161.200	247.10	246.95	-0.15	-0.06
166.800	295.90	295.34	-0.56	-0.19
172.000	348.70	347.05	-1.65	-0.47
176.500	398.50	397.62	-0.88	-0.22
184.000	496.10	495.29	-0.81	-0.16
187.600	548.90	548.70	-0.20	-0.04
190.600	597.70	596.73	-0.97	-0.16
193.500	646.50	646.37	-0.13	-0.02
196.400	696.40	699.32	2.92	0.42
199.200	753.10	753.74	0.64	0.08

STANDARD DEVIATION = 1.16

ETHYL NITRATE C2H5NO3

GRAY P.,PRATT M.W.T.,LARKIN M.J.:
J.CHEM.SOC. 1956,210.

A = 7.16369 B = 1338.813 C = 224.955

B.P.(100) = 34.319

T	P EXPTL.	P CALCD.	DEV.	PERCENT
0.000	16.30	16.30	0.00	0.01
10.000	29.20	29.21	0.01	0.03
20.000	49.90	49.91	0.01	0.01
30.000	81.80	81.76	-0.04	-0.05
40.000	129.00	129.03	0.03	0.03
50.000	197.00	197.01	0.01	0.00
60.000	292.00	291.98	-0.02	-0.01

STANDARD DEVIATION = 0.03

ETHANE C2H6

LOOMIS A.G.,WALTERS J.E.: J.AM.CHEM.SOC. 48,2051(1926).

A = 6.82477 B = 663.484 C = 256.893

B.P.(760) = -88.665

T	P EXPTL.	P CALCD.	DEV.	PERCENT
-137.414	18.62	18.69	0.07	0.38
-129.883	39.67	39.89	0.22	0.57
-125.826	57.91	57.89	-0.02	-0.03
-118.694	106.40	105.64	-0.76	-0.71
-114.765	143.34	143.41	0.07	0.05
-110.521	195.55	195.85	0.30	0.15
-107.621	240.16	239.87	-0.29	-0.12
-105.214	280.67	282.17	1.50	0.54
-103.975	306.51	306.16	-0.35	-0.11
-101.450	360.47	360.11	-0.36	-0.10
-102.548	336.68	335.79	-0.89	-0.26
-99.088	417.85	417.18	-0.67	-0.16
-97.442	461.55	461.02	-0.53	-0.11
-95.427	517.10	519.58	2.48	0.48
-94.529	547.96	547.49	-0.47	-0.09
-93.400	584.90	584.25	-0.65	-0.11
-91.644	645.92	645.24	-0.68	-0.11
-90.687	681.19	680.52	-0.67	-0.10
-89.372	732.18	731.43	-0.75	-0.10
-88.511	763.04	766.35	3.31	0.43
-88.013	787.82	787.13	-0.69	-0.09
-87.136	820.80	824.79	3.99	0.49
-86.541	851.81	851.13	-0.68	-0.08
-85.848	883.04	882.62	-0.42	-0.05
-85.324	902.96	907.03	4.07	0.45
-84.771	933.96	933.35	-0.61	-0.07
-84.036	969.84	969.25	-0.59	-0.06
-83.292	1008.06	1006.67	-1.39	-0.14
-82.459	1056.40	1049.87	-6.53	-0.62
-81.720	1090.37	1089.39	-0.98	-0.09
-80.864	1138.02	1136.58	-1.44	-0.13
-80.373	1165.84	1164.36	-1.48	-0.13
-76.906	1376.36	1375.58	-0.78	-0.06
-73.241	1629.52	1629.49	-0.03	0.00

STANDARD DEVIATION = 1.48

CARBORANE-4 C2H6B4

SHAPIRO I.ET AL.: J.AM.CHEM.SOC. 85,3167(1963).

A = 6.55896 B = 898.363 C = 221.471

B.P.(100) = -24.417

T	P EXPTL.	P CALCD.	DEV.	PERCENT
-32.630	62.52	63.35	0.83	1.32
-31.630	68.00	67.11	-0.89	-1.31
-22.840	108.40	108.69	0.29	0.27
-17.120	146.30	145.47	-0.83	-0.57
0.050	316.60	318.81	2.21	0.70
5.010	391.40	391.15	-0.25	-0.06
11.130	499.30	497.41	-1.89	-0.38
14.490	564.10	564.56	0.46	0.08

STANDARD DEVIATION = 1.48

DIMETHYL BERYLIUM C2H6BE

COATES G.E.,GLOCLING F.,HUCK N.D.: J.CHEM.SOC. 1952,4496.

A = 19.08986 B = 11535.450 C = 496.636

B.P.(100) = 178.353

T	P EXPTL.	P CALCD.	DEV.	PERCENT
100.200	0.62	0.58	-0.04	-6.72
115.000	1.72	1.70	-0.02	-1.29
125.400	3.39	3.51	0.12	3.51
130.200	4.98	4.87	-0.11	-2.28
135.100	6.81	6.76	-0.05	-0.74
140.600	9.82	9.72	-0.10	-1.05
145.300	13.10	13.19	0.09	0.65
151.500	19.80	19.59	-0.21	-1.07
155.000	24.40	24.41	0.01	0.04
160.000	33.10	33.29	0.19	0.59
165.000	45.10	45.20	0.10	0.22
170.000	61.00	61.08	0.08	0.13
175.000	82.40	82.17	-0.23	-0.28
180.000	110.00	110.06	0.06	0.05

STANDARD DEVIATION = 0.14

DIMETHYL CADMIUM C2H6CD

LI J.C.M.: J.AM.CHEM.SOC. 78,1081(1956).

A = 6.49055 B = 1126.363 C = 201.073

B.P.(10) = 4.072

T	P EXPTL.	P CALCD.	DEV.	PERCENT
-2.670	6.50	6.51	0.01	0.12
5.830	11.13	11.13	0.00	0.04
8.530	13.13	13.09	-0.05	-0.34
11.520	15.60	15.57	-0.03	-0.18
12.600	16.61	16.56	-0.05	-0.28
14.000	17.92	17.92	0.00	0.02
16.400	20.46	20.48	0.02	0.08
17.650	21.83	21.92	0.09	0.41
18.670	23.14	23.16	0.02	0.09
18.680	23.17	23.17	0.00	0.01
18.860	23.31	23.40	0.09	0.38
21.260	26.64	26.57	-0.07	-0.25
22.810	28.86	28.81	-0.05	-0.18

STANDARD DEVIATION = 0.05

DIMETHYLDICHLOROSILANE C2H6CL2SI

JENKINS A.C.,CHAMBERS G.F.: IND.ENG.CHEM. 46,2367(1954).

A = 7.06208 B = 1280.287 C = 235.649

B.P.(760) = 70.547

T	P EXPTL.	P CALCD.	DEV.	PERCENT
27.800	159.50	159.35	-0.15	-0.09
38.000	241.50	241.83	0.33	0.14
48.000	353.50	353.55	0.05	0.01
58.000	504.40	503.68	-0.72	-0.14
66.100	658.80	659.46	0.66	0.10
72.100	798.00	797.83	-0.17	-0.02

STANDARD DEVIATION = 0.61

ETHYLDICHLOROSILANE C2H6CL2SI

JENKINS A.C.,CHAMBERS G.F.:
IND.ENG.CHEM. 46,2367(1954).

A = 7.24650 B = 1429.233 C = 251.769

B.P.(760) = 75.610

T	P EXPTL.	P CALCD.	DEV.	PERCENT
6.000	50.00	50.33	0.33	0.67
21.300	103.90	102.92	-0.98	-0.94
31.100	155.40	156.26	0.86	0.55
35.500	187.50	186.74	-0.76	-0.40
45.600	274.30	275.57	1.27	0.46
49.300	315.60	315.71	0.11	0.04
53.200	363.10	363.08	-0.02	-0.01
61.400	482.50	481.63	-0.87	-0.18
73.100	703.20	703.19	-0.01	0.00

STANDARD DEVIATION = 0.89

DIMETHYLAMINOTRIFLUOROSILANE C2H6F3NSI

VONGROSSE-RUYKEN H.,KLEESAAT R.:
Z.ANORG.ALLG.CHEM. 308,122(1961).

A = 6.49051 B = 794.659 C = 200.016

B.P.(100) = -23.052

T	P EXPTL.	P CALCD.	DEV.	PERCENT
-48.000	20.00	18.33	-1.67	-8.37
-28.500	70.00	72.01	2.01	2.87
0.000	332.00	329.26	-2.74	-0.82
15.500	634.00	635.74	1.74	0.27

STANDARD DEVIATION = 4.17

DIMETHYL ETHER C_2H_6O

KENNEDY R.M.,SAGENKAHN M.,ASTON J.G.:
J.AM.CHEM.SOC. 63,2267(1941).

A = 6.97603 B = 889.264 C = 241.957

B.P.(760) = -24.810

T	P EXPTL.	P CALCD.	DEV.	PERCENT
-78.220	35.13	35.07	-0.06	-0.16
-70.660	60.91	60.91	0.00	0.00
-65.250	87.80	87.83	0.03	0.03
-60.030	122.37	122.47	0.10	0.08
-55.140	164.34	164.42	0.08	0.05
-49.900	221.91	221.73	-0.18	-0.08
-45.100	287.53	287.56	0.03	0.01
-40.020	373.61	373.57	-0.04	-0.01
-35.100	475.55	475.45	-0.10	-0.02
-31.180	571.59	571.56	-0.03	-0.01
-27.670	670.27	670.14	-0.13	-0.02
-24.910	756.42	756.72	0.30	0.04

STANDARD DEVIATION = 0.14

ETHANOL C_2H_6O

KRETSCHMER C.B.,WIEBE R.:
J.AM.CHEM.SOC. 71,1793(1949).

A = 8.16556 B = 1624.080 C = 228.993

B.P.(760) = 78.322

T	P EXPTL.	P CALCD.	DEV.	PERCENT
0.000	11.95	11.84	-0.11	-0.94
25.000	59.02	59.07	0.05	0.08
34.988	103.03	103.11	0.08	0.07
44.994	172.95	172.97	0.02	0.01
50.000	220.94	220.97	0.03	0.01
54.988	279.79	279.62	-0.17	-0.06
65.000	438.04	437.86	-0.18	-0.04
78.553	766.71	766.97	0.26	0.03

STANDARD DEVIATION = 0.17

ETHANOL C2H60

AMBROSE D.,SPRAKE C.H.S.: J.CHEM.THERMODYNAMICS 2,631(1970).

A = 8.11220 B = 1592.864 C = 226.184

B.P.(760) = 78.298

	T	P EXPTL.	P CALCD.	DEV.	PERCENT
	19.622	42.95	42.86	-0.09	-0.22
	23.633	54.52	54.46	-0.06	-0.11
	25.722	61.54	61.51	-0.03	-0.05
	28.157	70.73	70.71	-0.02	-0.03
	33.334	94.25	94.28	0.03	0.03
	36.606	112.37	112.42	0.05	0.04
	39.237	129.01	129.10	0.09	0.07
	43.228	158.33	158.42	0.09	0.06
	47.349	194.37	194.49	0.12	0.06
	51.081	232.87	232.95	0.08	0.04
	54.664	275.72	275.78	0.06	0.02
	58.873	334.41	334.43	0.02	0.01
	62.865	399.54	399.47	-0.07	-0.02
	66.578	469.33	469.22	-0.11	-0.02
	70.559	555.29	555.09	-0.20	-0.04
	74.980	665.78	665.53	-0.25	-0.04
	77.405	733.72	733.53	-0.19	-0.03
	77.989	750.97	750.74	-0.23	-0.03
	78.340	761.45	761.24	-0.21	-0.03
	78.497	766.20	765.98	-0.22	-0.03
	79.188	787.44	787.14	-0.30	-0.04
	82.362	890.47	890.65	0.18	0.02
	85.836	1016.48	1016.68	0.20	0.02
	89.606	1168.78	1169.85	1.07	0.09
*	93.481	1345.02	1346.72	1.70	0.13

STANDARD DEVIATION = 0.28

ETHYLENE GLYCOL C2H6O2

JONES W.S.,TAMPLIN W.S.:
GLYCOLS,AMERICAN SOCIETY'S SERIES OF CHEMICAL MONOGRAPHS1952

A = 8.09083 B = 2088.936 C = 203.454

B.P.(760) = 197.492

T	P EXPTL.	P CALCD.	DEV.	PERCENT
50.000	0.70	0.71	0.01	0.89
55.000	0.90	1.02	0.12	13.28
60.000	1.30	1.45	0.15	11.64
65.000	1.90	2.04	0.14	7.33
70.000	2.60	2.83	0.23	8.83
75.000	3.70	3.88	0.18	4.88
80.000	5.00	5.26	0.26	5.26
85.000	6.70	7.06	0.36	5.42
90.000	9.20	9.38	0.18	2.00
95.000	12.00	12.35	0.35	2.91
100.000	16.00	16.10	0.10	0.65
105.000	21.00	20.82	-0.18	-0.84
110.000	27.00	26.70	-0.30	-1.10
115.000	33.00	33.98	0.98	2.96
120.000	43.00	42.91	-0.09	-0.20
125.000	54.00	53.82	-0.18	-0.34
130.000	68.00	67.03	-0.97	-1.42
135.000	83.00	82.95	-0.05	-0.06
140.000	103.00	102.02	-0.98	-0.95
145.000	125.00	124.72	-0.28	-0.22
150.000	153.00	151.62	-1.38	-0.90
155.000	183.00	183.31	0.31	0.17
160.000	219.00	220.48	1.48	0.67
165.000	263.00	263.85	0.85	0.32
170.000	312.00	314.24	2.24	0.72
175.000	373.00	372.53	-0.47	-0.13
180.000	435.00	439.68	4.68	1.08
185.000	520.00	516.72	-3.28	-0.63
190.000	605.00	604.78	-0.22	-0.04
195.000	710.00	705.05	-4.95	-0.70
197.600	760.00	762.44	2.44	0.32
200.000	820.00	818.82	-1.18	-0.14

STANDARD DEVIATION = 1.67

METHYL SULFOXIDE C_2H_6OS

DOUGLAS T.B.:
J.AM.CHEM.SOC. 70,2001(1948).

A = 7.76374 B = 2048.739 C = 231.556

B.P.(10) = 71.345

T	P EXPTL.	P CALCD.	DEV.	PERCENT
20.000	0.42	0.42	0.00	-0.16
25.000	0.60	0.60	0.00	0.01
30.000	0.85	0.85	0.00	-0.03
35.000	1.20	1.20	0.00	0.09
40.000	1.66	1.66	0.00	0.05
45.000	2.27	2.27	0.00	-0.08
50.000	3.07	3.07	0.00	0.02

STANDARD DEVIATION = 0.00

ETHANETHIOL C_2H_6S

OSBORN A.N.,DOUSLIN D.R.:
J.CHEM.ENG.DATA 11,502(1966).

A = 6.95178 B = 1084.376 C = 231.366

B.P.(760) = 35.003

T	P EXPTL.	P CALCD.	DEV.	PERCENT
0.405	187.57	187.55	-0.02	-0.01
5.236	233.72	233.69	-0.03	-0.01
10.111	289.13	289.18	0.05	0.02
15.017	355.22	355.30	0.08	0.02
19.954	433.56	433.56	0.00	0.00
24.933	525.86	525.86	0.00	0.00
29.944	633.99	633.87	-0.12	-0.02
35.000	760.00	759.93	-0.07	-0.01
40.092	906.06	906.02	-0.04	-0.01
45.221	1074.60	1074.51	-0.09	-0.01
50.390	1268.00	1268.05	0.05	0.00
55.604	1489.10	1489.58	0.48	0.03
60.838	1740.80	1740.79	-0.01	0.00
66.115	2026.00	2025.71	-0.29	-0.01

STANDARD DEVIATION = 0.18

DIMETHYL SULFIDE C2H6S

OSBORNE D.W.,POESCHER R.N.,YOST D.M.:
J.AM.CHEM.SOC. 64,169(1942).

A = 7.15091 B = 1195.584 C = 242.677

B.P.(100) = -10.566

T	P EXPTL.	P CALCD.	DEV.	PERCENT
-22.547	52.46	52.44	-0.02	-0.04
-10.028	102.75	102.78	0.03	0.03
0.096	168.29	168.35	0.06	0.04
4.943	210.33	210.19	-0.14	-0.07
9.857	260.85	260.96	0.11	0.04
15.138	326.37	326.24	-0.13	-0.04
20.087	398.84	398.92	0.08	0.02

STANDARD DEVIATION = 0.12

2,3-DITHIABUTANE C2H6S2

SCOTT D.W.,ET AL.:
J.AM.CHEM.SOC. 72,2424(1950).

A = 7.05045 B = 1386.698 C = 222.952

B.P.(100) = 51.617

T	P EXPTL.	P CALCD.	DEV.	PERCENT
0.000	6.78	6.77	-0.01	-0.11
15.000	16.73	16.70	-0.03	-0.16
20.000	22.02	22.02	0.00	-0.02
25.000	28.69	28.70	0.01	0.03
30.000	36.98	37.02	0.04	0.10
35.000	47.26	47.28	0.02	0.03
40.000	59.82	59.82	0.00	0.00
45.000	75.06	75.04	-0.02	-0.03
50.000	93.37	93.34	-0.03	-0.03
55.000	115.20	115.20	0.00	0.00
60.000	141.11	141.13	0.02	0.01

STANDARD DEVIATION = 0.02

2,3-DITHIABUTANE C2H6S2

SCOTT D.W.,ET AL.: J.AM.CHEM.SOC. 72,2424(1950).

A = 6.96185 B = 1336.488 C = 217.746

B.P.(760) = 109.741

T	P EXPTL.	P CALCD.	DEV.	PERCENT
61.411	149.41	149.37	-0.04	-0.03
67.301	187.57	187.58	0.01	0.01
73.234	233.72	233.77	0.05	0.02
79.201	289.13	289.13	0.00	0.00
85.218	355.22	355.20	-0.02	0.00
91.283	433.56	433.57	0.01	0.00
97.393	525.86	525.90	0.04	0.01
103.540	633.99	633.93	-0.06	-0.01
109.739	760.00	759.95	-0.05	-0.01
115.984	906.06	906.05	-0.01	0.00
122.273	1074.60	1074.55	-0.05	0.00
128.611	1268.00	1268.09	0.09	0.01

STANDARD DEVIATION = 0.05

DIMETHYLAMINE C2H7N

ASTON J.G.,EIDINOFF M.L.,FORSTER W.S.: J.AM.CHEM.SOC. 61,1539(1939).

A = 7.08212 B = 960.242 C = 221.667

B.P.(760) = 6.891

T	P EXPTL.	P CALCD.	DEV.	PERCENT
-71.773	4.86	4.74	-0.12	-2.43
-59.358	14.69	14.65	-0.04	-0.24
-51.082	28.35	28.38	0.03	0.10
-41.023	58.32	58.41	0.09	0.15
-31.082	110.58	110.59	0.01	0.01
-23.530	172.13	172.09	-0.04	-0.02
-16.711	249.54	249.46	-0.08	-0.03
-10.183	348.06	348.03	-0.03	-0.01
-2.978	491.22	491.15	-0.07	-0.01
2.774	636.50	636.42	-0.08	-0.01
4.520	686.45	686.70	0.25	0.04
6.800	757.37	757.07	-0.30	-0.04
6.858	758.62	758.93	0.31	0.04

STANDARD DEVIATION = 0.17

ETHANOLAMINE C2H7NO

MATTHEWS J.B.,SUMNER J.F.,MOELWYN-HUGHES E.A.:
TRANS.FARADAY SOC. 46,797(1950).

A = 7.45680 B = 1577.670 C = 173.368

B.P.(760) = 171.403

T	P EXPTL.	P CALCD.	DEV.	PERCENT
65.400	6.70	7.07	0.37	5.48
65.500	7.10	7.11	0.01	0.18
69.500	8.70	9.14	0.44	5.02
70.800	9.40	9.89	0.49	5.26
71.000	9.80	10.02	0.22	2.20
72.100	10.80	10.71	-0.09	-0.87
73.300	11.20	11.50	0.30	2.72
75.400	13.20	13.03	-0.17	-1.30
76.900	14.00	14.22	0.22	1.57
78.400	15.40	15.50	0.10	0.67
81.100	17.40	18.07	0.67	3.84
84.000	20.90	21.22	0.32	1.54
85.000	21.70	22.41	0.71	3.29
86.400	24.60	24.18	-0.42	-1.72
86.800	24.90	24.70	-0.20	-0.79
90.000	28.60	29.27	0.67	2.35
92.200	32.40	32.81	0.41	1.28
93.200	34.80	34.54	-0.26	-0.74
94.600	37.50	37.09	-0.41	-1.09
96.400	40.80	40.60	-0.20	-0.48
98.700	45.00	45.50	0.50	1.11
99.500	47.80	47.31	-0.49	-1.02
101.700	53.60	52.63	-0.97	-1.81
104.100	59.60	59.00	-0.60	-1.01
105.000	63.30	61.55	-1.75	-2.76
105.500	64.20	63.01	-1.19	-1.86
107.500	70.60	69.13	-1.47	-2.08
108.900	76.00	73.71	-2.29	-3.01
112.100	84.80	85.15	0.35	0.41
114.200	93.40	93.44	0.04	0.05
117.300	105.10	106.92	1.82	1.73
121.000	122.90	125.11	2.21	1.80
125.000	145.90	147.62	1.72	1.18
126.900	159.40	159.44	0.04	0.02
128.800	171.40	172.04	0.64	0.37
132.000	193.30	195.14	1.84	0.95
133.300	205.50	205.23	-0.27	-0.13
136.000	227.20	227.59	0.39	0.17
137.900	242.00	244.50	2.50	1.03
141.500	280.00	279.40	-0.60	-0.21
144.600	316.80	312.66	-4.14	-1.31
147.100	340.70	341.81	1.11	0.33
* 150.800	406.50	389.02	-17.48	-4.30
154.700	445.30	444.45	-0.85	-0.19
* 158.500	518.00	504.54	-13.46	-2.60
161.400	554.00	554.72	0.72	0.13
* 167.000	652.00	663.15	11.15	1.71
170.900	751.00	748.37	-2.63	-0.35

O,N-DIMETHYLHYDROXYLAMINE C2H7NO

BISSOT T.C.,PARRY R.W.,CAMPBELL D.H.: J.AM.CHEM.SOC. 79,796(1957).

A = 7.40539 B = 1245.581 C = 233.063

B.P.(760) = 42.228

T	P EXPTL.	P CALCD.	DEV.	PERCENT
-45.200	6.10	5.96	-0.14	-2.32
-42.000	8.00	7.69	-0.31	-3.82
-36.500	11.80	11.71	-0.09	-0.75
-31.400	16.80	16.94	0.14	0.82
-27.000	23.10	22.95	-0.15	-0.66
-22.600	30.60	30.70	0.10	0.32
-18.000	40.80	41.09	0.29	0.70
-13.300	54.90	54.65	-0.25	-0.46
-7.300	77.80	77.30	-0.50	-0.64
-4.200	91.50	91.82	0.32	0.35
0.000	114.40	115.08	0.68	0.60
4.000	141.10	141.64	0.54	0.38
8.000	173.20	173.13	-0.07	-0.04
12.000	210.90	210.23	-0.67	-0.32
16.000	253.80	253.71	-0.09	-0.04
19.700	301.40	300.29	-1.11	-0.37
23.900	361.50	361.48	-0.02	-0.01
28.000	430.40	430.73	0.33	0.08
32.900	526.60	527.37	0.77	0.15
37.500	630.20	633.48	3.28	0.52
42.400	765.00	764.94	-0.06	-0.01

STANDARD DEVIATION = 0.47

N,N-DIMETHYLHYDROXYLAMINE C_2H_7NO

BISSOT T.C.,PARRY R.W.,CAMPBELL D.H.: J.AM.CHEM.SOC. 79,796(1957).

A = 7.56577 B = 1415.962 C = 201.932

B.P.(100) = 52.473

T	P EXPTL.	P CALCD.	DEV.	PERCENT
17.600	13.00	13.06	0.06	0.44
19.100	14.50	14.44	-0.06	-0.40
25.200	21.40	21.46	0.06	0.29
29.800	28.50	28.54	0.04	0.13
34.700	38.00	38.19	0.19	0.50
40.300	52.40	52.52	0.12	0.22
44.900	67.30	67.49	0.19	0.28
50.300	90.10	89.55	-0.55	-0.62
55.100	114.60	113.99	-0.61	-0.53
60.200	146.00	145.90	-0.10	-0.07
65.100	183.30	183.30	0.00	0.00
70.200	229.70	230.44	0.74	0.32
75.100	284.40	284.83	0.43	0.15
80.200	351.70	352.35	0.65	0.19
85.000	427.00	427.50	0.50	0.12
90.000	521.00	519.34	-1.66	-0.32

STANDARD DEVIATION = 0.62

1,2-ETHYLENEDIAMINE C2H8N2

HIEBER W.,WOERNER A.:
Z.ELEKTROCHEM. 40, 252(1934).

A = 7.16871 B = 1336.235 C = 194.366

B.P.(760) = 117.263

T	P EXPTL.	P CALCD.	DEV.	PERCENT
26.510	13.10	13.15	0.05	0.40
29.490	16.10	15.83	-0.27	-1.66
32.010	18.00	18.45	0.45	2.50
34.600	21.50	21.52	0.02	0.07
36.920	24.40	24.62	0.22	0.90
38.400	27.50	26.79	-0.71	-2.57
39.770	30.80	28.95	-1.85	-6.01
42.510	34.40	33.70	-0.70	-2.03
46.650	43.20	42.12	-1.08	-2.49
51.710	54.70	54.77	0.07	0.13
56.820	70.40	70.63	0.23	0.33
60.040	82.10	82.48	0.38	0.46
62.290	91.40	91.70	0.30	0.33
65.600	106.20	106.82	0.62	0.59
71.020	136.20	136.03	-0.17	-0.12
71.710	139.70	140.19	0.49	0.35
76.510	173.60	172.07	-1.53	-0.88
81.360	208.80	210.12	1.32	0.63
88.890	284.10	282.68	-1.42	-0.50
91.420	309.90	311.21	1.31	0.42
91.780	313.70	315.46	1.76	0.56
94.500	355.00	349.07	-5.93	-1.67
97.380	383.50	387.77	4.27	1.11
101.010	435.10	441.43	6.33	1.46
102.920	467.90	471.98	4.08	0.87
105.500	519.00	515.94	-3.06	-0.59
107.030	547.10	543.52	-3.58	-0.65
108.010	568.10	561.81	-6.29	-1.11
112.380	649.60	649.45	-0.15	-0.02
116.210	736.20	734.96	-1.24	-0.17
117.400	760.50	763.29	2.79	0.37

STANDARD DEVIATION = 2.69

1,1-DIMETHYLHYDRAZINE C2H8N2

ASTON J.G.,WOOD J.L.,ZOLKI T.P.: J.AM.CHEM.SOC. 75,6202(1953).

A = 7.40813 B = 1305.908 C = 225.531

B.P.(100) = 15.940

T	P EXPTL.	P CALCD.	DEV.	PERCENT
-35.410	3.51	3.46	-0.05	-1.37
-23.330	8.92	8.91	-0.01	-0.16
-17.370	13.62	13.63	0.01	0.10
-13.130	18.20	18.19	-0.01	-0.06
-4.120	32.29	32.36	0.07	0.22
-2.070	36.65	36.66	0.01	0.01
5.060	55.62	55.57	-0.05	-0.09
9.760	72.07	72.10	0.03	0.05
15.210	96.39	96.30	-0.10	-0.10
19.935	122.40	122.47	0.07	0.06

STANDARD DEVIATION = 0.06

1,2-DIMETHYLHYDRAZINE C2H8N2

ASTON J.G.,JANZ G.J.,RUSSELL: J.AM.CHEM.SOC.73,1943(1951).

A = 5.61188 B = 633.588 C = 143.168

B.P.(10) = -5.786

T	P EXPTL.	P CALCD.	DEV.	PERCENT
1.370	17.01	16.92	-0.09	-0.55
9.730	29.19	29.38	0.19	0.64
15.810	42.49	42.31	-0.18	-0.41
20.900	56.13	56.25	0.12	0.21
24.460	67.99	67.94	-0.05	-0.07

STANDARD DEVIATION = 0.21

DIMETHYLPHOSPHINE BORINE C2H10BP

BURG A.B.,WAGNER R.I.:
J.AM.CHEM.SOC. 75,3872(1953).

A = 7.63469 B = 2033.385 C = 250.651

B.P.(100) = 110.218

T	P EXPTL.	P CALCD.	DEV.	PERCENT
29.500	2.36	2.38	0.02	0.83
39.900	4.25	4.33	0.08	1.84
50.000	7.44	7.44	0.00	-0.04
60.700	12.67	12.70	0.03	0.25
75.200	25.00	24.80	-0.20	-0.80
90.000	46.20	46.30	0.10	0.22
100.100	68.70	68.78	0.08	0.12
110.200	100.00	99.94	-0.06	-0.06

STANDARD DEVIATION = 0.12

1,2-DIMETHYLDISILANE C2H10SI2

CRAIG A.D.,MACDIARMID A.G.:
J.INORG.NUCL.CHEM. 24,161(1962).

A = 4.02431 B = 255.441 C = 129.152

B.P.(100) = -2.965

T	P EXPTL.	P CALCD.	DEV.	PERCENT
-46.200	11.80	8.81	-2.99	-25.35
-41.800	15.10	12.59	-2.51	-16.62
-38.800	17.70	15.74	-1.96	-11.05
-25.200	35.60	36.90	1.30	3.65
-18.400	48.70	52.23	3.53	7.24
-13.800	60.70	64.54	3.84	6.33
1.200	120.20	116.06	-4.14	-3.45
0.000	114.50	111.29	-3.21	-2.80
0.000	107.00	111.29	4.29	4.01

STANDARD DEVIATION = 3.95

N-DIMETHYLAMINODIBORANE C2H11B2N

FURUKAVA G.T.,ET AL.:
J.RES.NATL.BUR.STANDARDS 55,201(1955).

A = 8.34005 B = 1917.352 C = 302.730

B.P.(100) = -0.311

T	P EXPTL.	P CALCD.	DEV.	PERCENT
-38.450	12.10	12.16	0.06	0.52
-35.650	14.80	14.49	-0.31	-2.09
-30.410	19.80	19.92	0.12	0.59
-27.590	23.80	23.52	-0.28	-1.18
-23.870	29.00	29.13	0.13	0.45
-16.680	43.20	43.37	0.17	0.39
-11.210	57.70	57.94	0.24	0.41
-8.300	67.30	67.29	-0.01	-0.01
-4.270	82.40	82.39	-0.01	-0.01
-1.830	92.80	92.89	0.09	0.10
0.340	103.50	103.18	-0.32	-0.31
10.060	162.50	162.26	-0.24	-0.15
14.020	193.30	193.57	0.27	0.14

STANDARD DEVIATION = 0.23

PENTAFLUOROCHLOROACETONE C3CLF5O

MURPHY K.P.:
J.CHEM.ENG.DATA 9,259(1964).

A = 6.84844 B = 925.269 C = 225.445

B.P.(760) = 7.759

T	P EXPTL.	P CALCD.	DEV.	PERCENT
-40.910	68.55	68.29	-0.26	-0.38
-34.310	101.23	101.75	0.52	0.51
-30.230	127.91	128.44	0.53	0.42
-22.040	200.18	199.32	-0.87	-0.43
-10.320	354.92	352.67	-2.25	-0.63
0.000	553.43	554.95	1.52	0.27
0.000	553.58	554.95	1.37	0.25
13.520	946.20	947.23	1.03	0.11
29.680	1664.40	1666.15	1.75	0.11
31.760	1786.00	1782.55	-3.45	-0.19

STANDARD DEVIATION = 1.95

PERFLUOROPROPENE C3F6

WHIPPLE G.H.:

IND.ENG.CHEM. 44, 1664(1952)

A = 7.35515 B = 1012.065 C = 256.509

B.P.(760) = -30.316

T	P EXPTL.	P CALCD.	DEV.	PERCENT
-40.700	463.00	462.93	-0.07	-0.01
-37.200	552.00	550.00	-2.00	-0.36
-35.560	597.00	595.13	-1.87	-0.31
-30.110	769.00	767.15	-1.85	-0.24
-24.830	963.00	969.97	6.97	0.72
-15.870	1403.00	1410.62	7.62	0.54
-8.300	1889.00	1895.30	6.30	0.33
-3.300	2273.00	2281.36	8.36	0.37
1.950	2771.00	2750.31	-20.69	-0.75
7.000	3293.00	3269.08	-23.92	-0.73
12.100	3867.00	3866.75	-0.25	-0.01
19.750	4896.00	4916.82	20.82	0.43

STANDARD DEVIATION =13.58

HEXAFLUOROACETONE C3F6O

PLAUSH A.C.,PACE E.L.:

J.CHEM.PHYS. 47,44(1967).

A = 6.65019 B = 725.898 C = 219.853

B.P.(760) = -27.275

T	P EXPTL.	P CALCD.	DEV.	PERCENT
-78.790	32.00	31.94	-0.06	-0.20
-73.500	48.70	49.01	0.31	0.64
-68.230	73.00	72.89	-0.11	-0.15
-62.990	105.40	105.34	-0.06	-0.06
-57.770	148.90	148.47	-0.43	-0.29
-52.600	204.30	204.21	-0.09	-0.05
-47.460	274.90	275.09	0.19	0.07
-42.360	362.90	363.47	0.57	0.16
-37.310	471.30	471.63	0.33	0.07
-32.300	602.10	602.32	0.22	0.04
-27.780	743.60	742.86	-0.74	-0.10
-27.110	765.80	765.68	-0.12	-0.02

STANDARD DEVIATION = 0.39

PERFLUOROPROPANE C3F8

PACE E.L.,PLAUSH A.C.:

J.CHEM.PHYS. 47,38(1967).

A = 6.91935 B = 825.829 C = 241.233

B.P.(760) = -36.746

T	P EXPTL.	P CALCD.	DEV.	PERCENT
-79.370	66.00	65.67	-0.34	-0.51
-73.530	98.80	98.86	0.06	0.06
-67.750	143.70	144.23	0.53	0.37
-62.010	204.20	204.89	0.69	0.34
-56.320	284.70	284.00	-0.70	-0.25
-50.680	385.80	385.03	-0.77	-0.20
-45.120	511.30	510.94	-0.36	-0.07
-41.820	600.00	599.86	-0.14	-0.02
-39.060	682.90	683.26	0.36	0.05
-36.340	773.50	774.13	0.63	0.08

STANDARD DEVIATION = 0.61

2,2,3,3,3-PENTAFLUOROPROPANOL C3H3F5O

MEEKS A.C.,GOLDFARB I.J.:

J.CHEM.ENG.DATA 12,196(1967).

A = 6.30865 B = 830.564 C = 153.824

B.P.(10) = 2.631

T	P EXPTL.	P CALCD.	DEV.	PERCENT
0.150	8.31	8.21	-0.10	-1.17
3.000	10.45	10.29	-0.16	-1.51
3.700	10.80	10.87	0.07	0.60
5.400	12.40	12.37	-0.03	-0.25
5.800	12.70	12.75	0.05	0.37
7.600	14.55	14.57	0.02	0.13
8.700	15.75	15.79	0.04	0.22
9.700	17.00	16.96	-0.04	-0.22
11.400	19.10	19.13	0.03	0.17
12.700	20.90	20.94	0.04	0.20
14.900	24.30	24.32	0.02	0.10
15.200	24.60	24.82	0.22	0.89
17.500	28.85	28.89	0.04	0.14
18.250	30.45	30.33	-0.12	-0.40
20.100	34.15	34.14	-0.01	-0.04
20.650	35.55	35.34	-0.21	-0.59
22.400	39.30	39.40	0.10	0.26
23.050	41.00	41.01	0.01	0.02

STANDARD DEVIATION = 0.11

THIAZOLE C_3H_3NS

SOULIE M.A.,GOURSOT P.,PENELOUX A.,METZGER J.:
J.CHIM.PHYS. 66,603(1969).

A = 7.14201 B = 1425.351 C = 216.255

B.P.(760) = 118.241

T	P EXPTL.	P CALCD.	DEV.	PERCENT
62.977	109.00	109.01	0.01	0.01
66.943	128.51	128.51	0.00	0.00
74.543	173.97	173.97	0.00	0.00
80.045	214.55	214.54	-0.01	-0.01
89.613	303.41	303.38	-0.03	-0.01
92.219	332.17	332.16	-0.01	0.00
94.844	363.39	363.36	-0.03	-0.01
97.852	402.00	401.99	-0.01	0.00
101.841	458.26	458.27	0.01	0.00
104.118	493.12	493.14	0.02	0.00
105.845	520.95	520.98	0.03	0.01
107.547	549.60	549.64	0.04	0.01
108.465	565.54	565.62	0.08	0.01
109.746	588.46	588.54	0.08	0.01
111.981	630.26	630.29	0.03	0.01
114.600	682.17	682.21	0.04	0.01
112.869	647.64	647.53	-0.11	-0.02
114.215	674.43	674.37	-0.06	-0.01
116.827	729.05	728.99	-0.06	-0.01
117.381	741.06	741.02	-0.04	-0.01
118.261	760.44	760.44	0.00	0.00

STANDARD DEVIATION = 0.05

TRIOXANE $C_3H_6O_3$

SEREBRYANNAYA I.I.,BYK S.SH.:
KHIM.PROM. 23,828(1965).

A = 7.81861 B = 1783.243 C = 247.107

B.P.(760) = 114.035

T	P EXPTL.	P CALCD.	DEV.	PERCENT
56.300	87.50	87.34	-0.16	-0.19
64.600	125.40	125.23	-0.17	-0.14
74.000	183.50	184.15	0.65	0.36
88.000	314.00	314.19	0.19	0.06
98.160	452.00	450.59	-1.41	-0.31
104.200	552.30	552.81	0.51	0.09
113.820	754.50	754.87	0.37	0.05

STANDARD DEVIATION = 0.85

PROPYNE C_3H_4

VAN HOOK W.A.:

J.CHEM.PHYS. 46,1907(1967).

A = 7.87500 B = 1247.062 C = 273.069

B.P.(760) = -23.366

T	P EXPTL.	P CALCD.	DEV.	PERCENT
-111.600	1.42	1.42	0.00	-0.12
-107.920	2.11	2.11	0.00	-0.09
-104.870	2.89	2.89	0.00	-0.02
-104.310	3.06	3.06	0.00	-0.07
-99.510	4.90	4.90	0.00	-0.10
* -98.900	4.72	5.19	0.47	9.90
-95.850	6.89	6.89	0.00	-0.01
-95.020	7.43	7.43	0.00	0.00
-92.100	9.64	9.64	0.00	-0.02
-91.950	9.77	9.77	0.00	-0.05
-91.420	10.23	10.23	0.00	-0.02
-86.950	14.95	14.95	0.00	0.00
-85.080	17.44	17.43	-0.01	-0.05
-84.770	17.88	17.87	-0.01	-0.03
-83.890	19.20	19.19	-0.01	-0.06
-78.180	29.93	29.93	0.00	0.02
-78.010	30.33	30.32	-0.01	-0.03
-71.550	48.60	48.61	0.01	0.01
-70.910	50.76	50.85	0.09	0.18
-64.960	76.33	76.32	-0.01	-0.01
-62.430	90.11	90.08	-0.03	-0.03
-61.860	93.49	93.46	-0.03	-0.04
-55.580	138.40	138.39	-0.01	-0.01
-52.430	167.08	167.09	0.01	0.01
-49.220	201.36	201.38	0.02	0.01
-43.990	269.97	269.90	-0.08	-0.03
-41.930	301.77	301.79	0.03	0.01
-38.300	365.66	365.71	0.05	0.01
* -32.440	492.74	492.60	-0.14	-0.03
-29.720	562.86	562.88	0.02	0.00
-23.330	761.23	761.27	0.04	0.01
-18.880	931.07	931.02	-0.05	-0.01

STANDARD DEVIATION = 0.03

PROPADIENE C_3H_4

LAMB A.B.,ROPER E.E.: J.AM.CHEM.SOC.62,806(1940).

A = 7.45500 B = 1033.494 C = 260.839

B.P.(10) = -100.732

T	P EXPTL.	P CALCD.	DEV.	PERCENT
-135.230	0.17	0.17	0.00	-0.75
-131.980	0.27	0.27	0.00	0.77
-128.590	0.44	0.44	0.00	-0.73
-127.560	0.50	0.50	0.00	0.39
-126.130	0.61	0.61	0.00	-0.54
-122.330	0.98	0.99	0.01	0.52
-115.500	2.21	2.21	0.00	-0.07
-84.980	37.86	37.86	0.00	0.00

STANDARD DEVIATION = 0.00

1,2,3-TRIBROMOPROPANE $C_3H_4BR_3$

DREISBACH R.R.,SHRADER S.A.: IND.ENG.CHEM. 41,2879(1949).

A = 7.03719 B = 1735.321 C = 195.417

B.P.(100) = 149.085

T	P EXPTL.	P CALCD.	DEV.	PERCENT
127.980	47.16	46.91	-0.25	-0.53
133.110	57.04	56.89	-0.15	-0.26
137.420	66.39	66.60	0.21	0.31
141.150	75.86	76.08	0.22	0.28
155.540	123.76	123.78	0.02	0.02
186.940	315.52	315.29	-0.23	-0.07
205.200	507.50	507.66	0.16	0.03

STANDARD DEVIATION = 0.25

ACRYLIC ACID C3H4O2

GUBKOV A.N.,FERMOR N.A.,SMIRNOV N.I.:
ZH.PRIKL.KHIM. 37,2204(1964).

A = 5.65204 B = 648.629 C = 154.683

B.P.(100) = 22.924

T	P EXPTL.	P CALCD.	DEV.	PERCENT
20.000	88.00	86.87	-1.13	-1.29
30.000	142.00	138.01	-3.99	-2.81
40.000	210.00	209.08	-0.92	-0.44
50.000	289.00	304.16	15.16	5.24
60.000	430.00	427.27	-2.73	-0.63
70.000	590.00	582.34	-7.66	-1.30

STANDARD DEVIATION =10.23

ALLYL CHLORIDE C3H5CL

IOFFE I.I.,YAMNOLSKAYA E.S.:
ZH.PRIKL.KHIM. 17,527(1944).

A = 5.29716 B = 418.375 C = 128.168

B.P.(760) = 44.976

T	P EXPTL.	P CALCD.	DEV.	PERCENT
13.200	216.50	217.61	1.11	0.51
14.300	235.50	229.37	-6.13	-2.60
14.500	239.50	231.55	-7.95	-3.32
19.900	297.50	296.21	-1.29	-0.43
21.100	306.50	312.11	5.61	1.83
21.700	320.50	320.28	-0.22	-0.07
25.400	365.00	373.93	8.93	2.45
26.300	383.50	387.85	4.35	1.14
31.900	476.00	482.42	6.42	1.35
34.600	531.50	533.07	1.57	0.29
36.600	571.00	572.77	1.77	0.31
37.600	588.50	593.33	4.83	0.82
39.800	636.00	640.26	4.26	0.67
41.200	683.00	671.34	-11.66	-1.71
43.900	746.50	734.02	-12.48	-1.67

STANDARD DEVIATION = 7.15

METHYL CHLOROACETATE C3H5CLO2

GOEBEL K.H.,SCHAFFENGER J.,OPEL G.: CHEM.TECH. 19,307(1967).

A = 7.00445 B = 1306.263 C = 187.301

B.P.(760) = 129.473

T	P EXPTL.	P CALCD.	DEV.	PERCENT
45.020	24.20	24.09	-0.11	-0.47
50.090	31.70	31.76	0.06	0.19
55.000	40.90	41.06	0.16	0.38
59.980	52.50	52.72	0.22	0.41
65.900	70.80	70.06	-0.74	-1.05
71.800	92.10	91.82	-0.28	-0.31
77.760	119.30	119.20	-0.10	-0.08
86.310	169.00	169.94	0.94	0.55
103.590	325.00	326.51	1.51	0.46
111.950	438.60	435.86	-2.74	-0.63
118.060	529.80	532.95	3.15	0.59
124.340	655.10	649.96	-5.14	-0.78
129.860	766.00	768.85	2.85	0.37

STANDARD DEVIATION = 2.36

ALLYLTRICHLOROSILANE C3H5CL3SI

JENKINS A.C.,CHAMBERS G.F.: IND.ENG.CHEM. 46,2367(1954).

A = 6.71253 B = 1179.265 C = 190.799

B.P.(760) = 116.965

T	P EXPTL.	P CALCD.	DEV.	PERCENT
45.700	53.00	53.23	0.23	0.44
57.900	93.50	93.50	0.00	0.00
76.300	199.50	198.35	-1.15	-0.58
92.000	348.90	348.78	-0.12	-0.03
104.900	525.20	530.24	5.04	0.96
115.100	724.30	720.19	-4.11	-0.57

STANDARD DEVIATION = 3.81

PROPIONITRILE C3H5N

DREISBACH R.R.,SHRADER S.A.:
IND.ENG.CHEM. 41,2879(1949).

A = 6.83945 B = 1221.990 C = 211.240

B.P.(760) = 97.449

T	P EXPTL.	P CALCD.	DEV.	PERCENT
35.500	75.86	77.08	1.22	1.60
45.760	123.76	121.52	-2.24	-1.81
70.450	315.52	317.24	1.72	0.55
84.440	507.50	508.91	1.41	0.28
97.350	760.00	757.78	-2.22	-0.29

STANDARD DEVIATION = 2.86

PROPIONITRILE C3H5N

MILAZZO G.:
ANNALI DI CHIMICA 46,1105(1956).

A = 5.27821 B = 665.523 C = 159.098

B.P.(10) = -3.537

T	P EXPTL.	P CALCD.	DEV.	PERCENT
-84.660	0.00	0.00	0.00	-95.27
-77.010	0.01	0.00	-0.01	-89.02
-67.420	0.04	0.01	-0.03	-74.95
-65.490	0.05	0.01	-0.04	-70.75
-59.720	0.10	0.04	-0.06	-61.08
-52.960	0.20	0.10	-0.10	-49.34
-46.190	0.37	0.24	-0.13	-34.59
-34.950	1.00	0.83	-0.17	-17.30
-22.850	2.67	2.48	-0.20	-7.30
-13.080	5.30	5.25	-0.05	-0.89
-2.950	10.10	10.38	0.28	2.75
6.560	18.00	18.23	0.23	1.27
16.420	31.10	30.65	-0.45	-1.44
22.050	40.00	40.21	0.21	0.52

STANDARD DEVIATION = 0.21

ETHYL ISOTHIOCYANATE C3H5NS

BAUER H.,BURSCHKIES K.: BER. 68,1243(1935).

A = 7.10597 B = 1567.473 C = 234.159

B.P.(10) = 22.553

T	P EXPTL.	P CALCD.	DEV.	PERCENT
10.000	4.90	4.85	-0.05	-0.94
12.000	5.40	5.47	0.07	1.35
14.000	6.20	6.16	-0.04	-0.65
16.000	7.00	6.92	-0.08	-1.15
18.000	7.80	7.76	-0.04	-0.54
20.000	8.70	8.68	-0.02	-0.20
22.000	9.60	9.70	0.10	1.05
24.000	10.80	10.82	0.02	0.18
26.000	12.00	12.05	0.05	0.40
28.000	13.40	13.39	-0.01	-0.06
30.000	15.00	14.86	-0.14	-0.91
32.000	16.50	16.47	-0.03	-0.17
34.000	18.40	18.22	-0.18	-0.96
37.000	21.00	21.15	0.15	0.72
40.000	24.40	24.47	0.07	0.27
42.000	26.80	26.91	0.11	0.43
46.000	32.30	32.44	0.14	0.42
48.000	35.50	35.54	0.04	0.10
50.000	39.10	38.88	-0.22	-0.55

STANDARD DEVIATION = 0.11

CYCLOPROPANE C_3H_6

RUEHRWEIN R.A.,POWELL T.M.: J.AM.CHEM.SOC. 68,1063(1946).

A = 6.88788 B = 856.012 C = 246.495

B.P.(760) = -32.870

T	P EXPTL.	P CALCD.	DEV.	PERCENT
-90.029	26.26	26.12	-0.14	-0.53
-81.810	48.95	48.98	0.03	0.06
-77.465	66.73	66.62	-0.11	-0.16
-73.304	88.24	88.16	-0.08	-0.09
-68.117	122.02	122.75	0.73	0.60
-63.234	164.91	164.77	-0.14	-0.08
-58.114	220.85	220.72	-0.13	-0.06
-53.107	289.59	289.39	-0.20	-0.07
-48.029	375.88	375.61	-0.27	-0.07
-43.283	473.90	473.67	-0.23	-0.05
-37.983	606.18	606.10	-0.08	-0.01
-34.809	698.39	698.40	0.01	0.00
-32.083	785.59	786.16	0.57	0.07

STANDARD DEVIATION = 0.33

PROPYLENE C3H6

POWELL T.M.,GIAUQUE W.F.:
J.AM.CHEM.SOC. 61,2366(1939).

A = 6.82359 B = 786.532 C = 247.243

B.P.(760) = -47.756

T	P EXPTL.	P CALCD.	DEV.	PERCENT
-107.336	15.98	15.91	-0.07	-0.42
-100.379	29.41	29.38	-0.03	-0.10
-96.635	39.92	39.92	0.00	0.00
-93.199	52.32	52.20	-0.12	-0.23
-88.614	73.19	73.33	0.14	0.19
-83.649	103.54	103.69	0.15	0.15
-77.517	154.52	154.69	0.17	0.11
-70.435	237.31	237.18	-0.13	-0.06
-65.705	309.96	309.73	-0.23	-0.08
-59.757	425.47	425.04	-0.43	-0.10
-52.872	598.30	598.46	0.16	0.03
-47.175	779.92	780.30	0.38	0.05

STANDARD DEVIATION = 0.24

1,2-DICHLOROPROPANE C3H6CL2

DREISBACH R.R.,SHRADER S.A.:
IND.ENG.CHEM. 41,2879(1949).

A = 6.98074 B = 1308.138 C = 222.845

B.P.(760) = 96.219

T	P EXPTL.	P CALCD.	DEV.	PERCENT
44.780	123.76	123.82	0.06	0.05
69.000	315.52	315.09	-0.43	-0.14
83.180	507.50	508.30	0.80	0.16
96.200	760.00	759.57	-0.43	-0.06

STANDARD DEVIATION = 1.01

GAMA-CHLOROPROPYLTRICHLOROSILANE C3H6CL4SI

JENKINS A.C.,CHAMBERS G.F.: IND.ENG.CHEM. 46,2367(1954).

A = 7.15644 B = 1679.069 C = 210.382

B.P.(760) = 182.325

T	P EXPTL.	P CALCD.	DEV.	PERCENT
87.100	32.50	32.52	0.02	0.06
121.800	126.40	126.41	0.01	0.00
149.300	308.00	307.77	-0.23	-0.07
159.100	409.20	409.30	0.10	0.03
171.600	575.80	576.44	0.64	0.11
179.400	706.40	705.86	-0.54	-0.08

STANDARD DEVIATION = 0.51

ACETONE C3H6O

BROWN I.,SMITH F.: AUSTRAL.J.CHEM.10,423(1957).

A = 7.15853 B = 1231.232 C = 231.766

B.P.(760) = 56.058

T	P EXPTL.	P CALCD.	DEV.	PERCENT
37.680	388.19	388.19	0.00	0.00
41.580	450.99	451.06	0.07	0.02
44.960	512.08	511.98	-0.10	-0.02
45.000	512.80	512.74	-0.06	-0.01
47.010	552.07	552.04	-0.03	0.00
49.310	599.82	599.94	0.12	0.02
51.910	658.01	658.05	0.04	0.01
56.020	759.06	759.01	-0.05	-0.01

STANDARD DEVIATION = 0.09

ACETONE C3H6O

BOUBLIK T.,AIM K.:
COLLECTION CZECHOSLOV.CHEM.COMMUN. 36(IN PRESS).

A = 7.11714 B = 1210.595 C = 229.664

B.P.(760) = 56.101

T	P EXPTL.	P CALCD.	DEV.	PERCENT
-12.949	33.93	33.96	0.03	0.10
-5.424	52.26	52.30	0.04	0.07
-0.103	69.80	69.76	-0.04	-0.06
4.882	90.35	90.30	-0.05	-0.05
8.666	109.13	109.05	-0.08	-0.07
13.019	134.42	134.51	0.09	0.07
16.731	159.87	159.92	0.05	0.03
20.939	193.37	193.38	0.01	0.00
24.330	224.44	224.34	-0.10	-0.05
28.351	266.22	266.18	-0.04	-0.01
32.138	311.05	311.20	0.15	0.05
36.649	372.81	372.71	-0.10	-0.03
42.094	459.75	459.68	-0.07	-0.02
48.344	578.49	578.90	0.41	0.07
55.285	739.35	739.04	-0.31	-0.04

STANDARD DEVIATION = 0.17

1-PROPEN-3-OL C3H6O

EWERT M.: BULL.SOC.CHIM.BELG.,45,509(1936).

A = 11.18698 B = 4068.457 C = 392.732

B.P.(760) = 97.079

T	P EXPTL.	P CALCD.	DEV.	PERCENT
21.000	22.80	22.56	-0.24	-1.03
25.000	28.10	28.03	-0.07	-0.26
30.000	36.30	36.54	0.24	0.67
35.000	47.10	47.35	0.25	0.52
40.000	61.20	60.98	-0.22	-0.36
97.080	760.00	760.01	0.01	0.00

STANDARD DEVIATION = 0.28

PROPYLENE OXIDE C3H6O

MCDONALD R.A.,SHRADER S.A.,STULL D.R.:
J.CHEM.ENG.DATA 4,311(1959).

A = 7.01443 B = 1086.369 C = 228.594

B.P.(760) = 34.220

T	P EXPTL.	P CALCD.	DEV.	PERCENT
-24.170	49.99	50.13	0.14	0.29
-13.660	91.28	91.20	-0.08	-0.09
0.770	190.10	189.66	-0.44	-0.23
17.050	389.80	390.74	0.94	0.24
33.110	730.40	729.93	-0.47	-0.06
34.750	774.80	774.70	-0.10	-0.01

STANDARD DEVIATION = 0.67

ETHYL FORMATE C3H6O2

MERTL I.,POLAK J.: VSCHT-UTZCHT PRAHA,1964.

A = 7.00902 B = 1123.943 C = 218.247

B.P.(760) = 54.012

T	P EXPTL.	P CALCD.	DEV.	PERCENT
3.893	89.03	89.00	-0.03	-0.03
5.883	98.67	98.70	0.03	0.03
8.963	115.28	115.43	0.15	0.13
12.395	136.84	136.74	-0.10	-0.07
16.185	164.03	163.94	-0.09	-0.05
19.290	189.33	189.39	0.06	0.03
23.715	231.35	231.15	-0.20	-0.09
28.260	281.40	281.54	0.14	0.05
32.775	340.12	340.05	-0.07	-0.02
38.220	423.30	423.26	-0.04	-0.01
43.630	521.10	521.36	0.26	0.05
49.435	645.55	645.97	0.42	0.07
53.560	748.60	748.06	-0.54	-0.07

STANDARD DEVIATION = 0.26

METHYL ACETATE C3H6O2

MERTL I.,POLAK J.:
COLLECTION CZECHOSLOV.CHEM.COMM. 30,3526(1965).

A = 7.06524 B = 1157.630 C = 219.726

B.P.(760) = 56.926

T	P EXPTL.	P CALCD.	DEV.	PERCENT
1.758	69.02	68.95	-0.07	-0.10
4.058	78.12	78.03	-0.09	-0.11
6.568	89.03	89.05	0.02	0.02
8.563	98.67	98.71	0.04	0.04
11.635	115.28	115.26	-0.02	-0.02
15.145	136.84	136.92	0.08	0.06
18.975	164.03	164.26	0.23	0.14
22.060	189.33	189.41	0.08	0.04
26.520	231.35	231.28	-0.07	-0.03
31.060	281.40	281.34	-0.06	-0.02
35.590	340.12	339.73	-0.39	-0.11
41.090	423.30	423.40	0.10	0.02
46.500	521.10	521.13	0.03	0.00
52.310	645.55	645.37	-0.18	-0.03
55.840	731.40	731.69	0.29	0.04

STANDARD DEVIATION = 0.17

PROPIONIC ACID C3H6O2

KAHLBAUM G.W.A.:

BER. 16,2476(1883).

A = 6.40330 B = 950.175 C = 130.343

B.P.(760) = 139.402

T	P EXPTL.	P CALCD.	DEV.	PERCENT
56.500	21.31	20.79	-0.52	-2.43
57.600	22.46	22.27	-0.19	-0.86
63.500	31.34	31.73	0.39	1.26
68.800	41.70	42.85	1.15	2.77
69.200	44.20	43.81	-0.39	-0.89
70.400	47.30	46.78	-0.52	-1.11
139.400	760.00	759.95	-0.05	-0.01

STANDARD DEVIATION = 0.75

PROPIONIC ACID C3H6O2

DREISBACH R.R.,SHRADER S.A.:

IND.ENG.CHEM. 41, 2879(1949).

A = 7.71423 B = 1733.418 C = 217.724

B.P.(100) = 85.627

T	P EXPTL.	P CALCD.	DEV.	PERCENT
72.390	57.04	54.86	-2.18	-3.82
76.750	66.39	67.26	0.87	1.30
79.680	75.86	76.86	1.00	1.32
90.730	123.76	124.32	0.56	0.45
114.620	315.52	315.13	-0.39	-0.12
128.340	507.50	507.30	-0.20	-0.04

STANDARD DEVIATION = 1.53

THIACYCLOBUTANE C3H6S

SCOTT D.W.,ET AL.: J.AM.CHEM.SOC.75,2795(1953).

A = 7.01616 B = 1321.007 C = 224.474

B.P.(760) = 94.968

T	P EXPTL.	P CALCD.	DEV.	PERCENT
48.357	149.41	149.39	-0.02	-0.02
54.044	187.57	187.58	0.01	0.00
59.771	233.72	233.74	0.02	0.01
65.534	289.13	289.13	0.00	0.00
71.341	355.22	355.23	0.01	0.00
77.187	433.56	433.57	0.01	0.00
83.073	525.86	525.85	-0.01	0.00
88.998	633.99	633.94	-0.05	-0.01
94.968	760.00	759.99	-0.01	0.00
100.977	906.06	906.07	0.01	0.00
107.027	1074.60	1074.58	-0.02	0.00
113.118	1268.00	1268.05	0.05	0.00
119.249	1489.10	1489.13	0.03	0.00
125.421	1740.80	1740.70	-0.10	-0.01
131.639	2026.00	2026.04	0.04	0.00

STANDARD DEVIATION = 0.04

2-BROMOPROPANE C3H7BR

REX A.: Z.PHYSIK.CHEM. 55,355(1906).

A = 6.92493 B = 1176.460 C = 231.274

B.P.(100) = 7.605

T	P EXPTL.	P CALCD.	DEV.	PERCENT
0.000	69.10	68.87	-0.23	-0.33
10.000	111.40	111.92	0.52	0.46
20.000	175.40	174.96	-0.44	-0.25
30.000	264.20	264.33	0.13	0.05

STANDARD DEVIATION = 0.72

1-CHLOROPROPANE C3H7CL

KEMME R.H.,KREPS S.I.:
J.CHEM.ENG.DATA 14,98(1969).

A = 6.92648 B = 1110.191 C = 227.944

B.P.(760) = 46.471

T	P EXPTL.	P CALCD.	DEV.	PERCENT
-25.100	28.80	28.40	-0.40	-1.38
-19.600	39.90	39.61	-0.29	-0.72
-13.700	55.40	55.54	0.14	0.25
-6.900	79.50	80.17	0.67	0.84
1.200	120.50	120.65	0.15	0.12
12.100	200.20	200.23	0.03	0.01
21.800	303.90	302.80	-1.10	-0.36
34.900	504.00	504.32	0.32	0.06
47.100	776.00	776.35	0.35	0.05

STANDARD DEVIATION = 0.60

2-CHLOROPROPANE C3H7CL

REX A.:
Z.PHYSIK.CHEM. 55,355(1906).

A = 7.77125 B = 1582.487 C = 288.286

B.P.(100) = -14.084

T	P EXPTL.	P CALCD.	DEV.	PERCENT
0.000	191.70	191.40	-0.30	-0.15
10.000	291.50	292.40	0.90	0.31
20.000	435.50	434.58	-0.92	-0.21
30.000	629.70	630.00	0.30	0.05

STANDARD DEVIATION = 1.36

N,N-DIMETHYLFORMAMIDE C3H7NO

RAM G.,SHARAF A.R.:
J.IND.CHEM.SOC. 45,13(1968).

A = 6.92796 B = 1400.869 C = 196.434

B.P.(100) = 87.836

T	P EXPTL.	P CALCD.	DEV.	PERCENT
30.000	5.50	5.51	0.01	0.21
40.000	10.00	10.07	0.07	0.68
50.000	17.50	17.51	0.01	0.08
60.000	29.50	29.18	-0.32	-1.09
70.000	46.50	46.78	0.28	0.61
80.000	72.50	72.49	-0.01	-0.01
90.000	109.00	108.95	-0.05	-0.05

STANDARD DEVIATION = 0.22

N-METHYLACETAMIDE C3H7NO

RAM G.,SHARAF A.R.:
J.IND.CHEM.SOC. 45,13(1968).

A = 2.63112 B = 121.682 C = -9.305

B.P.(10) = 83.906

T	P EXPTL.	P CALCD.	DEV.	PERCENT
40.000	0.50	0.05	-0.45	-90.71
50.000	0.50	0.44	-0.06	-12.50
60.000	1.50	1.70	0.20	13.43
70.000	3.50	4.23	0.73	20.84
80.000	8.00	8.13	0.13	1.58
90.000	13.50	13.28	-0.22	-1.63

STANDARD DEVIATION = 0.53

ISO-PROPYL NITRATE C3H7NO3

GRAY P.,PRATT M.W.T.: J.CHEM.SOC. 1957,2163.

A = 7.26632 B = 1434.396 C = 225.202

B.P.(100) = 47.170

T	P EXPTL.	P CALCD.	DEV.	PERCENT
0.000	7.30	7.89	0.59	8.05
10.000	14.60	14.71	0.11	0.78
20.000	26.50	26.09	-0.41	-1.55
30.000	44.50	44.23	-0.27	-0.62
40.000	72.10	72.05	-0.05	-0.07
50.000	113.00	113.28	0.28	0.25
60.000	172.00	172.55	0.55	0.32
70.000	256.00	255.43	-0.57	-0.22

STANDARD DEVIATION = 0.51

1-NITROPROPANE C3H7NO2

DREISBACH R.R.,SHRADER S.A.: IND.ENG.CHEM.41,2879(1949).

A = 7.11462 B = 1467.448 C = 215.232

B.P.(760) = 131.371

T	P EXPTL.	P CALCD.	DEV.	PERCENT
58.650	57.04	57.10	0.06	0.11
62.000	66.39	66.28	-0.11	-0.16
65.110	75.86	75.88	0.02	0.03
76.970	123.76	123.76	0.00	0.00
102.720	315.52	315.72	0.20	0.06
117.560	507.50	507.12	-0.38	-0.08
131.380	760.00	760.20	0.20	0.03

STANDARD DEVIATION = 0.25

PROPYL NITRATE C3H7NO3

GRAY P.,PRATT M.W.T.: J.CHEM.SOC. 1957,2163.

A = 6.95491 B = 1294.360 C = 206.749

B.P.(100) = 54.479

T	P EXPTL.	P CALCD.	DEV.	PERCENT
0.000	5.30	4.95	-0.35	-6.65
10.000	10.00	9.62	-0.38	-3.79
20.000	17.80	17.64	-0.16	-0.88
30.000	30.60	30.74	0.14	0.45
40.000	50.70	51.20	0.50	0.98
50.000	81.40	81.95	0.55	0.68
60.000	128.00	126.64	-1.37	-1.07
70.000	189.00	189.62	0.62	0.33

STANDARD DEVIATION = 0.79

PROPANE C3H8

KEMP J.D.,EGAN C.J.: J.AM.CHEM.SOC. 60,1521(1938).

A = 6.84343 B = 818.540 C = 248.677

B.P.(760) = -42.111

T	P EXPTL.	P CALCD.	DEV.	PERCENT
-107.130	11.61	11.50	-0.11	-0.97
-99.880	22.00	22.00	0.00	-0.01
-93.358	37.39	37.44	0.05	0.14
-85.853	65.44	65.51	0.07	0.10
-78.069	111.03	111.08	0.05	0.05
-70.290	179.92	179.83	-0.09	-0.05
-63.274	268.33	268.23	-0.10	-0.04
-58.218	351.66	351.32	-0.34	-0.10
-52.941	458.13	458.74	0.61	0.13
-48.102	578.89	578.70	-0.19	-0.03
-44.394	686.19	686.33	0.14	0.02
-41.738	772.71	772.61	-0.11	-0.01

STANDARD DEVIATION = 0.26

METHYL ETHYL ETHER C3H8O

ARONOVICH KH.A.,KASTORSKII L.P.,FEDOROVA K.F.:
ZH.FIZ.KHIM. 41,20(1967).

A = 5.51755 B = 434.530 C = 158.029

B.P.(760) = 6.770

T	P EXPTL.	P CALCD.	DEV.	PERCENT
5.000	712.30	711.53	-0.77	-0.11
6.700	758.70	758.05	-0.65	-0.09
6.900	760.00	763.65	3.65	0.48
7.250	773.70	773.53	-0.17	-0.02
7.700	788.40	786.35	-2.05	-0.26

STANDARD DEVIATION = 3.05

1-PROPANOL C3H8O

AMBROSE D.,SPRAKE C.H.S.:
J.CHEM.THERMODYNAMICS 2,631(1970).

A = 7.74416 B = 1437.686 C = 198.463

B.P.(760) = 97.153

T	P EXPTL.	P CALCD.	DEV.	PERCENT
60.168	153.30	153.23	-0.07	-0.05
65.599	199.36	199.37	0.01	0.01
70.255	247.70	247.74	0.04	0.02
74.507	300.03	300.15	0.12	0.04
78.022	350.10	350.18	0.08	0.02
81.174	400.80	400.78	-0.02	0.00
83.931	449.84	449.89	0.05	0.01
86.490	499.83	499.83	0.00	0.00
88.856	550.05	550.00	-0.05	-0.01
91.026	599.69	599.62	-0.07	-0.01
93.143	651.63	651.52	-0.11	-0.02
94.955	699.00	698.84	-0.16	-0.02
96.837	751.04	750.94	-0.10	-0.01
97.595	772.80	772.80	0.00	0.00
98.513	800.06	799.98	-0.08	-0.01
100.155	850.63	850.55	-0.08	-0.01
101.666	899.35	899.37	0.02	0.00
103.166	949.94	950.08	0.14	0.01
104.575	999.53	999.81	0.28	0.03

STANDARD DEVIATION = 0.11

2-PROPANOL C3H8O

AMBROSE D.,SPRAKE C.H.S.: J.CHEM.THERMODYNAMICS 2,631(1970).

A = 7.74021 B = 1359.517 C = 197.527

B.P.(760) = 82.243

T	P EXPTL.	P CALCD.	DEV.	PERCENT
52.323	199.07	199.01	-0.06	-0.03
56.779	247.85	247.87	0.02	0.01
60.798	300.15	300.19	0.04	0.01
64.091	349.62	349.65	0.03	0.01
67.087	400.32	400.38	0.06	0.02
69.704	449.52	449.56	0.04	0.01
72.131	499.55	499.55	0.00	0.00
74.372	549.69	549.71	0.02	0.00
76.454	600.03	599.97	-0.06	-0.01
78.431	651.22	651.14	-0.08	-0.01
80.160	698.87	698.80	-0.07	-0.01
81.931	750.65	750.55	-0.10	-0.01
82.958	782.06	781.96	-0.10	-0.01
85.090	850.69	850.65	-0.04	0.00
86.550	900.42	900.48	0.06	0.01
87.922	949.38	949.47	0.09	0.01
89.261	999.22	999.35	0.13	0.01

STANDARD DEVIATION = 0.07

ETHYLENEGLYCOL MONOMETHYL ETHER C3H8O2

PICK J.,FRIED V.,HALA E.,VILIM O.:
COLLECTION CZECH.CHEM.COMM.21,260(1956).

A = 7.84980 B = 1793.982 C = 236.877

B.P.(760) = 124.159

T	P EXPTL.	P CALCD.	DEV.	PERCENT
56.100	53.80	53.27	-0.53	-0.98
59.600	63.30	62.92	-0.38	-0.60
61.600	68.80	69.08	0.28	0.40
65.300	81.70	81.83	0.13	0.16
67.200	88.70	89.13	0.43	0.49
69.700	99.60	99.57	-0.03	-0.03
72.800	114.10	113.95	-0.15	-0.13
75.600	128.00	128.42	0.42	0.33
77.800	140.50	140.85	0.35	0.25
81.300	162.48	162.73	0.25	0.16
88.200	215.20	214.36	-0.84	-0.39
91.400	243.10	242.63	-0.47	-0.19
94.400	271.50	271.92	0.42	0.15
102.200	363.70	362.25	-1.45	-0.40
105.300	404.00	404.52	0.52	0.13
112.600	519.60	520.54	0.94	0.18
117.200	605.80	606.94	1.14	0.19
120.200	668.60	669.44	0.84	0.13
123.500	746.30	744.25	-2.05	-0.27

STANDARD DEVIATION = 0.85

METHYLAL C3H8O2

NICOLINI E.,LAFFITTE P.:
COMPTES RENDUS 229,757(1949).

A = 6.87215 B = 1049.221 C = 220.570

B.P.(100) = -5.219

T	P EXPTL.	P CALCD.	DEV.	PERCENT
0.000	130.70	130.40	-0.30	-0.23
5.000	166.00	166.24	0.24	0.14
10.000	209.40	209.70	0.30	0.14
15.000	261.50	261.92	0.42	0.16
20.000	325.10	324.15	-0.95	-0.29
25.000	397.80	397.69	-0.11	-0.03
30.000	483.50	483.96	0.46	0.09
35.000	584.50	584.42	-0.08	-0.01

STANDARD DEVIATION = 0.55

GLYCEROL C3H8O3

RICHARDSON A.: J.CHEM.SOC.(LONDON),49,761(1886).

A = 6.16501 B = 1036.056 C = 28.097

B.P.(100) = 220.656

T	P EXPTL.	P CALCD.	DEV.	PERCENT
183.250	20.46	18.32	-2.14	-10.48
183.400	20.51	18.46	-2.05	-9.98
192.000	30.62	28.69	-1.93	-6.30
195.300	34.37	33.67	-0.70	-2.03
199.800	41.81	41.58	-0.23	-0.56
200.800	44.87	43.52	-1.35	-3.01
201.300	45.61	44.52	-1.09	-2.39
205.800	52.77	54.38	1.61	3.06
211.500	65.61	69.32	3.71	5.65
217.300	86.73	87.71	0.98	1.13
220.300	100.81	98.63	-2.18	-2.16
224.300	115.25	114.85	-0.40	-0.35
227.000	130.54	126.93	-3.61	-2.76
229.500	137.95	138.99	1.04	0.76
237.100	183.50	181.24	-2.26	-1.23
241.800	201.23	211.98	10.75	5.34
246.400	239.95	245.82	5.87	2.45
248.500	258.63	262.59	3.96	1.53
257.300	347.09	342.59	-4.50	-1.30
260.400	385.33	374.78	-10.55	-2.74

STANDARD DEVIATION = 4.57

1-PROPANETHIOL C3H8S

PENNINGTON R.E.,ET AL.:
J.AM.CHEM.SOC. 78,3266(1956).

A = 6.92825 B = 1183.175 C = 224.607

B.P.(760) = 67.719

T	P EXPTL.	P CALCD.	DEV.	PERCENT
24.275	149.41	149.38	-0.03	-0.02
29.563	187.57	187.59	0.02	0.01
34.891	233.72	233.76	0.04	0.02
40.254	289.13	289.13	0.00	0.00
45.663	355.22	355.22	0.00	0.00
51.113	433.56	433.54	-0.02	0.00
56.605	525.86	525.83	-0.03	-0.01
62.139	633.99	633.93	-0.06	-0.01
67.719	760.00	759.98	-0.02	0.00
73.341	906.06	906.09	0.03	0.00
79.004	1074.60	1074.59	-0.01	0.00
84.710	1268.00	1268.04	0.04	0.00
90.464	1489.20	1489.32	0.12	0.01
96.255	1740.80	1740.86	0.06	0.00
102.088	2026.00	2025.83	-0.17	-0.01

STANDARD DEVIATION = 0.07

2-PROPANETHIOL C3H8S

MCCULLOUGH J.P.,ET AL.:
J.AM.CHEM.SOC. 76,4796(1954).

A = 6.87791 B = 1114.206 C = 226.195

B.P.(760) = 52.559

T	P EXPTL.	P CALCD.	DEV.	PERCENT
10.697	149.41	149.44	0.03	0.02
15.770	187.57	187.54	-0.03	-0.02
20.899	233.72	233.71	-0.01	-0.01
26.071	289.13	289.15	0.02	0.01
31.282	355.22	355.23	0.01	0.00
36.536	433.56	433.55	-0.01	0.00
41.833	525.86	525.84	-0.02	0.00
47.175	633.99	634.00	0.01	0.00
52.558	760.00	759.98	-0.03	0.00
57.985	906.06	906.00	-0.06	-0.01
63.461	1074.60	1074.62	0.02	0.00
68.979	1268.00	1268.13	0.13	0.01
74.540	1489.10	1489.23	0.13	0.01
80.143	1740.80	1740.72	-0.08	0.00
85.795	2026.00	2025.90	-0.10	-0.01

STANDARD DEVIATION = 0.07

2-THIABUTANE C3H8S

SCOTT D.W.,ET AL.:
J.AM.CHEM.SOC. 73,261(1951).

A = 6.93773 B = 1182.117 C = 224.730

B.P.(760) = 66.653

T	P EXPTL.	P CALCD.	DEV.	PERCENT
23.435	149.41	149.38	-0.03	-0.02
28.695	187.57	187.57	0.00	0.00
33.997	233.72	233.75	0.03	0.01
39.339	289.13	289.18	0.05	0.02
44.717	355.22	355.24	0.02	0.01
50.136	433.56	433.53	-0.03	-0.01
55.600	525.86	525.83	-0.03	-0.01
61.104	633.99	633.94	-0.05	-0.01
66.655	760.00	760.04	0.04	0.00
72.241	906.06	906.03	-0.03	0.00
77.870	1074.60	1074.46	-0.14	-0.01
83.551	1268.00	1268.18	0.18	0.01
89.265	1489.10	1489.22	0.12	0.01
95.020	1740.80	1740.69	-0.11	-0.01
100.825	2026.00	2026.01	0.01	0.00

STANDARD DEVIATION = 0.09

2-THIABUTANE C3H8S

OSBORN A.N.,DOUSLIN D.R.:
J.CHEM.ENG.DATA 11,502(1966).

A = 6.93773 B = 1182.117 C = 224.730

B.P.(760) = 66.653

T	P EXPTL.	P CALCD.	DEV.	PERCENT
23.435	149.41	149.38	-0.03	-0.02
28.695	187.57	187.57	0.00	0.00
33.997	233.72	233.75	0.03	0.01
39.339	289.13	289.18	0.05	0.02
44.717	355.22	355.24	0.02	0.01
50.136	433.56	433.53	-0.03	-0.01
55.600	525.86	525.83	-0.03	-0.01
61.104	633.99	633.94	-0.05	-0.01
66.655	760.00	760.04	0.04	0.00
72.241	906.06	906.03	-0.03	0.00
77.870	1074.60	1074.46	-0.14	-0.01
83.551	1268.00	1268.18	0.18	0.01
89.265	1489.10	1489.22	0.12	0.01
95.020	1740.80	1740.69	-0.11	-0.01
100.825	2026.00	2026.01	0.01	0.00

STANDARD DEVIATION = 0.09

TRIMETHYLALUMINIUM C3H9AL

MCCULLOUGH J.P.,ET AL.:
J.PHYS.CHEM. 67,677(1963).

A = 7.57029 B = 1734.715 C = 242.778

B.P.(760) = 127.139

T	P EXPTL.	P CALCD.	DEV.	PERCENT
63.818	81.64	81.72	0.08	0.09
66.779	92.52	92.56	0.04	0.04
69.750	104.63	104.64	0.01	0.01
72.722	118.06	118.03	-0.03	-0.03
75.707	132.95	132.90	-0.05	-0.04
78.701	149.41	149.36	-0.05	-0.03
84.696	187.57	187.51	-0.06	-0.03
90.712	233.72	233.66	-0.06	-0.02
96.748	289.13	289.11	-0.02	-0.01
102.802	355.22	355.28	0.06	0.02
108.869	433.56	433.68	0.12	0.03
114.949	525.86	526.04	0.18	0.03
121.039	633.99	634.15	0.16	0.02
127.122	760.00	759.63	-0.37	-0.05

STANDARD DEVIATION = 0.15

METHYL BORATE C3H9BO3

CHRISTOPHER P.M.,SHILMAN A.:
J.CHEM.ENG.DATA 12,333(1967).

A = 7.64600 B = 1491.512 C = 245.547

B.P.(760) = 67.455

T	P EXPTL.	P CALCD.	DEV.	PERCENT
31.200	180.40	180.53	0.13	0.07
38.000	244.90	243.11	-1.79	-0.73
45.000	325.70	325.49	-0.21	-0.06
50.200	397.10	400.68	3.58	0.90
54.100	465.00	466.05	1.05	0.23
56.700	513.20	514.35	1.15	0.22
60.400	591.10	590.11	-0.99	-0.17
63.400	662.70	658.07	-4.63	-0.70
67.400	757.10	758.54	1.44	0.19

STANDARD DEVIATION = 2.68

METHYLBORIC ANHYDRIDE C3H9B3O3

BURG A.B.:
J.AM.CHEM.SOC. 62,2228(1940).

A = 8.00409 B = 1726.125 C = 257.902

B.P.(100) = 29.590

T	P EXPTL.	P CALCD.	DEV.	PERCENT
0.000	20.50	20.47	-0.03	-0.14
14.300	46.50	46.00	-0.50	-1.08
18.900	58.00	58.63	0.63	1.09
21.700	67.80	67.70	-0.10	-0.15
24.800	79.20	79.12	-0.08	-0.11
32.200	113.00	113.24	0.24	0.22
39.900	161.50	161.38	-0.12	-0.07
47.700	227.10	226.88	-0.22	-0.10
54.800	304.70	304.82	0.12	0.04

STANDARD DEVIATION = 0.37

TRIMETHYLCHLOROSILANE C3H9CLSI

JENKINS A.C.,CHAMBERS G.F.:
IND.ENG.CHEM. 46,2367(1954).

A = 7.05581 B = 1245.506 C = 240.727

B.P.(760) = 57.598

T	P EXPTL.	P CALCD.	DEV.	PERCENT
2.600	86.70	86.53	-0.17	-0.20
27.000	252.30	253.32	1.02	0.40
38.700	397.80	396.69	-1.11	-0.28
43.000	464.10	463.46	-0.64	-0.14
48.500	561.20	561.68	0.48	0.08
55.600	711.90	712.31	0.41	0.06

STANDARD DEVIATION = 1.02

TRIMETHYLCHLORSILANE C3H9CLSI

CAPKOVA A.,FRIED V.:

COLLECTION 29,336(1964).

A = 6.75981 B = 1082.730 C = 221.300

B.P.(100) = 6.173

T	P EXPTL.	P CALCD.	DEV.	PERCENT
1.100	78.20	77.88	-0.32	-0.41
5.860	98.70	98.50	-0.20	-0.20
12.960	136.90	137.37	0.47	0.34
20.340	189.50	190.13	0.63	0.33
29.920	281.60	281.79	0.19	0.07
34.740	340.20	339.67	-0.53	-0.16
40.630	423.40	422.81	-0.59	-0.14
46.450	521.20	520.00	-1.20	-0.23
52.880	645.40	646.90	1.50	0.23

STANDARD DEVIATION = 0.92

1-PROPYLAMINE C3H9N

OSBORN A.G.,DOUSLIN D.R.:

J.CHEM.ENG.DATA13,534(1968).

A = 6.92651 B = 1044.051 C = 210.836

B.P.(760) = 47.229

T	P EXPTL.	P CALCD.	DEV.	PERCENT
22.973	289.13	289.13	0.00	0.00
27.750	355.22	355.22	0.00	0.00
32.564	433.56	433.56	0.00	0.00
37.414	525.86	525.84	-0.02	0.00
42.304	633.99	634.01	0.02	0.00
47.229	760.00	759.99	-0.01	0.00
52.193	906.06	906.06	0.00	0.00
57.195	1074.60	1074.57	-0.03	0.00
62.235	1268.00	1268.02	0.02	0.00
67.314	1489.10	1489.17	0.07	0.00
72.430	1740.80	1740.74	-0.06	0.00
77.587	2026.00	2025.98	-0.02	0.00

STANDARD DEVIATION = 0.03

2-PROPYLAMINE C3H9N

OSBORN A.G.,DOUSLIN D.R.: J.CHEM.ENG.DATA13,534(1968).

A = 6.89025 B = 985.685 C = 214.074

B.P.(760) = 31.767

T	P EXPTL.	P CALCD.	DEV.	PERCENT
3.922	233.72	233.71	-0.01	0.00
8.471	289.13	289.14	0.01	0.00
13.055	355.22	355.22	0.00	0.00
17.677	433.56	433.56	0.00	0.00
22.336	525.86	525.86	0.00	0.00
27.032	633.99	633.98	-0.01	0.00
31.766	760.00	759.97	-0.03	0.00
36.539	906.06	906.07	0.01	0.00
41.349	1074.60	1074.55	-0.05	-0.01
46.199	1268.00	1268.05	0.05	0.00
51.087	1489.10	1489.17	0.08	0.01
56.014	1740.80	1740.84	0.04	0.00
60.978	2026.00	2025.91	-0.09	0.00

STANDARD DEVIATION = 0.05

TRIMETHYLAMINE C_3H_9N

ASTON J.G.,ET AL.:

J.AM.CHEM.SOC. 66,1171(1944).

A = 6.85755 B = 955.944 C = 237.515

B.P.(760) = 2.869

T	P EXPTL.	P CALCD.	DEV.	PERCENT
-80.315	6.04	5.98	-0.06	-1.04
-74.081	10.25	10.20	-0.05	-0.52
-62.339	25.16	25.15	-0.01	-0.05
-51.938	50.83	50.86	0.03	0.06
-46.842	69.79	69.83	0.04	0.06
-41.774	94.12	94.16	0.04	0.04
-35.617	132.64	132.67	0.03	0.02
-28.507	192.20	192.24	0.04	0.02
-24.155	238.31	238.30	-0.01	0.00
-23.067	251.23	251.11	-0.12	-0.05
-20.164	288.03	288.01	-0.02	-0.01
-15.974	348.82	348.80	-0.02	-0.01
-11.422	426.05	426.04	-0.01	0.00
-8.985	472.83	472.65	-0.18	-0.04
-7.399	505.14	505.10	-0.04	-0.01
-3.113	601.61	601.63	0.02	0.00
0.780	701.27	701.39	0.12	0.02
2.928	761.51	761.72	0.21	0.03
3.454	777.15	777.09	-0.06	-0.01

STANDARD DEVIATION = 0.09

O,N,N,-TRIMETHYLHYDROXYLAMINE C3H9NO

BISSOT T.C.,PARRY R.W.,CAMPBELL D.H.:
J.AM.CHEM.SOC. 79,796(1957).

A = 6.76581 B = 979.551 C = 222.179

B.P.(100) = -16.642

T	P EXPTL.	P CALCD.	DEV.	PERCENT
-78.500	0.90	0.89	-0.01	-1.38
-63.500	3.60	3.91	0.31	8.73
-45.200	17.30	17.02	-0.28	-1.61
-39.000	26.50	26.20	-0.30	-1.13
-35.000	35.50	34.09	-1.41	-3.98
-29.800	47.50	47.21	-0.29	-0.61
-25.200	61.70	62.08	0.38	0.62
-20.000	81.50	83.34	1.84	2.26
-14.100	113.60	114.35	0.75	0.66
-9.300	144.40	146.01	1.61	1.11
-4.800	186.00	181.81	-4.19	-2.25
0.000	226.90	227.50	0.60	0.26
3.900	270.70	271.04	0.34	0.13
8.300	328.00	327.91	-0.09	-0.03
11.800	379.10	379.60	0.50	0.13
15.900	448.00	448.15	0.15	0.03
20.000	525.40	526.11	0.71	0.14
23.300	597.50	596.28	-1.22	-0.20

STANDARD DEVIATION = 1.40

DIMETHYL METHYLPHOSPHONATE C3H9O3P

KOSOLAPOFF G.M.:
J.CHEM.SOC. 1955,2964.

A = 5.10641 B = 573.682 C = 68.153

B.P.(100) = 116.523

T	P EXPTL.	P CALCD.	DEV.	PERCENT
63.000	4.90	5.40	0.50	10.18
70.300	9.70	9.18	-0.52	-5.35
80.500	18.50	17.67	-0.83	-4.49
90.000	30.50	30.13	-0.37	-1.20
98.000	44.50	45.05	0.55	1.23
106.000	64.50	64.91	0.41	0.63
117.000	100.50	101.86	1.36	1.35
128.000	154.50	151.97	-2.53	-1.64
135.000	190.50	191.66	1.16	0.61

STANDARD DEVIATION = 1.38

TRIMETHYLHYDRAZINE C3H10N2

ASTON J.G.,ZOLKI T.P.,WOOD J.L.:
J.AM.CHEM.SOC. 77,281(1955).

A = 7.10680 B = 1189.876 C = 222.056

B.P.(100) = 10.943

T	P EXPTL.	P CALCD.	DEV.	PERCENT
-16.221	21.20	21.19	-0.01	-0.06
-9.369	32.52	32.53	0.01	0.03
-4.102	44.41	44.41	0.00	0.00
1.062	59.38	59.41	0.03	0.05
6.408	79.19	79.18	-0.01	-0.01
6.416	79.25	79.22	-0.03	-0.04
13.826	115.44	115.46	0.02	0.01
13.835	115.51	115.51	0.00	0.00

STANDARD DEVIATION = 0.02

TRIMETHYLSILANOL C3H10OSI

GRUBB W.T.,OSTHOFF R.C.:
J.AM.CHEM.SOC. 75,2230(1953).

A = 8.12659 B = 1657.645 C = 219.193

B.P.(100) = 51.373

T	P EXPTL.	P CALCD.	DEV.	PERCENT
18.000	13.90	13.74	-0.16	-1.15
22.900	19.30	19.03	-0.27	-1.40
24.500	21.20	21.11	-0.09	-0.44
24.600	21.40	21.24	-0.16	-0.74
29.700	29.50	29.28	-0.22	-0.76
32.700	34.90	35.14	0.24	0.70
37.700	47.00	47.20	0.20	0.42
41.600	59.30	58.94	-0.36	-0.61
46.600	77.00	77.62	0.62	0.81
54.100	115.40	115.12	-0.28	-0.25
61.100	161.80	163.16	1.36	0.84
64.500	193.90	192.08	-1.82	-0.94
68.900	234.40	235.90	1.50	0.64
76.400	329.20	330.16	0.96	0.29
79.100	374.90	371.09	-3.81	-1.02
84.700	468.00	469.77	1.77	0.38

STANDARD DEVIATION = 1.44

BORINE TRIMETHYLAMINE C3H12BN

BURG A.B.,SCHLESINGER H.I.: J.AM.CHEM.SOC. 59,780(1937).

A = 7.35084 B = 870.229 C = 258.027

B.P.(100) = -95.393

T	P EXPTL.	P CALCD.	DEV.	PERCENT
-137.000	1.40	1.45	0.05	3.36
-122.000	9.00	8.98	-0.02	-0.20
-121.200	9.70	9.79	0.09	0.93
-120.300	10.70	10.77	0.07	0.68
-112.800	22.50	22.84	0.34	1.50
-111.800	25.40	25.10	-0.30	-1.19
-103.500	52.80	52.39	-0.41	-0.77
-95.600	98.60	98.44	-0.16	-0.16
-87.500	175.00	176.88	1.88	1.07
-87.200	181.00	180.56	-0.44	-0.24
-78.800	314.00	312.89	-1.11	-0.35

STANDARD DEVIATION = 0.82

CARBON SUBOXIDE C302

MCDOUGALL L.A.,KILPATRICK J.E.:
J.CHEM.PHYS. 42,2311(1965).

A = 7.19004 B = 1101.326 C = 249.182

B.P.(100) = -36.982

T	P EXPTL.	P CALCD.	DEV.	PERCENT
-112.188	0.16	0.14	-0.02	-11.55
-111.039	0.19	0.17	-0.03	-13.12
-108.467	0.24	0.23	-0.01	-3.80
-105.477	0.34	0.34	0.00	-1.20
-102.428	0.50	0.48	-0.02	-3.06
-102.033	0.54	0.51	-0.03	-5.98
-99.547	0.73	0.68	-0.05	-7.40
-98.124	0.76	0.79	0.03	4.34
-95.741	1.08	1.03	-0.05	-4.70
-91.938	1.58	1.54	-0.05	-2.85
-87.903	2.32	2.30	-0.02	-0.95
-82.999	3.60	3.65	0.05	1.52
-77.876	5.69	5.77	0.08	1.37
-73.217	8.51	8.54	0.03	0.31
-67.935	13.11	12.99	-0.12	-0.91
-62.491	19.54	19.54	0.00	-0.02
-57.113	28.52	28.58	0.06	0.20
-51.776	40.80	40.84	0.04	0.09
-46.561	56.81	56.84	0.03	0.05
-41.412	77.56	77.51	-0.05	-0.07
-38.666	90.91	90.88	-0.03	-0.03
-35.653	107.78	107.72	-0.06	-0.05
-32.680	126.76	126.80	0.04	0.03
-29.686	148.78	148.77	-0.01	-0.01
-26.985	171.10	171.20	0.10	0.06
-24.373	195.53	195.48	-0.05	-0.03

STANDARD DEVIATION = 0.05

TRIFLUOROACETIC ANHYDRIDE C4F6O3

KREGLEWSKI A.: BULL.ACAD.POL.SCI. 10,629(1962).

A = 6.13577 B = 1026.129 C = 201.976

B.P.(10) = -2.176

T	P EXPTL.	P CALCD.	DEV.	PERCENT
-1.830	10.17	10.21	0.04	0.36
8.000	17.72	17.74	0.02	0.10
16.730	27.89	27.79	-0.10	-0.34
23.220	37.99	37.95	-0.04	-0.12
28.380	47.94	48.00	0.06	0.12
32.810	58.14	58.25	0.11	0.19
38.650	74.36	74.36	0.00	0.00
38.870	75.11	75.03	-0.08	-0.10

STANDARD DEVIATION = 0.08

PERFLUOROCYCLOBUTANE C4F8

FURUKAVA G.T.,MCCOSKEY R.E.,REILLY M.L.: J.RESEARCH NATL.BUR.STANDARDS 52,11(1954).

A = 6.81529 B = 842.487 C = 225.189

B.P.(760) = -5.976

T	P EXPTL.	P CALCD.	DEV.	PERCENT
-31.830	226.30	226.33	0.03	0.01
-26.160	303.30	303.26	-0.04	-0.01
-20.140	406.50	406.48	-0.02	-0.01
-13.640	547.30	547.36	0.06	0.01
-9.270	661.90	661.90	0.00	0.00
-6.500	743.70	743.68	-0.02	0.00
-5.260	782.60	782.75	0.15	0.02
-2.200	886.20	886.01	-0.19	-0.02
0.480	984.80	984.85	0.05	0.01

STANDARD DEVIATION = 0.11

PERFLUOROBUTENE C_4F_8

FOWLER R.D.,ET AL.: IND.ENG.CHEM.39,375(1947).

A = 9.22238 B = 2401.651 C = 381.507

B.P.(760) = -2.792

T	P EXPTL.	P CALCD.	DEV.	PERCENT
-28.300	264.48	264.74	0.26	0.10
-20.000	380.00	379.26	-0.74	-0.19
0.000	643.60	645.69	2.09	0.25
10.100	1231.20	1229.05	-2.15	-0.17
20.000	1740.40	1740.97	0.57	0.03

STANDARD DEVIATION = 2.23

PERFLUOROBUTANE C_4F_{10}

BROWN J.A.,MEARS W.H.: J.PHYS.CHEM. 62,960(1958).

A = 7.03513 B = 990.269 C = 240.376

B.P.(760) = -2.005

T	P EXPTL.	P CALCD.	DEV.	PERCENT
-39.880	124.64	124.75	0.11	0.09
-31.910	193.04	192.70	-0.34	-0.18
-24.320	282.72	282.98	0.26	0.09
-15.270	432.44	432.53	0.09	0.02
-3.910	703.76	703.64	-0.12	-0.02

STANDARD DEVIATION = 0.33

PERFLUOROBUTANE C4F10

SIMONS J.H., MAUSTELLER J.W.: J.CHEM.PHYS. 20, 1516(1952).

A = 6.67232 B = 822.316 C = 219.103

B.P.(100) = -43.105

T	P EXPTL.	P CALCD.	DEV.	PERCENT
-39.930	121.00	121.01	0.01	0.00
-34.730	163.00	163.02	0.02	0.01
-26.810	249.00	248.85	-0.15	-0.06
-19.540	356.00	356.23	0.23	0.06
-13.210	477.00	476.89	-0.11	-0.02

STANDARD DEVIATION = 0.21

ENNEAFLUORO-2-(FLUOROSULFATO)BUTANE C4F10O3S

DELFINO J.J., SHREEVE J.M.: INORG.CHEM. 5,308(1966).

A = 4.05331 B = 153.833 C = 55.315

B.P.(760) = 75.887

T	P EXPTL.	P CALCD.	DEV.	PERCENT
-7.150	26.20	7.23	-18.97	-72.39
6.850	55.50	37.91	-17.60	-31.70
21.350	106.80	111.36	4.56	4.27
27.950	155.40	160.62	5.22	3.36
34.750	198.60	221.45	22.85	11.51
37.950	231.20	253.44	22.24	9.62
43.150	283.60	309.74	26.14	9.22
47.650	338.50	362.47	23.97	7.08
52.650	408.80	425.07	16.27	3.98
54.950	446.50	455.18	8.68	1.94
55.550	454.00	463.16	9.16	2.02
59.150	515.60	512.12	-3.48	-0.67
62.550	582.40	559.94	-22.46	-3.86
67.050	678.80	625.37	-53.43	-7.87
68.550	714.90	647.69	-67.21	-9.40
81.250	803.50	845.01	41.51	5.17

STANDARD DEVIATION =31.66

DIACETYLENE C4H2

STRAUS F.:

BER. 59,1664(1926).

A = 4.99079 B = 326.359 C = 143.223

B.P.(10C) = -34.101

T	P EXPTL.	P CALCD.	DEV.	PERCENT
-78.200	1.60	0.94	-0.66	-41.45
-70.000	5.00	3.42	-1.58	-31.65
-60.000	12.00	11.73	-0.27	-2.26
-53.000	21.80	23.63	1.83	8.41
-50.000	28.70	30.90	2.20	7.66
-40.000	68.60	67.47	-1.13	-1.65
-37.000	85.00	82.87	-2.13	-2.51
-35.500	93.50	91.45	-2.05	-2.20
-30.000	129.40	128.33	-1.07	-0.83
-20.000	214.90	219.92	5.02	2.33
-14.000	287.50	291.90	4.40	1.53
-10.000	345.50	347.59	2.09	0.60
-5.000	431.50	426.26	-5.24	-1.21
0.000	517.60	515.35	-2.25	-0.43

STANDARD DEVIATION = 3.07

DIACETYLENE C4H2

TANNEBERGER H.:

BER. 66,484(1933).

A = 6.17072 B = 624.381 C = 185.131

B.P.(10) = -64.378

T	P EXPTL.	P CALCD.	DEV.	PERCENT
-84.800	1.50	0.89	-0.61	-40.92
-77.000	3.50	2.49	-1.01	-28.82
-73.600	4.00	3.74	-0.26	-6.59
-68.000	6.50	6.92	0.42	6.46
-62.500	11.50	12.00	0.50	4.35
-58.800	16.50	16.92	0.42	2.53
-53.600	26.50	26.53	0.03	0.11
-48.500	40.00	39.89	-0.11	-0.27
-43.500	58.00	57.84	-0.16	-0.28
-39.200	78.00	78.01	0.01	0.01

STANDARD DEVIATION = 0.55

BIS(TRIFLUOROMETHYL)ACETOXYPHOSPHINE C4H3F6O2P

PETERSON L.K.,BURG A.B.:
J.AM.CHEM.SOC. 86,2587(1964).

A = 7.39131 B = 1426.254 C = 220.371

B.P.(10) = 2.784

T	P EXPTL.	P CALCD.	DEV.	PERCENT
0.000	8.32	8.30	-0.02	-0.20
9.500	15.30	15.37	0.07	0.47
17.200	24.50	24.42	-0.08	-0.31
21.350	31.00	30.97	-0.03	-0.11
28.300	45.20	45.27	0.07	0.15
39.500	80.00	79.98	-0.02	-0.02

STANDARD DEVIATION = 0.08

2,2,3,3,4,4,4-HEPTAFLUOROBUTANOL C4H3F7O

MEEKS A.C.,GOLDFARB I.J.:
J.CHEM.ENG.DATA 12,196(1967).

A = 2.95417 B = 102.245 C = 46.265

B.P.(10) = 6.057

T	P EXPTL.	P CALCD.	DEV.	PERCENT
0.000	6.73	5.55	-1.18	-17.56
5.100	9.75	9.20	-0.55	-5.68
7.150	11.40	10.96	-0.44	-3.82
9.350	12.90	13.05	0.15	1.19
9.800	13.10	13.50	0.40	3.09
11.000	14.75	14.75	0.00	-0.02
13.200	17.30	17.17	-0.13	-0.76
15.200	18.80	19.53	0.73	3.88
15.350	19.55	19.71	0.16	0.83
16.900	21.25	21.65	0.40	1.89
18.400	23.25	23.61	0.36	1.53
19.900	24.15	25.64	1.49	6.16
21.400	27.40	27.74	0.34	1.24
23.900	31.90	31.40	-0.50	-1.56
24.800	34.15	32.77	-1.38	-4.06

STANDARD DEVIATION = 0.79

FURAN C4H4O

GUTHRIE G.B.,ET AL.: J.AM.CHEM.SOC. 74,4662(1952).

A = 6.97527 B = 1060.870 C = 227.742

B.P.(760) = 31.357

T	P EXPTL.	P CALCD.	DEV.	PERCENT
2.552	233.72	233.71	-0.01	0.00
7.267	289.13	289.14	0.01	0.00
12.018	355.22	355.27	0.05	0.01
16.797	433.56	433.54	-0.02	-0.01
21.614	525.86	525.81	-0.05	-0.01
26.469	633.99	633.99	0.00	0.00
31.357	760.00	759.99	-0.01	0.00
36.279	906.06	906.02	-0.04	0.00
41.241	1074.60	1074.65	0.05	0.00
46.232	1268.00	1267.98	-0.02	0.00
51.265	1489.10	1489.23	0.13	0.01
56.329	1740.80	1740.78	-0.02	0.00
61.430	2026.00	2025.91	-0.09	0.00

STANDARD DEVIATION = 0.06

THIOPHENE C4H4S

WADDINGTON G.,ET AL.: J.AM.CHEM.SOC. 71,797(1949).

A = 6.94710 B = 1238.618 C = 220.448

B.P.(760) = 84.158

T	P EXPTL.	P CALCD.	DEV.	PERCENT
39.061	149.41	149.34	-0.07	-0.05
44.560	187.57	187.59	0.02	0.01
50.094	233.72	233.79	0.07	0.03
55.663	289.13	289.17	0.04	0.02
61.276	355.22	355.25	0.03	0.01
66.931	433.56	433.56	0.00	0.00
72.629	525.86	525.83	-0.03	-0.01
78.370	633.99	633.93	-0.06	-0.01
84.155	760.00	759.92	-0.08	-0.01
89.985	906.06	906.00	-0.06	-0.01
95.860	1074.60	1074.58	-0.02	0.00
101.776	1268.00	1268.04	0.04	0.00
107.738	1489.10	1489.24	0.14	0.01
113.741	1740.80	1740.84	0.04	0.00
119.787	2026.00	2025.91	-0.09	0.00

STANDARD DEVIATION = 0.07

CHLOROPRENE C4H5CL

GUBKOV A.N.,FERMOR N.A.,SMIRNOV N.I.: ZH.PRIKL.KHIM. 37,2204(1964).

A = 6.16150 B = 783.451 C = 179.661

B.P.(760) = 59.146

T	P EXPTL.	P CALCD.	DEV.	PERCENT
20.000	174.00	172.82	-1.18	-0.68
30.000	265.00	265.92	0.92	0.35
40.000	389.60	393.42	3.82	0.98
50.000	568.00	562.55	-5.45	-0.96
60.000	779.00	780.73	1.73	0.22

STANDARD DEVIATION = 4.98

ETHYL TRICHLORACETATE C4H5CL3O2

USANOVICH M.,DEMBICKIJ A.: ZH.OBSHCH.KHIM. 29,1781(1959).

A = 7.72537 B = 1927.007 C = 233.734

B.P.(10) = 52.794

T	P EXPTL.	P CALCD.	DEV.	PERCENT
44.200	6.60	6.20	-0.41	-6.14
57.300	13.20	12.71	-0.49	-3.72
65.500	18.70	19.30	0.60	3.21
72.400	26.70	26.96	0.26	0.97
77.800	34.70	34.66	-0.04	-0.12
84.300	46.50	46.37	-0.13	-0.28
88.600	55.80	55.86	0.06	0.10
95.200	73.70	73.62	-0.08	-0.10

STANDARD DEVIATION = 0.41

PYRROLE

C4H5N

SCOTT D.W.,ET AL.:

J.PHYS.CHEM. 71,2263(1967).

A = 7.29469 B = 1501.563 C = 210.417

B.P.(760) = 129.774

T	P EXPTL.	P CALCD.	DEV.	PERCENT
65.671	71.87	71.78	-0.09	-0.13
68.522	81.64	81.58	-0.06	-0.08
71.374	92.52	92.48	-0.04	-0.04
74.233	104.63	104.61	-0.02	-0.02
77.098	118.06	118.07	0.01	0.01
79.970	132.95	132.98	0.03	0.02
82.847	149.41	149.46	0.05	0.03
88.622	187.57	187.67	0.10	0.05
94.422	233.72	233.85	0.13	0.06
100.244	289.13	289.23	0.10	0.04
106.096	355.22	355.31	0.09	0.03
111.972	433.56	433.59	0.03	0.01
117.875	525.86	525.81	-0.05	-0.01
123.806	633.99	633.86	-0.13	-0.02
129.764	760.00	759.77	-0.23	-0.03
135.753	906.06	905.82	-0.24	-0.03
141.768	1074.60	1074.31	-0.29	-0.03
147.812	1268.00	1267.83	-0.17	-0.01
153.884	1489.10	1489.11	0.01	0.00
159.984	1740.80	1741.03	0.23	0.01
166.109	2026.00	2026.52	0.52	0.03

STANDARD DEVIATION = 0.19

ALLYL ISOTHIOCYANATE C4H5NS

BAUER H.,BURSCHKIES K.:
BER. 68,1243(1935).

A = 5.12658 B = 791.434 C = 154.019

B.P.(10) = 37.771

T	P EXPTL.	P CALCD.	DEV.	PERCENT
10.000	2.40	2.00	-0.40	-16.61
14.000	2.70	2.61	-0.09	-3.44
18.000	3.40	3.36	-0.04	-1.32
20.000	3.70	3.79	0.09	2.42
24.000	4.80	4.80	-0.01	-0.11
28.000	6.00	6.00	0.00	0.08
30.000	6.60	6.69	0.09	1.44
34.000	8.30	8.27	-0.04	-0.42
37.000	9.60	9.62	0.02	0.25
40.000	10.90	11.15	0.25	2.33
44.000	13.30	13.48	0.18	1.38
48.000	16.10	16.18	0.08	0.49
50.000	18.00	17.67	-0.33	-1.81

STANDARD DEVIATION = 0.20

METHYL-2-THIAZOLE C4H5NS

SOULIE M.A.,GOURSOT P.,PENELOUX A.,METZGER J.:
J.CHIM.PHYS. 66,607(1969).

A = 7.04212 B = 1407.052 C = 209.326

B.P.(760) = 128.801

T	P EXPTL.	P CALCD.	DEV.	PERCENT
80.503	153.90	153.94	0.04	0.03
84.719	180.76	180.70	-0.06	-0.03
99.978	311.16	311.19	0.03	0.01
102.195	335.25	335.28	0.03	0.01
106.281	383.67	383.60	-0.07	-0.02
108.314	409.65	409.65	0.00	0.00
110.735	442.54	442.51	-0.03	-0.01
112.233	463.88	463.88	0.00	0.00
114.066	491.11	491.14	0.03	0.01
116.062	522.21	522.27	0.06	0.01
119.692	582.84	582.91	0.07	0.01
119.724	583.50	583.47	-0.03	-0.01
121.925	622.89	622.91	0.02	0.00
124.157	665.07	665.06	-0.01	0.00
124.785	677.37	677.31	-0.06	-0.01
125.730	696.13	696.10	-0.03	-0.01
126.821	718.28	718.28	0.00	0.00
127.240	726.96	726.96	0.00	0.00
128.458	752.62	752.63	0.01	0.00

1,3-BUTADIENE C4H6

HEISIG G.B.:

J.AM.CHEM.SOC. 55,2304(1933).

A = 6.85364 B = 933.586 C = 239.511

B.P.(760) = -4.518

T	P EXPTL.	P CALCD.	DEV.	PERCENT
-75.500	14.60	14.50	-0.10	-0.67
-63.400	35.40	35.69	0.29	0.81
-51.600	77.10	76.81	-0.29	-0.38
-39.400	153.40	154.28	0.88	0.57
-38.600	161.70	161.02	-0.68	-0.42
-32.700	219.00	218.50	-0.50	-0.23
-26.100	301.40	301.34	-0.06	-0.02
-19.900	400.50	400.45	-0.05	-0.01
-15.500	484.20	485.35	1.15	0.24
-10.400	599.90	600.93	1.03	0.17
-5.600	729.70	728.52	-1.18	-0.16
-1.500	854.00	853.48	-0.52	-0.06

STANDARD DEVIATION = 0.80

1,3-BUTADIENE C4H6

VAVGHAN W.E.:

J.AM.CHEM.SOC. 54,3863(1932).

A = 7.73810 B = 1313.362 C = 275.463

B.P.(100) = -46.578

T	P EXPTL.	P CALCD.	DEV.	PERCENT
-81.900	9.20	8.97	-0.23	-2.48
-79.100	11.00	11.21	0.21	1.92
-38.500	156.90	156.90	0.00	0.00
-24.000	327.50	327.50	0.00	0.00

STANDARD DEVIATION = 0.31

1-BUTYNE C4H6

ASTON J.G.,MASRANGELO V.R.,MOESSEN G.W.:
J.AM.CHEM.SOC. 72,5287(1950).

A = 7.02038 B = 1005.739 C = 234.895

B.P.(760) = 8.062

T	P EXPTL.	P CALCD.	DEV.	PERCENT
-78.758	3.92	3.79	-0.13	-3.24
-68.911	9.24	9.14	-0.10	-1.04
-58.723	20.54	20.49	-0.05	-0.24
-48.679	41.64	41.63	-0.01	-0.01
-41.225	67.08	67.19	0.11	0.17
-36.414	89.65	89.78	0.13	0.15
-29.761	131.04	131.08	0.04	0.03
-23.857	179.82	179.77	-0.05	-0.03
-18.745	233.07	233.03	-0.04	-0.02
-14.437	287.38	287.30	-0.08	-0.03
-10.244	349.64	349.53	-0.11	-0.03
-5.838	426.39	426.19	-0.20	-0.05
-1.564	513.10	512.90	-0.20	-0.04
2.725	613.55	613.52	-0.03	0.00
6.161	704.87	704.95	0.08	0.01
8.063	759.84	760.02	0.18	0.02
9.553	805.16	805.48	0.32	0.04

STANDARD DEVIATION = 0.15

CYCLOBUTENE C4H6

HEISIG G.B.:
J.AM.CHEM.SOC. 63,1698(1941).

A = 7.30570 B = 1165.999 C = 261.059

B.P.(760) = 2.450

T	P EXPTL.	P CALCD.	DEV.	PERCENT
-77.100	8.70	9.28	0.58	6.61
-62.900	26.50	26.40	-0.10	-0.39
-55.000	44.50	44.37	-0.13	-0.28
-46.800	74.00	73.06	-0.94	-1.27
-37.700	120.90	121.73	0.83	0.69
-30.700	175.20	175.40	0.20	0.12
-17.900	322.20	323.95	1.75	0.54
-13.000	404.00	402.91	-1.09	-0.27
-5.100	564.60	562.71	-1.89	-0.34
-2.100	635.90	635.41	-0.49	-0.08
-0.200	685.70	685.25	-0.45	-0.07
1.900	742.10	743.96	1.86	0.25

STANDARD DEVIATION = 1.23

CYCLOBUTANONE C4H6O

BENSON S.W.,KISTIAKOWSKI G.B.:
J.AM.CHEM.SOC. 64,80(1942).

A = 6.11668 B = 933.951 C = 183.188

B.P.(10) = -0.657

T	P EXPTL.	P CALCD.	DEV.	PERCENT
-24.060	2.00	1.77	-0.23	-11.60
-14.060	4.10	3.93	-0.17	-4.12
-7.860	6.30	6.16	-0.14	-2.17
-0.060	10.20	10.39	0.19	1.88
5.440	14.30	14.64	0.34	2.34
14.640	24.80	24.87	0.07	0.28
14.940	25.50	25.28	-0.22	-0.86
22.740	38.30	38.14	-0.16	-0.43
23.640	40.00	39.91	-0.09	-0.22
24.940	42.40	42.59	0.19	0.44
25.240	43.20	43.23	0.03	0.06

STANDARD DEVIATION = 0.22

VINYL ACETATE C4H6O2

CAPKOVA A.,FRIED V.:
COLL.CZECH.CHEM.COMM. 28,2235(1963).

A = 7.21010 B = 1296.130 C = 226.655

B.P.(760) = 72.731

T	P EXPTL.	P CALCD.	DEV.	PERCENT
21.830	98.70	98.62	-0.08	-0.08
25.120	115.30	115.38	0.08	0.07
28.800	136.90	136.86	-0.04	-0.03
32.850	164.20	164.24	0.04	0.02
36.130	189.50	189.59	0.09	0.05
40.820	231.50	231.36	-0.14	-0.06
45.640	281.60	281.89	0.29	0.10
50.370	340.20	339.90	-0.30	-0.09
56.120	423.40	423.14	-0.26	-0.06
61.830	521.20	521.45	0.25	0.05
67.920	645.40	645.80	0.40	0.06
72.040	743.00	742.66	-0.34	-0.05

STANDARD DEVIATION = 0.26

ACETIC ANHYDRIDE $C_4H_6O_3$

MCDONALD R.A.,SHRADER S.A.,STULL D.R.:
J.CHEM.ENG.DATA 4,311(1959).

A = 7.14948 B = 1444.718 C = 199.817

B.P.(760) = 138.629

T	P EXPTL.	P CALCD.	DEV.	PERCENT
62.840	44.62	44.57	-0.05	-0.10
67.260	54.95	54.97	0.02	0.03
70.730	64.23	64.49	0.26	0.41
74.100	74.99	75.02	0.03	0.04
85.810	123.89	123.43	-0.46	-0.37
111.030	317.29	317.54	0.25	0.08
125.560	511.62	512.09	0.47	0.09
136.120	706.23	706.19	-0.04	-0.01
137.520	736.37	735.81	-0.56	-0.08
138.020	746.94	746.63	-0.31	-0.04
138.480	756.82	756.69	-0.13	-0.02
138.980	767.45	767.76	0.31	0.04
139.420	777.41	777.60	0.19	0.02

STANDARD DEVIATION = 0.33

ETHYL CHLOROACETATE C4H7CLO2

NELSON O.A.: IND.ENG.CHEM.20,1380(1928).

A = 6.96683 B = 1355.873 C = 188.181

B.P.(760) = 143.652

T	P EXPTL.	P CALCD.	DEV.	PERCENT
25.000	5.00	4.04	-0.96	-19.15
32.000	7.00	6.44	-0.56	-8.01
41.000	12.50	11.24	-1.26	-10.10
47.400	17.00	16.27	-0.73	-4.30
54.500	24.40	23.98	-0.42	-1.74
59.900	31.90	31.72	-0.18	-0.55
65.700	43.30	42.29	-1.01	-2.33
72.500	60.30	58.28	-2.02	-3.34
79.300	77.70	79.03	1.33	1.71
85.200	98.60	101.67	3.07	3.11
92.000	130.50	134.14	3.64	2.79
98.500	172.80	172.69	-0.11	-0.06
104.600	217.00	216.68	-0.32	-0.15
110.000	263.40	262.83	-0.57	-0.22
114.900	312.80	311.31	-1.49	-0.48
120.200	373.50	371.60	-1.90	-0.51
124.900	435.30	432.60	-2.70	-0.62
130.000	509.50	507.58	-1.92	-0.38
134.700	586.70	585.50	-1.20	-0.20
139.800	679.80	680.50	0.70	0.10
142.400	730.00	733.40	3.40	0.47
144.200	766.00	771.88	5.88	0.77
144.600	787.90	780.65	-7.25	-0.92
145.900	806.40	809.67	3.27	0.41

STANDARD DEVIATION = 2.77

ETHYL IODOACETATE C4H7IO2

GOULD C.,HOLZMAN G.,NIEMANN C.: ANAL.CHEM. 19,204(1947).

A = 4.07372 B = 374.640 C = 54.767

B.P.(10) = 67.118

T	P EXPTL.	P CALCD.	DEV.	PERCENT
28.800	1.00	0.39	-0.61	-61.04
40.800	2.00	1.42	-0.58	-28.80
51.400	3.00	3.51	0.51	16.90
60.000	6.00	6.45	0.45	7.45
72.000	13.00	13.13	0.13	1.02
88.600	29.00	28.88	-0.12	-0.42

STANDARD DEVIATION = 0.63

1-BUTENE C4H8

COFFIN C.C.,MAASS O.:
J.AM.CHEM.SOC. 50,1427(1928).

A = 6.53101 B = 810.261 C = 228.066

B.P.(760) = -6.089

	T	P EXPTL.	P CALCD.	DEV.	PERCENT
*	-77.500	23.50	14.11	-9.39	-39.95
	-60.100	50.50	50.94	0.44	0.87
	-57.100	62.00	61.90	-0.10	-0.16
	-52.300	83.30	83.39	0.09	0.11
	-44.600	130.80	130.20	-0.60	-0.46
	-39.200	175.30	174.13	-1.17	-0.67
	-33.700	230.00	230.29	0.29	0.13
	-32.600	242.20	243.07	0.87	0.36
	-29.900	276.90	276.83	-0.07	-0.03
	-28.000	302.90	302.72	-0.18	-0.06
	-25.600	338.00	338.10	0.10	0.03
	-23.100	377.70	378.32	0.62	0.17
	-19.400	445.10	444.59	-0.51	-0.11
	-17.000	491.30	492.17	0.87	0.18
	-14.300	549.60	550.30	0.70	0.13
	-11.300	621.70	620.95	-0.75	-0.12
	-8.500	691.30	692.99	1.69	0.24
	-6.700	740.20	742.56	2.36	0.32
	-6.100	760.20	759.67	-0.53	-0.07
	-4.400	811.40	809.79	-1.61	-0.20
	-3.700	833.70	831.14	-2.56	-0.31

STANDARD DEVIATION = 1.17

CIS-2-BUTENE

C4H8

SCOTT R.B.,FERBUSON W.J.,BRICKWEDDE F.G.:
J.RESEARCH NATL.BUR.STANDARDS 33,1(1944).

A = 6.86102 B = 956.398 C = 236.572

B.P.(760) = 3.717

T	P EXPTL.	P CALCD.	DEV.	PERCENT
-70.090	13.09	13.07	-0.02	-0.16
-62.548	23.15	23.19	0.04	0.16
-55.370	38.29	38.28	-0.01	-0.03
-48.103	61.12	61.16	0.04	0.06
-40.075	98.58	98.58	0.00	0.00
-31.547	157.18	157.12	-0.06	-0.04
-23.129	240.04	240.00	-0.04	-0.02
-14.822	353.25	353.23	-0.02	-0.01
-6.716	501.46	501.37	-0.09	-0.02
-6.706	501.51	501.58	0.07	0.01
0.560	672.53	672.71	0.18	0.03
3.689	759.14	759.20	0.06	0.01
6.702	850.50	850.46	-0.04	0.00
11.144	1000.27	1000.35	0.08	0.01
15.520	1167.07	1167.27	0.20	0.02
19.885	1354.53	1354.39	-0.14	-0.01
22.759	1489.85	1489.61	-0.24	-0.02

STANDARD DEVIATION = 0.11

TRANS-2-BUTENE

C4H8

GUTTMAN L.,PITZER K.S.:

J.AM.CHEM.SOC. 67,324(1945).

A = 6.87825 B = 964.897 C = 240.498

B.P.(760) = 0.881

T	P EXPTL.	P CALCD.	DEV.	PERCENT
-71.454	14.92	14.80	-0.12	-0.80
-60.543	32.76	32.84	0.08	0.24
-51.118	60.64	60.70	0.06	0.10
-42.899	98.83	98.89	0.06	0.06
-37.359	134.51	134.37	-0.14	-0.10
-28.303	214.37	214.30	-0.07	-0.03
-21.001	303.58	303.60	0.02	0.01
-14.259	410.47	410.49	0.02	0.00
-9.958	492.86	493.02	0.16	0.03
-3.927	630.38	630.33	-0.05	-0.01
0.982	762.97	762.93	-0.04	-0.01

STANDARD DEVIATION = 0.10

CYCLOBUTANE C4H8

RATHJENS G.W.,GWINN W.D.: J.AM.CHEM.SOC. 75,5629(1953).

A = 6.91631   B = 1024.539   C = 241.373

B.P.(760) = 12.508

T	P EXPTL.	P CALCD.	DEV.	PERCENT
-59.930	18.73	18.61	-0.12	-0.65
-56.129	24.32	24.30	-0.02	-0.09
-49.001	38.79	38.95	0.16	0.41
-45.921	47.27	47.25	-0.02	-0.04
-33.859	95.34	95.30	-0.04	-0.04
-27.261	135.20	135.29	0.09	0.06
-20.421	190.38	190.28	-0.10	-0.05
-15.817	236.58	236.61	0.03	0.01
-10.312	303.59	303.56	-0.03	-0.01
-3.057	414.25	414.23	-0.03	-0.01
5.300	579.31	579.27	-0.04	-0.01
12.195	751.22	751.32	0.10	0.01

STANDARD DEVIATION = 0.09

2-METHYLPROPENE C4H8

LAMB A.B.,ROPER E.E.: J.AM.CHEM.SOC. 62,806(1940).

A = 6.53103   B = 802.620   C = 226.979

B.P.(760) = -7.097

T	P EXPTL.	P CALCD.	DEV.	PERCENT
-56.750	65.44	65.48	0.04	0.06
-25.300	357.80	355.92	-1.88	-0.53
-20.880	431.00	433.21	2.21	0.51
-7.470	748.50	749.21	0.71	0.09
-0.670	966.00	964.87	-1.13	-0.12
0.000	987.80	988.41	0.61	0.06
0.000	989.00	988.41	-0.59	-0.06

STANDARD DEVIATION = 1.65

ETHYLVINYLDICHLOROSILANE C4H8CL2SI

JENKINS A.C.,CHAMBERS G.F.:
IND.ENG.CHEM. 46,2367(1954).

A = 6.85924 B = 1330.948 C = 210.833

B.P.(760) = 123.708

T	P EXPTL.	P CALCD.	DEV.	PERCENT
44.900	45.50	45.16	-0.34	-0.74
73.500	149.90	150.76	0.86	0.58
93.400	306.10	305.12	-0.98	-0.32
106.700	465.80	465.28	-0.52	-0.11
117.000	628.40	630.09	1.69	0.27
121.700	719.90	719.09	-0.81	-0.11

STANDARD DEVIATION = 1.37

BUTYRALDEHYDE C4H8O

SEPRAKOVA M.,PAULECH J.,DYKYJ J.:
CHEM.ZVESTI 13,313(1959).

A = 6.38544 B = 913.590 C = 185.483

B.P.(760) = 75.199

T	P EXPTL.	P CALCD.	DEV.	PERCENT
30.710	145.30	144.42	-0.88	-0.61
38.360	200.80	201.39	0.59	0.29
43.930	251.90	253.01	1.11	0.44
48.330	300.00	300.66	0.66	0.22
52.360	350.00	350.17	0.17	0.05
55.900	400.00	398.67	-1.33	-0.33
62.290	500.00	499.14	-0.86	-0.17
67.730	600.00	599.02	-0.98	-0.16
74.030	731.40	732.87	1.47	0.20

STANDARD DEVIATION = 1.19

ISOBUTYRALDEHYDE C4H8O

SEPRAKOVA M.,PAULECH J.,DYKYJ J.: CHEM.ZVESTI 13,313(1959).

A = 6.73513 B = 1053.178 C = 209.134

B.P.(760) = 64.113

T	P EXPTL.	P CALCD.	DEV.	PERCENT
12.930	98.30	98.27	-0.03	-0.03
18.950	131.30	131.10	-0.20	-0.15
23.980	164.90	164.91	0.01	0.01
29.680	210.80	211.39	0.59	0.28
37.080	286.90	286.83	-0.07	-0.02
44.630	384.90	384.50	-0.40	-0.11
48.310	441.00	440.77	-0.23	-0.05
54.530	550.70	550.46	-0.24	-0.04
56.910	597.10	597.66	0.56	0.09
62.850	729.30	729.31	0.01	0.00

STANDARD DEVIATION = 0.37

METHYL ETHYL KETONE C4H8O

COLLERSON R.R., ET AL.: J.CHEM.SOC. 3697(1965).

A = 7.06356 B = 1261.339 C = 221.969

B.P.(760) = 79.589

T	P EXPTL.	P CALCD.	DEV.	PERCENT
42.788	199.28	199.26	-0.02	-0.01
48.148	247.70	247.72	0.02	0.01
53.026	299.75	299.77	0.02	0.01
57.080	349.46	349.48	0.02	0.01
60.821	401.05	401.07	0.02	0.00
64.005	449.69	449.65	-0.04	-0.01
67.009	499.73	499.72	-0.01	0.00
69.734	548.93	548.91	-0.02	0.00
72.343	599.55	599.56	0.01	0.00
74.839	651.42	651.44	0.02	0.00
76.950	698.03	698.05	0.02	0.00
79.221	751.08	751.11	0.03	0.00
81.268	801.72	801.63	-0.09	-0.01
83.161	850.78	850.71	-0.07	-0.01
85.013	901.02	900.99	-0.03	0.00
86.715	949.18	949.23	0.05	0.01
88.444	1000.23	1000.31	0.08	0.01

STANDARD DEVIATION = 0.05

TETRAHYDROFURAN C4H8O

SCOTT D.W.:
J.CHEM.THERMODYN. 2,833(1970).

A = 6.99515 B = 1202.290 C = 226.254

B.P.(760) = 65.965

T	P EXPTL.	P CALCD.	DEV.	PERCENT
23.139	149.41	149.38	-0.03	-0.02
28.362	187.57	187.58	0.01	0.01
33.620	233.72	233.73	0.01	0.01
38.917	289.13	289.16	0.03	0.01
44.251	355.22	355.26	0.04	0.01
49.620	433.56	433.55	-0.01	0.00
55.029	525.86	525.83	-0.03	0.00
60.475	633.99	633.91	-0.08	-0.01
65.965	760.00	759.99	-0.01	0.00
71.489	906.06	906.03	-0.03	0.00
77.054	1074.60	1074.56	-0.04	0.00
82.659	1268.00	1268.10	0.10	0.01
88.300	1489.10	1489.19	0.09	0.01
93.980	1740.80	1740.78	-0.02	0.00
99.700	2026.00	2025.94	-0.06	0.00

STANDARD DEVIATION = 0.05

BUTYRIC ACID C4H8O2

DREISBACH R.R.,SHRADER S.A.:
IND.ENG.CHEM. 41,2879(1949).

A = 7.73990 B = 1764.680 C = 199.892

B.P.(760) = 163.279

T	P EXPTL.	P CALCD.	DEV.	PERCENT
90.920	47.16	46.97	-0.19	-0.41
95.010	57.04	57.01	-0.03	-0.05
98.350	66.39	66.52	0.13	0.20
101.320	75.86	76.09	0.23	0.30
112.570	123.76	123.66	-0.10	-0.08
136.770	315.52	314.92	-0.60	-0.19
150.700	507.50	508.71	1.21	0.24
163.250	760.00	759.32	-0.68	-0.09

STANDARD DEVIATION = 0.69

BUTYRIC ACID C4H802

JASPER J.J.,MILLER G.B.:
J.PHYS.CHEM. 59, 441(1955).

A = 8.71019 B = 2433.014 C = 255.189

B.P.(100) = 107.396

T	P EXPTL.	P CALCD.	DEV.	PERCENT
20.000	0.77	0.74	-0.03	-3.96
30.000	1.60	1.51	-0.09	-5.63
40.000	3.00	2.94	-0.06	-2.09
50.000	5.50	5.47	-0.03	-0.54
60.000	9.80	9.79	-0.01	-0.06
70.000	16.80	16.92	0.12	0.70
80.000	28.00	28.28	0.28	1.02
90.000	45.30	45.90	0.60	1.33
100.000	73.80	72.49	-1.31	-1.78
110.000	112.40	111.64	-0.76	-0.67
120.000	166.70	168.04	1.34	0.80
130.000	247.00	247.61	0.61	0.25
140.000	357.20	357.76	0.56	0.16
150.000	509.10	507.62	-1.48	-0.29

STANDARD DEVIATION = 0.82

1,4-DIOXANE C4H802

CRENSHAW J.L.,COPE A.C.,FINKELSTEIN N.,ROGAN R.
J.AM.CHEM.SOC.60,2310(1938).

A = 7.43155 B = 1554.679 C = 240.337

B.P.(760) = 101.296

T	P EXPTL.	P CALCD.	DEV.	PERCENT
20.000	28.90	28.82	-0.08	-0.27
25.000	37.40	37.35	-0.05	-0.14
30.000	47.90	47.93	0.03	0.07
40.000	76.80	76.87	0.07	0.10
50.000	119.30	119.34	0.04	0.04
60.000	179.90	179.92	0.02	0.01
70.000	264.20	264.17	-0.03	-0.01
80.000	378.80	378.68	-0.12	-0.03
90.000	531.20	531.12	-0.08	-0.01
100.000	730.30	730.26	-0.04	-0.01
105.000	850.20	850.39	0.19	0.02

STANDARD DEVIATION = 0.10

ETHYL ACETATE C4H8O2

MERTL I.,POLAK J.:
COLLECTION CZECHOSLOV.CHEM.COMM. 30,3526(1965).

A = 7.10179 B = 1244.951 C = 217.881

B.P.(760) = 77.063

T	P EXPTL.	P CALCD.	DEV.	PERCENT
15.580	58.79	58.77	-0.02	-0.03
18.680	69.02	69.04	0.02	0.02
21.115	78.12	78.11	-0.01	-0.02
23.755	89.03	89.04	0.01	0.02
25.860	98.67	98.65	-0.02	-0.02
29.135	115.28	115.30	0.02	0.01
32.830	136.84	136.80	-0.04	-0.03
36.875	164.03	164.04	0.01	0.00
40.175	189.33	189.42	0.09	0.05
44.905	231.35	231.35	0.00	0.00
49.715	281.40	281.47	0.07	0.02
54.505	340.12	339.81	-0.31	-0.09
60.320	423.30	423.42	0.12	0.03
66.045	521.10	521.21	0.11	0.02
72.190	645.55	645.50	-0.05	-0.01
75.830	729.60	729.60	0.00	0.00

STANDARD DEVIATION = 0.10

ISOBUTYRIC ACID C4H8O2

KAHLBAUM G.W.A.:
BER. 16,2476(1883).

A = 4.89380 B = 382.571 C = 38.033

B.P.(760) = 152.018

T	P EXPTL.	P CALCD.	DEV.	PERCENT
57.500	11.36	7.75	-3.61	-31.79
69.200	18.14	21.19	3.05	16.82
72.800	28.84	27.67	-1.17	-4.05
78.800	39.36	41.62	2.26	5.74
85.000	61.40	60.86	-0.54	-0.89
152.000	760.00	759.66	-0.34	-0.04

STANDARD DEVIATION = 3.12

METHYL PROPIONATE C4H8O2

MERTL I.,POLAK J.: COLLECTION CZECHOSLOV.CHEM.COMM. 30,3526(1965).

A = 6.94244 B = 1170.236 C = 208.751

B.P.(760) = 79.369

T	P EXPTL.	P CALCD.	DEV.	PERCENT
20.590	69.02	69.16	0.14	0.20
23.015	78.12	78.20	0.08	0.11
25.650	89.03	89.12	0.09	0.10
27.725	98.67	98.58	-0.09	-0.09
31.000	115.28	115.18	-0.10	-0.08
34.695	136.84	136.61	-0.23	-0.17
38.770	164.03	163.91	-0.12	-0.07
42.090	189.33	189.32	-0.01	-0.01
46.855	231.35	231.29	-0.06	-0.03
51.720	281.40	281.62	0.22	0.08
56.560	340.12	340.12	0.00	0.00
62.420	423.30	423.59	0.29	0.07
68.195	521.10	521.12	0.02	0.00
74.460	645.55	646.26	0.71	0.11
78.755	745.90	744.96	-0.94	-0.13

STANDARD DEVIATION = 0.37

PROPYL FORMATE C4H8O2

NELSON O.A.: IND.ENG.CHEM.20,1382(1928).

A = 6.84755 B = 1126.539 C = 203.459

B.P.(760) = 80.538

T	P EXPTL.	P CALCD.	DEV.	PERCENT
26.200	88.20	87.55	-0.65	-0.73
30.800	109.10	109.29	0.19	0.18
35.200	134.50	134.05	-0.45	-0.34
40.500	170.40	169.75	-0.65	-0.38
46.200	216.50	216.39	-0.11	-0.05
50.000	252.90	252.86	-0.04	-0.02
58.400	348.90	351.12	2.22	0.64
63.700	421.80	427.37	5.57	1.32
63.900	430.70	430.48	-0.22	-0.05
70.500	540.30	543.84	3.54	0.66
71.000	554.80	553.30	-1.50	-0.27
73.500	600.00	602.58	2.58	0.43
75.100	648.90	635.88	-13.02	-2.01
78.900	724.00	720.78	-3.23	-0.45
80.500	759.40	759.07	-0.33	-0.04
82.300	798.80	804.03	5.23	0.65

STANDARD DEVIATION = 4.52

THIACYCLOPENTANE

C4H8S

OSBORN A.N., DOUSLIN D.R.:

J.CHEM.ENG.DATA 11,502(1966).

A = 6.99427 B = 1401.169 C = 219.515

B.P.(760) = 121.116

T	P EXPTL.	P CALCD.	DEV.	PERCENT
71.182	149.41	149.36	-0.05	-0.03
77.278	187.57	187.60	0.03	0.02
83.405	233.72	233.74	0.02	0.01
89.580	289.13	289.15	0.02	0.01
95.803	355.22	355.30	0.08	0.02
102.056	433.56	433.51	-0.05	-0.01
108.365	525.86	525.83	-0.03	-0.01
114.716	633.99	633.94	-0.05	-0.01
121.114	760.00	759.96	-0.04	-0.01
127.558	906.06	906.07	0.01	0.00
134.046	1074.60	1074.60	0.00	0.00
140.574	1268.00	1267.92	-0.08	-0.01
147.163	1489.10	1489.40	0.30	0.02
153.775	1740.80	1740.59	-0.21	-0.01
160.451	2026.00	2026.04	0.04	0.00

STANDARD DEVIATION = 0.12

1-BROMOBUTANE

C4H9BR

SMYTH C.P., ENGEL E.W.:

J.AM.CHEM.SOC.51,2646(1929).

A = 6.64814 B = 1070.971 C = 206.251

B.P.(100) = 24.157

T	P EXPTL.	P CALCD.	DEV.	PERCENT
20.000	82.30	82.15	-0.15	-0.19
30.000	130.20	130.30	0.10	0.08
40.000	198.50	199.08	0.58	0.29
50.000	295.80	294.28	-1.52	-0.52
60.000	420.70	422.40	1.70	0.40
70.000	591.40	590.65	-0.75	-0.13

STANDARD DEVIATION = 1.43

1-BROMOBUTANE C4H9BR

MILAZZO G.:

ANNALI DI CHIMICA 46,1105(1956).

A = 5.28138 B = 685.001 C = 160.880

B.P.(10) = -0.884

T	P EXPTL.	P CALCD.	DEV.	PERCENT
-77.970	0.01	0.00	0.00	-82.28
-73.180	0.01	0.00	-0.01	-76.36
-58.750	0.08	0.04	-0.04	-50.64
-49.580	0.22	0.13	-0.08	-37.71
-32.950	1.00	0.85	-0.16	-15.50
-15.970	3.75	3.58	-0.17	-4.44
-11.150	5.00	5.09	0.09	1.74
-0.690	9.85	10.12	0.27	2.75
2.630	12.35	12.36	0.01	0.08
11.050	20.00	19.82	-0.18	-0.88
22.670	35.40	35.43	0.03	0.10

STANDARD DEVIATION = 0.15

2-BROMO-2-METHYLPROPANE C4H9BR

BRYCE-SMITH D.,HOWLETT K.E.:

J.CHEM.SOC. (1951)1141.

A = 7.39592 B = 1512.672 C = 262.221

B.P.(760) = 72.803

T	P EXPTL.	P CALCD.	DEV.	PERCENT
0.000	42.80	42.39	-0.41	-0.97
6.800	59.00	59.30	0.30	0.51
10.600	70.70	71.02	0.32	0.45
15.700	90.00	89.77	-0.23	-0.26
20.800	112.90	112.51	-0.39	-0.35
22.700	121.90	122.13	0.23	0.19
25.000	135.30	134.69	-0.61	-0.45
28.400	155.40	155.22	-0.18	-0.11
31.300	174.50	174.74	0.24	0.13
35.700	209.30	208.21	-1.09	-0.52
40.000	243.40	245.89	2.49	1.02
49.300	344.50	346.86	2.36	0.69
56.600	450.40	448.06	-2.34	-0.52
63.500	567.60	564.74	-2.86	-0.50
72.800	758.00	759.91	1.91	0.25

STANDARD DEVIATION = 1.62

1-CHLOROBUTANE C4H9CL

KEMME R.H.,KREPS S.I.:
J.CHEM.ENG.DATA 14,98(1969).

A = 6.83694 B = 1173.790 C = 218.126

B.P.(760) = 78.576

T	P EXPTL.	P CALCD.	DEV.	PERCENT
-16.700	10.10	10.22	0.12	1.21
-10.500	15.20	15.26	0.06	0.39
-6.200	19.80	19.87	0.07	0.37
0.800	29.60	29.88	0.28	0.94
6.200	40.20	40.22	0.02	0.04
12.000	54.80	54.49	-0.31	-0.57
19.700	79.80	79.70	-0.10	-0.13
28.800	121.60	121.15	-0.45	-0.37
41.100	203.10	203.65	0.55	0.27
51.300	302.00	302.21	0.21	0.07
65.500	499.10	499.37	0.27	0.05
78.500	758.80	758.22	-0.58	-0.08

STANDARD DEVIATION = 0.36

2-CHLOROBUTANE C4H9CL

ROLAND M.:
BULL.SOC.CHIM.BELG. 37,117(1928).

A = 6.79923 B = 1149.124 C = 224.682

B.P.(100) = 14.757

T	P EXPTL.	P CALCD.	DEV.	PERCENT
0.320	49.30	49.21	-0.09	-0.18
10.020	79.95	80.01	0.06	0.07
20.180	127.50	127.73	0.23	0.18
29.960	193.80	193.43	-0.37	-0.19
39.850	285.10	285.26	0.16	0.06

STANDARD DEVIATION = 0.34

2-CHLORO-2-METHYLPROPANE C4H9CL

G.CHOLIZ CALERO, ET.AL.:
REV.ACAD.CIENC.EX.FIS.QUIM. Y NAT. ZARAGOZA 24(2),137(1969).

A = 4.89600 B = 334.986 C = 114.047

B.P.(100) = 1.625

T	P EXPTL.	P CALCD.	DEV.	PERCENT
22.000	277.50	271.47	-6.03	-2.17
22.420	281.77	276.25	-5.52	-1.96
25.990	318.18	319.06	0.88	0.28
30.000	365.50	371.94	6.44	1.76
30.060	369.37	372.77	3.40	0.92
32.070	395.09	401.25	6.16	1.56
37.370	478.01	482.68	4.67	0.98
40.000	524.00	526.54	2.54	0.48
41.040	544.52	544.52	0.00	0.00
42.530	573.90	570.91	-2.99	-0.52
46.560	655.99	646.03	-9.96	-1.52

STANDARD DEVIATION = 6.07

PYRROLIDINE C4H9N

MCCULLOUGH J.P.,ET AL.:
J.AM.CHEM.SOC. 81,5884(1959).

A = 6.92455 B = 1179.991 C = 205.248

B.P.(760) = 86.559

T	P EXPTL.	P CALCD.	DEV.	PERCENT
43.158	149.41	149.38	-0.03	-0.02
48.439	187.57	187.58	0.01	0.00
53.760	233.72	233.74	0.02	0.01
59.122	289.13	289.16	0.03	0.01
64.523	355.22	355.22	0.00	0.00
69.969	433.56	433.56	0.00	0.00
75.456	525.86	525.85	-0.01	0.00
80.987	633.99	634.00	0.01	0.00
86.558	760.00	759.96	-0.04	0.00
92.173	906.06	906.01	-0.05	-0.01
97.833	1074.60	1074.55	-0.05	0.00
103.537	1268.00	1268.07	0.07	0.01
109.280	1489.10	1489.09	-0.01	0.00
115.072	1740.80	1740.84	0.04	0.00
120.905	2026.00	2026.00	0.00	0.00

STANDARD DEVIATION = 0.04

N,N-DIMETHYLACETAMIDE C4H9NO

RAM G.,SHARAF A.R.:

J.IND.CHEM.SOC. 45,13(1968).

A = 9.72090 B = 3273.800 C = 334.490

B.P.(100) = 89.528

T	P EXPTL.	P CALCD.	DEV.	PERCENT
30.000	5.50	5.48	-0.02	-0.30
40.000	10.00	9.53	-0.47	-4.75
50.000	16.00	16.08	0.08	0.49
60.000	26.00	26.43	0.43	1.65
70.000	42.50	42.39	-0.11	-0.26
80.000	66.50	66.45	-0.05	-0.07
90.000	102.00	102.00	0.00	0.00

STANDARD DEVIATION = 0.33

N-METHYLPROPIONAMIDE C4H9NO

RAM G.,SHARAF A.R.:

J.IND.CHEM.SOC. 45,13(1968).

A = -0.91032 B = 119.427 C = -148.023

B.P.(10) = 85.507

T	P EXPTL.	P CALCD.	DEV.	PERCENT
30.000	0.50	1.26	0.76	152.70
40.000	1.00	1.57	0.57	56.76
50.000	1.50	2.03	0.53	35.50
60.000	3.00	2.80	-0.20	-6.82
70.000	4.50	4.17	-0.33	-7.29
80.000	7.00	7.00	0.00	0.06
90.000	14.00	14.06	0.06	0.42

STANDARD DEVIATION = 0.58

BUTYL NITRATE $C_4H_9NO_3$

GRAY P.,PRATT M.W.T.:

J.CHEM.SOC. 1957,2163.

A = 8.05427 B = 1992.830 C = 254.299

B.P.(10) = 28.201

T	P EXPTL.	P CALCD.	DEV.	PERCENT
0.000	1.70	1.65	-0.05	-2.89
10.000	3.30	3.27	-0.03	-0.99
20.000	6.30	6.15	-0.15	-2.33
30.000	11.10	11.08	-0.02	-0.16
40.000	18.90	19.18	0.28	1.47
50.000	32.10	32.01	-0.09	-0.27
60.000	51.80	51.73	-0.07	-0.14
70.000	81.10	81.14	0.04	0.05

STANDARD DEVIATION = 0.15

ISOBUTYL NITRATE $C_4H_9NO_3$

GRAY P.,PRATT M.W.T.:

J.CHEM.SOC. 1957,2163.

A = 8.16435 B = 2022.666 C = 262.395

B.P.(100) = 65.729

T	P EXPTL.	P CALCD.	DEV.	PERCENT
0.000	2.90	2.86	-0.04	-1.49
10.000	5.40	5.48	0.08	1.50
20.000	10.00	10.04	0.04	0.42
30.000	17.80	17.65	-0.15	-0.84
40.000	30.10	29.89	-0.21	-0.70
50.000	48.50	48.94	0.44	0.90
60.000	77.90	77.71	-0.19	-0.25
70.000	120.00	120.01	0.01	0.01

STANDARD DEVIATION = 0.25

BUTANE C4H10

ASTON J.G.,MESSERLY G.H.:
J.AM.CHEM.SOC. 62,1917(1940).

A = 6.82485 B = 943.453 C = 239.711

B.P.(760) = -0.501

T	P EXPTL.	P CALCD.	DEV.	PERCENT
-78.043	9.90	9.75	-0.15	-1.49
-60.482	36.26	36.38	0.12	0.34
-46.874	85.59	85.58	-0.01	-0.02
-37.328	145.58	145.59	0.01	0.01
-26.639	249.53	249.45	-0.08	-0.03
-16.946	388.77	388.73	-0.04	-0.01
-10.883	503.34	503.34	0.00	0.00
-6.361	604.97	605.01	0.04	0.01
-2.753	697.13	697.15	0.02	0.00
-1.123	742.20	742.21	0.01	0.00
-0.344	764.50	764.53	0.03	0.00

STANDARD DEVIATION = 0.08

2-METHYLPROPANE C4H10

ASTON J.G.,KENNEDY R.M.,SCHUMANN S.C.:
J.AM.CHEM.SOC. 62,2059(1940).

A = 6.78866 B = 899.617 C = 241.942

B.P.(760) = -11.734

T	P EXPTL.	P CALCD.	DEV.	PERCENT
-85.090	11.37	11.30	-0.07	-0.58
-71.704	31.96	31.93	-0.03	-0.09
-56.431	86.85	86.95	0.10	0.11
-44.107	174.26	174.32	0.06	0.04
-27.576	391.02	390.86	-0.16	-0.04
-22.071	498.08	497.84	-0.24	-0.05
-18.761	572.67	572.50	-0.17	-0.03
-13.238	716.13	716.33	0.20	0.03
-11.609	763.44	763.72	0.28	0.04

STANDARD DEVIATION = 0.20

DIETHYLALUMINIUMCHLORIDE C4H10ALCL

FIC V.,DVORAK J.: CHEM.PRUM. 15,732(1965).

A = 8.22970 B = 2484.531 C = 255.447

B.P.(10) = 88.209

T	P EXPTL.	P CALCD.	DEV.	PERCENT
44.000	0.90	0.86	-0.04	-4.86
46.000	1.00	0.97	-0.03	-2.80
47.500	1.10	1.07	-0.03	-2.93
49.000	1.20	1.17	-0.03	-2.34
51.500	1.40	1.37	-0.03	-2.45
58.500	2.10	2.07	-0.03	-1.46
92.000	12.00	11.99	-0.01	-0.07
93.500	13.00	12.87	-0.13	-0.99
95.500	14.00	14.13	0.13	0.94
98.500	16.00	16.23	0.23	1.41
103.000	20.00	19.88	-0.12	-0.62
110.000	27.00	26.98	-0.02	-0.06
117.500	37.00	36.97	-0.03	-0.09
125.000	50.00	50.02	0.02	0.04

STANDARD DEVIATION = 0.10

DIETHYLDICHLOROSILANE C4H10CL2SI

JENKINS A.C.,CHAMBERS G.F.: IND.ENG.CHEM. 46,2367(1954).

A = 6.86287 B = 1346.279 C = 207.654

B.P.(760) = 130.433

T	P EXPTL.	P CALCD.	DEV.	PERCENT
48.100	40.40	39.71	-0.69	-1.71
65.300	83.90	85.23	1.33	1.59
83.700	176.20	174.62	-1.58	-0.90
100.000	305.90	306.84	0.94	0.31
116.500	511.70	512.45	0.75	0.15
127.700	706.20	705.28	-0.92	-0.13

STANDARD DEVIATION = 1.53

1-BUTANOL $C_4H_{10}O$

BIDDISCOMBE D.P.,ET AL.:

J.CHEM.SOC. 1963,1954.

A = 7.36366 B = 1305.198 C = 173.427

B.P.(760) = 117.727

T	P EXPTL.	P CALCD.	DEV.	PERCENT
89.212	247.86	247.80	-0.06	-0.03
93.662	299.83	299.84	0.01	0.00
97.357	349.54	349.60	0.06	0.02
100.742	400.91	400.95	0.04	0.01
103.636	449.53	449.59	0.06	0.01
106.367	499.78	499.80	0.02	0.00
108.885	550.03	550.05	0.02	0.00
111.161	598.97	598.93	-0.04	-0.01
113.429	651.11	651.08	-0.03	0.00
115.324	697.52	697.42	-0.10	-0.01
117.393	751.13	751.03	-0.10	-0.01
119.193	800.35	800.32	-0.03	0.00
120.940	850.66	850.62	-0.04	-0.01
122.564	899.66	899.63	-0.04	0.00
124.155	949.79	949.81	0.02	0.00
125.686	1000.00	1000.20	0.20	0.02

STANDARD DEVIATION = 0.08

2-BUTANOL $C_4H_{10}O$

BIDDISCOMBE D.P.,ET AL.:

J.CHEM.SOC. 1963,1954.

A = 7.20131 B = 1157.000 C = 168.279

B.P.(760) = 99.514

T	P EXPTL.	P CALCD.	DEV.	PERCENT
72.392	247.75	247.70	-0.05	-0.02
76.600	299.59	299.59	0.00	0.00
80.131	349.65	349.70	0.05	0.01
83.340	400.89	400.95	0.06	0.02
86.112	449.91	449.98	0.07	0.02
88.693	499.85	499.89	0.04	0.01
91.068	549.69	549.68	-0.01	0.00
93.303	600.13	600.10	-0.03	-0.01
95.394	650.76	650.58	-0.18	-0.03
97.252	698.26	698.24	-0.02	0.00
99.201	751.23	751.19	-0.04	-0.01
100.931	800.87	800.85	-0.02	0.00
102.611	851.57	851.54	-0.03	0.00
104.186	901.34	901.35	0.01	0.00
105.647	949.55	949.60	0.05	0.01
107.146	1001.12	1001.22	0.10	0.01

STANDARD DEVIATION = 0.07

DIETHYL ETHER C4H10O

TAYLOR R.S.,SMITH L.B.: J.AM.CHEM.SOC. 44,2450(1922).

A = 6.92032 B = 1064.066 C = 228.799

B.P.(100) = -12.539

T	P EXPTL.	P CALCD.	DEV.	PERCENT
-60.799	3.95	3.86	-0.09	-2.28
-55.748	5.93	5.91	-0.02	-0.37
-50.873	8.77	8.71	-0.06	-0.70
-45.998	12.62	12.57	-0.05	-0.38
-41.125	17.78	17.81	0.03	0.14
-36.231	24.77	24.81	0.04	0.17
-31.329	34.03	34.03	0.00	-0.01
-26.421	45.81	45.97	0.16	0.36
-21.502	61.31	61.27	-0.04	-0.06
-16.578	80.67	80.61	-0.06	-0.08
-11.637	104.79	104.82	0.03	0.03
-6.698	134.76	134.71	-0.05	-0.04
0.009	186.13	186.14	0.01	0.01
4.975	233.73	233.69	-0.04	-0.02
9.937	290.62	290.56	-0.06	-0.02
14.903	358.15	358.15	0.00	0.00
19.871	437.70	437.81	0.11	0.03

STANDARD DEVIATION = 0.07

2-METHYL-1-PROPANOL C4H10O

BIDDISCOMBE D.P.,ET AL.:

J.CHEM.SOC. 1963,1954.

A = 7.29491 B = 1230.810 C = 170.947

B.P.(760) = 107.889

T	P EXPTL.	P CALCD.	DEV.	PERCENT
80.207	247.96	247.91	-0.05	-0.02
84.524	299.99	299.99	0.00	0.00
88.063	349.02	349.08	0.06	0.02
91.381	400.86	400.90	0.04	0.01
94.205	449.73	449.79	0.06	0.01
96.815	499.15	499.17	0.02	0.00
99.289	549.99	549.97	-0.02	0.00
101.546	599.92	599.88	-0.04	-0.01
103.696	650.79	650.76	-0.03	-0.01
105.589	698.46	698.39	-0.07	-0.01
107.568	751.21	751.14	-0.07	-0.01
109.326	800.68	800.65	-0.03	0.00
111.017	850.70	850.71	0.01	0.00
112.624	900.61	900.56	-0.05	-0.01
114.101	948.40	948.43	0.03	0.00
115.623	999.73	999.86	0.13	0.01

STANDARD DEVIATION = 0.06

2-METHYL-2-PROPANOL C4H10O

BIDDISCOMBE D.P.,ET AL.:

J.CHEM.SOC. 1963,1954.

A = 7.20340 B = 1092.971 C = 170.503

B.P.(760) = 82.348

T	P EXPTL.	P CALCD.	DEV.	PERCENT
60.781	300.46	300.42	-0.04	-0.01
64.054	349.69	349.68	-0.01	0.00
67.076	400.77	400.81	0.04	0.01
69.707	450.08	450.12	0.04	0.01
72.138	499.89	499.94	0.05	0.01
74.338	548.75	548.77	0.02	0.00
76.439	598.90	598.93	0.03	0.00
78.509	651.90	651.88	-0.02	0.00
80.198	697.88	697.81	-0.07	-0.01
81.996	749.56	749.52	-0.04	-0.01
83.625	799.03	798.97	-0.06	-0.01
85.272	851.67	851.57	-0.10	-0.01
86.730	900.41	900.41	0.00	0.00
88.127	949.27	949.27	0.00	0.00
89.560	1001.39	1001.56	0.17	0.02

STANDARD DEVIATION = 0.07

METHYL PROPYL ETHER C4H10O

BINGHAM E.C.:

AM.CHEM.J. 43,287(1910).

A = 6.11856 B = 708.691 C = 179.873

B.P.(760) = 39.011

T	P EXPTL.	P CALCD.	DEV.	PERCENT
-0.600	151.80	146.36	-5.44	-3.58
3.420	180.00	178.70	-1.30	-0.72
7.010	212.60	212.03	-0.57	-0.27
8.810	230.90	230.45	-0.45	-0.19
12.580	272.20	272.99	0.80	0.29
14.300	293.30	294.29	0.99	0.34
17.600	336.90	338.66	1.76	0.52
18.670	352.60	354.08	1.48	0.42
20.020	376.00	374.29	-1.71	-0.45
21.010	388.30	389.66	1.36	0.35
23.000	420.50	421.98	1.48	0.35
23.910	436.00	437.41	1.41	0.32
24.850	453.30	453.79	0.49	0.11
26.800	487.50	489.24	1.74	0.36
28.250	517.60	516.90	-0.70	-0.13
29.000	533.40	531.66	-1.74	-0.33
30.120	555.90	554.28	-1.62	-0.29
31.010	574.50	572.76	-1.74	-0.30
31.870	594.00	591.05	-2.95	-0.50
32.720	612.50	609.55	-2.95	-0.48
33.800	633.90	633.66	-0.24	-0.04
34.840	658.10	657.54	-0.56	-0.09
35.580	675.20	674.93	-0.27	-0.04
36.730	703.80	702.62	-1.18	-0.17
38.340	739.50	742.78	3.28	0.44
39.340	765.60	768.56	2.96	0.39

STANDARD DEVIATION = 1.77

DIETHYLENE GLYCOL $C_4H_{10}O_2$

RINKENBACH W.H.: IND.ENG.CHEM.19,474(1927).

A = 7.63666 B = 1939.359 C = 162.714

B.P.(760) = 245.070

T	P EXPTL.	P CALCD.	DEV.	PERCENT
130.000	8.00	10.26	2.26	28.27
135.000	12.00	13.26	1.26	10.49
140.000	16.00	16.99	0.99	6.16
145.000	21.00	21.59	0.59	2.79
148.500	25.00	25.41	0.41	1.65
150.000	27.00	27.22	0.22	0.83
155.000	34.00	34.08	0.08	0.25
160.000	43.00	42.38	-0.62	-1.45
165.000	54.00	52.34	-1.66	-3.08
170.000	66.00	64.23	-1.77	-2.68
175.000	80.00	78.35	-1.65	-2.06
180.000	96.00	95.02	-0.98	-1.02
185.000	116.00	114.60	-1.40	-1.20
190.000	135.00	137.49	2.49	1.84
195.000	165.00	164.10	-0.90	-0.54
200.000	195.00	194.92	-0.08	-0.04
205.000	228.00	230.44	2.44	1.07
210.000	268.00	271.21	3.21	1.20
215.000	316.00	317.83	1.83	0.58
220.000	370.00	370.91	0.91	0.25
225.000	430.00	431.14	1.14	0.27
230.000	499.00	499.24	0.24	0.05
235.000	577.00	575.96	-1.04	-0.18
240.000	669.00	662.12	-6.88	-1.03
* 243.000	734.00	718.69	-15.31	-2.09

STANDARD DEVIATION = 2.15

1,1-DIMETHOXYETHANE C4H10O2

NICOLINI E.,LAFFITTE P.:
COMPTES RENDUS 229,757(1949).

A = 6.71892 B = 1050.532 C = 209.237

B.P.(100) = 13.385

T	P EXPTL.	P CALCD.	DEV.	PERCENT
0.000	50.30	49.90	-0.40	-0.79
5.000	65.40	65.36	-0.04	-0.06
10.000	84.40	84.56	0.16	0.19
15.000	107.90	108.14	0.24	0.22
20.000	136.80	136.83	0.03	0.02
25.000	171.20	171.40	0.20	0.11
30.000	212.20	212.68	0.48	0.23
35.000	261.30	261.60	0.30	0.11
40.000	320.00	319.10	-0.90	-0.28
45.000	387.00	386.20	-0.80	-0.21
50.000	464.40	463.99	-0.41	-0.09
55.000	553.50	553.59	0.09	0.02
60.000	655.20	656.18	0.98	0.15

STANDARD DEVIATION = 0.57

ETHYLENEGLYCOL MONOETHYL ETHER C4H10O2

PICK J.,FRIED V.,HALA E.,VILIM O.:
COLLECTION CZECH.CHEM.COMM.21,260(1956).

A = 7.87457 B = 1843.463 C = 234.170

B.P.(760) = 134.983

T	P EXPTL.	P CALCD.	DEV.	PERCENT
63.000	46.80	46.90	0.10	0.22
63.800	48.30	48.73	0.43	0.90
76.400	87.20	86.86	-0.34	-0.39
86.800	136.60	135.26	-1.34	-0.98
92.200	169.00	168.34	-0.66	-0.39
96.800	200.80	201.69	0.89	0.45
102.500	249.70	250.61	0.91	0.36
107.400	298.00	300.29	2.29	0.77
112.400	360.80	359.26	-1.54	-0.43
117.800	432.60	433.53	0.93	0.21
122.400	507.80	506.51	-1.29	-0.25
127.600	601.00	601.03	0.03	0.01
134.000	737.60	737.02	-0.58	-0.08

STANDARD DEVIATION = 1.21

2-METHYL-2-PROPANETHIOL $C_4H_{10}S$

MCCULLOUGH J.P.,ET AL.:
J.AM.CHEM.SOC. 75,1818(1953).

A = 6.78871 B = 1116.076 C = 221.378

B.P.(760) = 64.217

T	P EXPTL.	P CALCD.	DEV.	PERCENT
20.496	149.41	149.43	0.02	0.01
25.785	187.57	187.57	0.00	0.00
31.127	233.72	233.72	0.00	0.00
36.519	289.13	289.14	0.01	0.00
41.959	355.22	355.23	0.01	0.00
47.446	433.56	433.52	-0.04	-0.01
52.983	525.86	525.78	-0.08	-0.02
58.573	633.99	633.91	-0.08	-0.01
64.217	760.00	760.00	0.00	0.00
69.908	906.06	906.07	0.01	0.00
75.654	1074.60	1074.69	0.09	0.01
81.449	1268.00	1268.20	0.20	0.02
87.294	1489.10	1489.28	0.18	0.01
93.188	1740.80	1740.70	-0.10	-0.01
99.138	2026.00	2025.76	-0.24	-0.01

STANDARD DEVIATION = 0.12

3,4-DITHIAHEXANE $C_4H_{10}S_2$

OSBORN A.G.,DOUSLIN D.R.: J.CHEM.ENG.DATA 11,502(1966).

A = 6.96592 B = 1479.991 C = 208.505

B.P.(760) = 153.985

T	P EXPTL.	P CALCD.	DEV.	PERCENT
100.567	149.41	149.39	-0.02	-0.02
107.079	187.57	187.61	0.04	0.02
113.627	233.72	233.72	0.00	0.00
120.230	289.13	289.13	0.00	0.00
126.884	355.22	355.24	0.02	0.00
133.579	433.56	433.50	-0.06	-0.01
140.336	525.86	525.87	0.01	0.00
147.136	633.99	634.01	0.02	0.00
153.986	760.00	760.02	0.02	0.00
160.884	906.06	906.04	-0.02	0.00

STANDARD DEVIATION = 0.03

3,4-DITHIAHEXANE $C_4H_{10}S_2$

SCOTT D.W.,ET AL.: J.AM.CHEM.SOC. 74,2478(1952).

A = 7.02014 B = 1512.784 C = 211.598

B.P.(10) = 39.689

T	P EXPTL.	P CALCD.	DEV.	PERCENT
0.000	0.76	0.74	-0.02	-2.27
15.000	2.22	2.21	-0.01	-0.52
20.000	3.10	3.08	-0.02	-0.72
25.000	4.23	4.23	0.00	-0.02
30.000	5.74	5.74	0.00	-0.08
35.000	7.66	7.68	0.02	0.30
40.000	10.14	10.17	0.03	0.33
45.000	13.32	13.32	0.00	0.02
50.000	17.28	17.27	-0.01	-0.06
55.000	22.18	22.17	-0.01	-0.05
60.000	28.20	28.20	0.00	-0.01
65.000	35.56	35.55	-0.01	-0.02
70.000	44.47	44.46	-0.01	-0.01
75.000	55.16	55.17	0.01	0.02
80.000	67.96	67.96	0.00	0.00

STANDARD DEVIATION = 0.02

DIETHYLAMINE C4H11N

BITTRICH H.J.,KAUER E.:

Z.PHYS.CHEM. 219,226(1962).

A = 5.80159 B = 583.297 C = 144.145

B.P.(760) = 55.561

T	P EXPTL.	P CALCD.	DEV.	PERCENT
31.450	300.00	301.83	1.83	0.61
34.750	350.00	347.56	-2.44	-0.70
38.050	400.00	398.19	-1.81	-0.45
41.100	450.00	449.57	-0.43	-0.09
43.850	500.00	499.88	-0.12	-0.02
46.500	550.00	552.07	2.07	0.38
48.850	600.00	601.52	1.52	0.25
51.100	650.00	651.74	1.74	0.27
53.200	700.00	701.24	1.24	0.18
55.550	760.00	759.71	-0.29	-0.04
57.050	800.00	798.78	-1.22	-0.15
59.000	850.00	851.64	1.64	0.19
60.580	900.00	896.22	-3.78	-0.42

STANDARD DEVIATION = 2.06

2-METHYL-2-PROPYLAMINE C4H11N

OSBORN A.G.,DOUSLIN D.R.:

J.CHEM.ENG.DATA13,534(1968).

A = 6.78322 B = 993.328 C = 210.501

B.P.(760) = 44.041

T	P EXPTL.	P CALCD.	DEV.	PERCENT
19.325	289.13	289.16	0.03	0.01
24.177	355.22	355.22	0.00	0.00
29.075	433.56	433.55	-0.01	0.00
34.016	525.86	525.80	-0.06	-0.01
39.006	633.99	633.97	-0.02	0.00
44.040	760.00	759.98	-0.02	0.00
49.120	906.06	906.06	0.00	0.00
54.246	1074.60	1074.58	-0.02	0.00
59.417	1268.00	1267.99	-0.01	0.00
64.642	1489.10	1489.37	0.27	0.02
69.908	1740.80	1741.00	0.21	0.01
75.212	2026.00	2025.63	-0.37	-0.02

STANDARD DEVIATION = 0.17

DIETHANOLAMINE $C_4H_{11}NO_2$

MCDONALD R.A.,SHRADER S.A.,STULL D.R.: J.CHEM.ENG.DATA 4,311(1959).

A = 8.13880 B = 2327.938 C = 174.352

B.P.(100) = 204.865

T	P EXPTL.	P CALCD.	DEV.	PERCENT
193.780	65.31	65.34	0.03	0.04
197.440	75.44	75.41	-0.03	-0.05
200.780	85.71	85.73	0.02	0.03
203.640	95.27	95.52	0.25	0.27
206.160	105.38	104.93	-0.45	-0.43
216.750	153.42	153.66	0.24	0.15
241.320	345.51	345.45	-0.06	-0.02

STANDARD DEVIATION = 0.29

DIMETHYL ETHYLPHOSPHONATE $C_4H_{11}O_3P$

KOSOLAPOFF G.M.: J.CHEM.SOC. 1955,2964.

A = 5.03986 B = 559.815 C = 61.907

B.P.(100) = 122.251

T	P EXPTL.	P CALCD.	DEV.	PERCENT
60.200	1.60	2.85	1.25	78.29
69.400	5.60	5.98	0.38	6.73
78.200	11.10	11.07	-0.03	-0.25
88.500	21.10	20.79	-0.31	-1.47
98.000	35.60	34.59	-1.01	-2.83
108.000	55.60	55.59	-0.01	-0.01
118.500	85.10	86.46	1.36	1.59
128.000	123.60	123.60	0.00	0.00
137.000	168.60	168.04	-0.56	-0.33

STANDARD DEVIATION = 0.91

BIS-DIMETHYLAMINODIFLUOROSILANE C4H12F2N2SI

VONGROSSE-RUYKEN H.,KLEESAAT R.:
Z.ANORG.ALLG.CHEM. 308,122(1961).

A = 5.95180 B = 748.732 C = 146.942

B.P.(100) = 42.524

T	P EXPTL.	P CALCD.	DEV.	PERCENT
23.800	40.00	36.87	-3.13	-7.83
36.600	74.00	74.55	0.55	0.74
57.800	189.00	197.17	8.17	4.32
70.100	326.00	317.75	-8.25	-2.53
77.900	418.00	418.57	0.57	0.14
88.300	587.00	587.47	0.47	0.08

STANDARD DEVIATION = 6.96

DIMETHYL-N-DIMETHYLAMINOFLUOROSILANE C4H12FNSI

VONGROSSE-RUYKEN H.,KLEESAAT R.:
Z.ANORG.ALLG.CHEM. 308,122(1961).

A = 7.53902 B = 1530.262 C = 249.388

B.P.(100) = 26.881

T	P EXPTL.	P CALCD.	DEV.	PERCENT
11.400	47.00	46.90	-0.10	-0.21
17.800	65.00	64.82	-0.18	-0.27
23.300	84.00	84.58	0.58	0.69
30.800	120.00	119.53	-0.47	-0.40
36.700	155.00	154.92	-0.08	-0.05
43.300	204.00	204.51	0.51	0.25
51.300	282.00	281.72	-0.28	-0.10

STANDARD DEVIATION = 0.49

TETRAMETHYLLEAD C4H12PB

GOOD W.D.,ET AL.: J.PHYS.CHEM. 63,1136(1959).

A = 6.93769 B = 1335.345 C = 219.088

B.P.(100) = 51.351

T	P EXPTL.	P CALCD.	DEV.	PERCENT
0.000	7.00	6.96	-0.04	-0.55
15.000	17.09	17.11	0.02	0.12
20.000	22.52	22.52	0.00	-0.01
25.000	29.29	29.31	0.02	0.05
30.000	37.72	37.74	0.02	0.04
35.000	48.12	48.11	-0.01	-0.02
40.000	60.78	60.77	-0.01	-0.02
45.000	76.06	76.08	0.02	0.02
50.000	94.51	94.45	-0.06	-0.06
55.000	116.30	116.34	0.04	0.04
60.000	142.24	142.24	0.00	0.00

STANDARD DEVIATION = 0.03

TETRAMETHYLSILANE C4H12SI

ASTON J.G.,KENNEDY R.M.,MESSERLY G.H.: J.AM.CHEM.SOC.63,2343(1941).

A = 6.82239 B = 1033.724 C = 235.623

B.P.(100) = -21.264

T	P EXPTL.	P CALCD.	DEV.	PERCENT
-64.220	6.24	6.19	-0.05	-0.86
-55.980	11.71	11.70	-0.01	-0.11
-36.990	41.50	41.51	0.01	0.03
-30.310	61.28	61.31	0.03	0.05
-23.110	90.79	90.80	0.01	0.02
-17.080	123.68	123.69	0.01	0.00
-9.430	178.77	178.77	0.00	0.00
-3.970	229.12	229.09	-0.03	-0.01
1.410	289.35	289.26	-0.09	-0.03
8.830	392.31	392.34	0.03	0.01
9.790	407.57	407.57	0.00	0.00
14.810	494.99	495.04	0.05	0.01
15.250	503.41	503.36	-0.05	-0.01
19.460	588.54	588.69	0.15	0.03
20.490	611.29	611.20	-0.09	-0.01

STANDARD DEVIATION = 0.06

BIS-DIMETHYLAMINOBORINE C4H13BN2

WIBERG E.,BOLTZ A.:

Z.ANORG.CHEM. 257,131(1948).

A = 5.58452 B = 774.371 C = 170.642

B.P.(100) = 45.390

T	P EXPTL.	P CALCD.	DEV.	PERCENT
-25.000	2.00	1.85	-0.15	-7.42
-10.000	5.00	5.81	0.81	16.16
0.000	11.00	11.13	0.13	1.19
5.000	15.00	14.99	-0.01	-0.09
15.000	26.00	25.89	-0.11	-0.41
20.000	34.00	33.31	-0.69	-2.02
29.500	52.00	51.93	-0.07	-0.14
37.100	72.00	71.94	-0.06	-0.09
45.800	101.00	101.57	0.57	0.57
54.700	140.00	140.63	0.63	0.45
62.500	184.00	183.26	-0.74	-0.40

STANDARD DEVIATION = 0.56

TETRAIODOTHIOPHENE C4I4S

MILAZZO G.:

ANNALI DI CHIMICA 46,1105(1956).

A = 5.58544 B = 871.248 C = 175.594

B.P.(10) = 14.409

T	P EXPTL.	P CALCD.	DEV.	PERCENT
-64.830	0.01	0.01	0.00	-47.03
-55.860	0.03	0.02	-0.01	-37.71
-50.060	0.06	0.04	-0.02	-27.59
-37.460	0.22	0.19	-0.03	-13.35
-32.010	0.38	0.33	-0.05	-12.42
-22.850	0.85	0.76	-0.09	-10.45
-13.170	1.65	1.67	0.02	0.91
-6.120	2.90	2.78	-0.12	-4.02
2.950	4.95	5.08	0.13	2.59
8.240	7.05	7.02	-0.03	-0.47
21.250	14.20	14.43	0.23	1.64
23.570	16.45	16.25	-0.20	-1.20

STANDARD DEVIATION = 0.12

PERFLUOROCYCLOPENTANE C5F10

BARBER E.J.,CADY G.H.: J.PHYS.CHEM.60,504(1956).

A = 7.03963 B = 1069.251 C = 234.603

B.P.(760) = 22.502

T	P EXPTL.	P CALCD.	DEV.	PERCENT
17.060	618.29	617.86	-0.43	-0.07
19.950	689.82	690.44	0.62	0.09
22.570	761.55	761.93	0.38	0.05
26.100	866.80	867.39	0.59	0.07
29.120	967.30	966.45	-0.85	-0.09
32.250	1079.90	1078.29	-1.61	-0.15
35.960	1224.70	1223.71	-0.99	-0.08
40.010	1397.50	1399.46	1.96	0.14
42.820	1531.70	1532.50	0.80	0.05
46.290	1709.50	1710.07	0.57	0.03
49.420	1883.60	1883.50	-0.10	-0.01
52.750	2082.30	2082.52	0.22	0.01
56.180	2305.20	2304.00	-1.20	-0.05

STANDARD DEVIATION = 1.08

PERFLUOROPIPERIDINE C5F11N

GOOD W.D.,ET AL.: J.PHYS.CHEM. 67.1306(1963).

A = 6.85339 B = 1059.951 C = 217.201

B.P.(760) = 49.616

T	P EXPTL.	P CALCD.	DEV.	PERCENT
29.138	355.22	355.28	0.06	0.02
34.191	433.56	433.57	0.01	0.00
39.290	525.86	525.87	0.01	0.00
44.421	633.99	633.75	-0.24	-0.04
54.847	906.06	906.13	0.07	0.01
60.117	1074.60	1074.56	-0.04	0.00
65.437	1268.00	1268.17	0.17	0.01
70.798	1489.10	1489.30	0.20	0.01
76.201	1740.80	1740.84	0.04	0.00
81.646	2026.00	2025.73	-0.27	-0.01

STANDARD DEVIATION = 0.17

PERFLUORO-2-METHYLBUTANE C5F12

CROWDER G.A.,TAYLOR Z.L.,REED T.M.,YOUNG J.A.: J.CHEM.ENG.DATA 12,481(1967).

A = 6.90092 B = 1013.305 C = 222.024

B.P.(760) = 30.035

T	P EXPTL.	P CALCD.	DEV.	PERCENT
-44.450	15.80	15.65	-0.15	-0.94
-37.220	26.56	26.17	-0.39	-1.47
-20.100	76.14	76.33	0.19	0.25
-11.900	118.86	119.82	0.96	0.80
0.440	222.54	221.83	-0.72	-0.32
9.660	337.47	336.73	-0.74	-0.22
28.690	723.52	723.18	-0.34	-0.05
32.130	819.23	820.26	1.03	0.13

STANDARD DEVIATION = 0.82

PERFLUOROPENTANE C5F12

BARBER J.E.,CADY G.H.: J.PHYS.CHEM.60,504(1956).

A = 7.01785 B = 1072.850 C = 230.002

B.P.(760) = 29.326

	T	P EXPTL.	P CALCD.	DEV.	PERCENT
	9.670	347.56	347.95	0.39	0.11
	12.700	395.99	395.73	-0.26	-0.06
	15.630	447.94	447.74	-0.20	-0.05
	18.250	497.16	496.85	-0.31	-0.06
	20.690	547.68	547.38	-0.30	-0.05
	24.100	624.20	624.77	0.57	0.09
	28.740	743.95	743.76	-0.19	-0.03
	35.230	940.50	941.15	0.65	0.07
	41.120	1149.00	1150.19	1.19	0.10
	46.090	1357.40	1355.20	-2.20	-0.16
	50.800	1574.50	1574.64	0.14	0.01
	55.310	1809.60	1809.57	-0.03	0.00
*	54.940	1801.00	1789.33	-11.67	-0.65
	54.930	1789.50	1788.79	-0.71	-0.04
	58.230	1976.20	1978.40	2.20	0.11
*	60.760	2175.30	2128.42	-46.88	-2.16
	64.790	2391.50	2390.55	-0.95	-0.04

STANDARD DEVIATION = 1.07

NONAFLUOROCYCLOPENTANE C_5HF_9

BARBER E.J.,CADY G.H.: J.PHYS.CHEM.60,504(1956).

A = 6.94533 B = 1051.724 C = 220.110

B.P.(760) = 38.647

T	P EXPTL.	P CALCD.	DEV.	PERCENT
17.000	322.87	323.40	0.53	0.16
20.150	369.97	369.74	-0.23	-0.06
23.490	424.50	424.54	0.04	0.01
26.670	482.53	482.56	0.03	0.01
29.730	544.98	544.18	-0.80	-0.15
33.070	618.48	618.41	-0.07	-0.01
36.180	695.36	694.52	-0.84	-0.12
38.860	765.92	765.87	-0.05	-0.01
42.080	858.60	859.07	0.47	0.05
45.850	980.40	979.24	-1.16	-0.12
49.860	1118.80	1121.05	2.25	0.20
53.050	1243.00	1244.86	1.86	0.15
56.430	1387.50	1387.33	-0.17	-0.01
59.480	1526.40	1526.40	0.00	0.00
63.090	1704.60	1704.58	-0.02	0.00
67.050	1919.50	1917.92	-1.58	-0.08
70.390	2114.00	2113.20	-0.80	-0.04
73.500	2308.30	2308.29	-0.01	0.00
74.780	2391.90	2392.43	0.53	0.02

STANDARD DEVIATION = 0.98

2-FURFURALDEHYDE C5H4O2

MATTHEWS J.B.,SUMNER J.F.,MOELWYN-HUGHES E.A.: TRANS.FARADAY SOC. 46,797(1950).

A = 6.57589 B = 1198.693 C = 162.837

B.P.(760) = 161.566

T	P EXPTL.	P CALCD.	DEV.	PERCENT
55.870	13.25	12.45	-0.80	-6.06
56.070	13.45	12.59	-0.86	-6.39
64.580	20.90	20.18	-0.72	-3.43
67.700	24.91	23.79	-1.12	-4.52
75.050	35.10	34.43	-0.67	-1.90
76.470	37.76	36.89	-0.87	-2.31
84.950	54.41	54.74	0.33	0.60
87.380	61.80	60.99	-0.81	-1.31
92.150	74.70	74.97	0.27	0.36
93.050	77.32	77.88	0.56	0.72
96.770	90.46	90.90	0.44	0.48
102.130	113.10	112.71	-0.40	-0.35
103.310	117.80	118.03	0.23	0.20
105.090	127.60	126.45	-1.15	-0.90
107.270	137.20	137.42	0.22	0.16
111.650	162.40	161.75	-0.65	-0.40
116.250	185.50	190.91	5.41	2.92
118.400	203.90	205.90	2.00	0.98
120.850	223.10	224.11	1.01	0.45
124.750	256.10	255.73	-0.37	-0.15
126.050	263.00	267.01	4.01	1.53
126.350	271.00	269.67	-1.33	-0.49
132.150	326.90	325.34	-1.56	-0.48
133.800	339.80	342.72	2.92	0.86
135.950	367.10	366.45	-0.65	-0.18
138.350	399.50	394.44	-5.06	-1.27
139.950	422.80	414.01	-8.79	-2.08
143.800	462.30	464.22	1.92	0.41
146.850	517.80	507.25	-10.55	-2.04
149.650	538.00	549.42	11.42	2.12
151.700	579.40	581.97	2.57	0.44
153.950	618.60	619.40	0.80	0.13
155.100	641.00	639.23	-1.77	-0.28
* 160.750	764.20	743.86	-20.34	-2.66

STANDARD DEVIATION = 3.79

1-METHYLPYRROLE C5H7N

OSBORN A.G.,DOUSLIN D.R.: J.CHEM.ENG.DATA13,534(1968).

A = 7.08498 B = 1368.661 C = 212.800

B.P.(760) = 112.748

T	P EXPTL.	P CALCD.	DEV.	PERCENT
48.962	71.87	71.84	-0.03	-0.05
51.766	81.64	81.61	-0.03	-0.03
54.577	92.52	92.50	-0.02	-0.02
57.400	104.63	104.62	-0.01	-0.01
60.228	118.06	118.06	0.00	0.00
63.067	132.95	132.95	0.00	0.00
65.926	149.41	149.47	0.06	0.04
71.649	187.57	187.66	0.09	0.05
77.400	233.72	233.73	0.01	0.00
83.202	289.13	289.18	0.05	0.02
89.033	355.22	355.21	-0.01	0.00
94.908	433.56	433.57	0.01	0.00
100.815	525.86	525.82	-0.04	-0.01
106.761	633.99	633.93	-0.06	-0.01
112.743	760.00	759.88	-0.12	-0.02
118.767	906.06	906.00	-0.06	-0.01
124.827	1074.60	1074.53	-0.07	-0.01
130.925	1268.00	1268.04	0.04	0.00
137.060	1489.10	1489.22	0.12	0.01
143.230	1740.80	1740.81	0.01	0.00
149.440	2026.00	2026.08	0.08	0.00

STANDARD DEVIATION = 0.06

PYRIDINE C_5H_5N

MCCULLOUGH J.P.,ET AL.:
J.AM.CHEM.SOC. 79,4289(1957).

A = 7.04115 B = 1373.799 C = 214.979

B.P.(760) = 115.235

T	P EXPTL.	P CALCD.	DEV.	PERCENT
67.299	149.41	149.39	-0.02	-0.02
73.154	187.57	187.59	0.02	0.01
79.045	233.72	233.74	0.02	0.01
84.974	289.13	289.13	0.00	0.00
90.946	355.22	355.22	0.00	0.00
96.958	433.56	433.56	0.00	0.00
103.008	525.86	525.83	-0.03	-0.01
109.101	633.99	633.97	-0.02	0.00
115.234	760.00	759.98	-0.02	0.00
121.408	906.06	906.07	0.01	0.00
127.622	1074.60	1074.57	-0.03	0.00
133.878	1268.00	1268.07	0.07	0.01
140.174	1489.10	1489.20	0.10	0.01
146.509	1740.80	1740.78	-0.02	0.00
152.886	2026.00	2025.93	-0.07	0.00

STANDARD DEVIATION = 0.05

DIMETHYLMALONONITRILE C5H7NO2

RIBNER A.,WESTRUM E.F.: J.PHYS.CHEM. 71,1208(1967).

A = 7.03551 B = 1546.991 C = 202.002

B.P.(100) = 105.214

T	P EXPTL.	P CALCD.	DEV.	PERCENT
49.090	7.50	7.49	-0.01	-0.14
56.610	11.35	11.31	-0.04	-0.32
60.750	14.14	14.06	-0.08	-0.60
64.140	16.75	16.71	-0.04	-0.27
67.730	19.97	19.96	-0.01	-0.04
70.950	23.49	23.33	-0.16	-0.69
73.520	26.26	26.35	0.09	0.33
78.070	32.41	32.51	0.10	0.30
81.180	37.92	37.38	-0.54	-1.43
84.230	43.14	42.74	-0.40	-0.93
86.540	47.08	47.22	0.14	0.29
87.440	49.01	49.06	0.05	0.11
89.890	54.15	54.40	0.25	0.47
94.160	64.62	64.87	0.25	0.39
96.020	69.47	69.93	0.46	0.66
97.900	75.46	75.37	-0.09	-0.12
100.040	81.76	81.98	0.22	0.27
110.930	123.31	123.59	0.28	0.22
118.310	161.06	160.65	-0.42	-0.26
120.790	175.56	174.97	-0.59	-0.33
132.040	253.38	253.73	0.35	0.14
132.320	255.62	256.01	0.39	0.15
132.800	260.28	259.95	-0.33	-0.13
134.010	270.60	270.10	-0.50	-0.18
136.290	290.02	290.11	0.09	0.03
139.480	319.74	320.09	0.35	0.11

STANDARD DEVIATION = 0.31

2-METHYL-1,3-BUTADIENE C5H8

BOUBLIKOVA L.:

THESIS,UTZCHT,PRAGUE(1972).

A = 6.92587 B = 1091.747 C = 235.860

B.P.(760) = 34.036

T	P EXPTL.	P CALCD.	DEV.	PERCENT
-16.204	90.35	90.29	-0.06	-0.07
-12.508	109.13	109.11	-0.02	-0.02
-8.291	134.42	134.41	-0.01	0.00
-4.649	159.87	159.96	0.09	0.06
-0.541	193.37	193.39	0.02	0.01
2.793	224.44	224.52	0.08	0.04
6.725	266.22	266.32	0.10	0.04
10.416	311.05	311.07	0.02	0.01
14.852	372.81	372.64	-0.17	-0.05
20.211	459.75	459.64	-0.11	-0.02
26.339	578.49	578.18	-0.31	-0.05
33.257	739.40	739.77	0.37	0.05

STANDARD DEVIATION = 0.18

2-METHYL-1,3-BUTADIENE C5H8

OSBORN A.G.,DOUSLIN D.R.:

J.CHEM.ENG.DATA 14,208(1969).

A = 6.92883 B = 1095.050 C = 236.291

B.P.(10) = -51.592

T	P EXPTL.	P CALCD.	DEV.	PERCENT
-57.598	6.32	6.32	0.00	0.00
-55.186	7.63	7.63	0.00	-0.04
-52.770	9.16	9.16	0.00	0.01
-50.351	10.95	10.95	0.00	0.03
-47.930	13.04	13.04	0.00	-0.01
-45.507	15.45	15.46	0.01	0.04
-43.083	18.25	18.24	-0.01	-0.04
-40.656	21.45	21.45	0.00	0.00
-38.227	25.12	25.12	0.00	0.01

STANDARD DEVIATION = 0.00

3-METHYL-1,2-BUTADIENE C_5H_8

OSBORN A.G.,DOUSLIN D.R.: J.CHEM.ENG.DATA 14,208(1969).

A = 6.94717 B = 1105.979 C = 231.146

B.P.(760) = 40.837

T	P EXPTL.	P CALCD.	DEV.	PERCENT
0.578	149.41	149.39	-0.02	-0.01
5.479	187.57	187.58	0.01	0.01
10.416	233.72	233.73	0.01	0.00
15.391	289.13	289.14	0.01	0.00
20.404	355.22	355.23	0.01	0.00
25.454	433.56	433.55	-0.01	0.00
30.543	525.86	525.84	-0.02	0.00
35.671	633.99	633.99	0.00	0.00
40.837	760.00	759.99	-0.01	0.00
46.042	906.06	906.08	0.02	0.00

STANDARD DEVIATION = 0.02

3-METHYL-1,2-BUTADIENE C_5H_8

OSBORN A.G.,DOUSLIN D.R.: J.CHEM.ENG.DATA 14,208(1969).

A = 7.17072 B = 1201.676 C = 240.061

B.P.(10) = -45.322

T	P EXPTL.	P CALCD.	DEV.	PERCENT
-60.010	3.15	3.14	-0.01	-0.39
-57.598	3.84	3.84	0.00	0.12
-55.186	4.69	4.69	0.00	-0.09
-52.770	5.68	5.68	0.00	0.06
-50.351	6.86	6.86	0.00	0.02
-47.930	8.24	8.25	0.01	0.07
-45.507	9.86	9.87	0.01	0.06
-43.083	11.75	11.75	0.00	0.03
-40.656	13.95	13.94	-0.01	-0.04
-38.227	16.48	16.48	0.00	-0.01
-35.797	19.40	19.40	0.00	-0.01
-33.364	22.76	22.75	-0.01	-0.04
-30.929	26.58	26.59	0.01	0.03

STANDARD DEVIATION = 0.01

1,2-PENTADIENE C5H8

OSBORN A.G.,DOUSLIN D.R.:
J.CHEM.ENG.DATA 14,208(1969).

A = 6.95032 B = 1130.912 C = 232.336

B.P.(10) = -42.277

T	P EXPTL.	P CALCD.	DEV.	PERCENT
-60.010	2.44	2.44	0.00	0.07
-57.598	3.01	3.01	0.00	-0.07
-55.186	3.68	3.68	0.00	0.13
-52.770	4.49	4.49	0.00	0.01
-50.351	5.44	5.45	0.01	0.09
-47.930	6.57	6.57	0.00	0.01
-45.507	7.90	7.89	-0.01	-0.12
-43.083	9.44	9.43	-0.01	-0.07
-40.656	11.23	11.23	0.00	-0.01
-38.227	13.31	13.31	0.00	-0.01
-35.797	15.70	15.71	0.01	0.07
-33.364	18.47	18.47	0.00	0.02
-30.929	21.64	21.64	0.00	0.00
-28.493	25.26	25.26	0.00	-0.02

STANDARD DEVIATION = 0.01

CIS-1,3-PENTADIENE C5H8

OSBORN A.G.,DOUSLIN D.R.:
J.CHEM.ENG.DATA 14,208(1969).

A = 6.95889 B = 1133.203 C = 233.281

B.P.(10) = -43.111

T	P EXPTL.	P CALCD.	DEV.	PERCENT
-60.010	2.62	2.62	0.00	0.12
-57.598	3.23	3.23	0.00	-0.14
-55.186	3.95	3.94	-0.01	-0.14
-52.770	4.79	4.80	0.01	0.18
-50.351	5.81	5.81	0.00	-0.01
-47.930	7.01	7.00	-0.01	-0.15
-45.507	8.39	8.39	0.00	0.04
-43.083	10.01	10.02	0.01	0.10
-40.656	11.91	11.91	0.00	0.01
-38.227	14.10	14.10	0.00	-0.01
-35.797	16.63	16.62	-0.01	-0.05
-33.364	19.52	19.52	0.00	0.01
-30.929	22.84	22.84	0.00	0.01

STANDARD DEVIATION = 0.01

TRANS-1,3-PENTADIENE C5H8

OSBORN A.G.,DOUSLIN D.R.: J.CHEM.ENG.DATA 14,208(1969).

A = 6.97466 B = 1136.229 C = 235.356

B.P.(10) = -45.181

T	P EXPTL.	P CALCD.	DEV.	PERCENT
-60.010	3.13	3.12	-0.01	-0.19
-57.598	3.83	3.83	0.00	-0.13
-55.186	4.66	4.66	0.00	-0.04
-52.770	5.64	5.65	0.01	0.09
-50.351	6.81	6.81	0.00	-0.02
-47.930	8.17	8.17	0.00	0.03
-45.507	9.76	9.77	0.01	0.07
-43.083	11.61	11.62	0.01	0.08
-40.656	13.76	13.77	0.01	0.06
-38.227	16.27	16.25	-0.02	-0.14
-35.797	19.10	19.10	0.00	-0.02
-33.364	22.36	22.36	0.00	0.02
-30.929	26.09	26.09	0.00	0.02

STANDARD DEVIATION = 0.01

1,4-PENTADIENE C5H8

LAMB A.B.,ROPER E.E.: J.AM.CHEM.SOC.62,806(1940).

A = 6.93249 B = 1062.191 C = 236.307

B.P.(100) = -20.961

T	P EXPTL.	P CALCD.	DEV.	PERCENT
-78.840	1.60	1.54	-0.06	-3.87
-54.770	12.08	12.06	-0.02	-0.16
-37.970	37.80	37.76	-0.04	-0.11
-29.830	61.27	61.39	0.12	0.20
-21.050	99.67	99.53	-0.14	-0.14
-12.700	152.00	152.13	0.13	0.09
-3.970	229.20	229.46	0.26	0.11
0.010	273.70	273.97	0.27	0.10
0.010	275.00	273.97	-1.03	-0.37
2.450	304.50	304.54	0.04	0.01
4.520	332.30	332.57	0.27	0.08
16.320	533.10	534.44	1.34	0.25
18.080	572.70	571.46	-1.24	-0.22

STANDARD DEVIATION = 0.68

1,4-PENTADIENE C5H8

OSBORN A.G.,DOUSLIN D.R.: J.CHEM.ENG.DATA 14,208(1969).

A = 7.02593 B = 1098.676 C = 239.507

B.P.(10) = -57.183

T	P EXPTL.	P CALCD.	DEV.	PERCENT
-60.010	8.04	8.04	0.00	-0.04
-57.598	9.68	9.69	0.01	0.08
-55.186	11.62	11.62	0.00	0.02
-52.770	13.89	13.88	-0.01	-0.07
-50.351	16.51	16.51	0.00	-0.03
-47.930	19.54	19.55	0.01	0.03
-45.507	23.04	23.05	0.01	0.04
-43.083	27.08	27.07	-0.01	-0.02

STANDARD DEVIATION = 0.01

2,3-PENTADIENE C5H8

OSBORN A.G.,DOUSLIN D.R.: J.CHEM.ENG.DATA 14,208(1969).

A = 7.22100 B = 1239.539 C = 238.245

B.P.(10) = -38.995

T	P EXPTL.	P CALCD.	DEV.	PERCENT
-60.010	1.85	1.85	0.00	-0.16
-57.598	2.29	2.29	0.00	-0.11
-55.186	2.82	2.82	0.00	-0.11
-52.770	3.45	3.45	0.00	0.03
-50.351	4.21	4.21	0.00	-0.07
-47.930	5.10	5.10	0.00	0.08
-45.507	6.16	6.16	0.00	0.05
-43.083	7.40	7.41	0.01	0.10
-40.656	8.87	8.87	0.00	-0.05
-38.227	10.56	10.57	0.01	0.05
-35.797	12.54	12.54	0.00	-0.01
-33.364	14.83	14.82	-0.01	-0.04
-30.929	17.46	17.46	0.00	0.00
-28.493	20.49	20.49	0.00	-0.02
-26.054	23.95	23.95	0.00	0.02

STANDARD DEVIATION = 0.00

SPIROPENTANE C5H8

SCOTT D.W.,ET AL.:
J.AM.CHEM.SOC. 72,4664(1950).

A = 6.91700 B = 1090.084 C = 231.103

B.P.(760) = 38.974

T	P EXPTL.	P CALCD.	DEV.	PERCENT
3.632	187.57	187.55	-0.02	-0.01
8.565	233.72	233.72	0.00	0.00
13.537	289.13	289.16	0.03	0.01
18.546	355.22	355.25	0.03	0.01
23.592	433.56	433.55	-0.01	0.00
28.679	525.86	525.84	-0.02	0.00
33.808	633.99	634.01	0.02	0.00
38.977	760.00	760.06	0.06	0.01
44.181	906.06	906.04	-0.02	0.00
49.429	1074.60	1074.55	-0.05	-0.01
54.714	1268.00	1267.88	-0.12	-0.01
60.045	1489.10	1489.05	-0.05	0.00
65.421	1740.80	1740.97	0.17	0.01
70.829	2026.00	2026.00	0.00	0.00

STANDARD DEVIATION = 0.07

CYCLOPENTANONE C5H8O

BENSON S.W.,KISTIAKOWSKI G.B.:
J.AM.CHEM.SOC. 64,80(1942).

A = 2.90242 B = 162.903 C = 63.215

B.P.(10) = 22.415

T	P EXPTL.	P CALCD.	DEV.	PERCENT
-0.060	2.30	2.10	-0.20	-8.53
13.840	6.00	6.14	0.14	2.37
23.740	10.40	10.69	0.29	2.79
25.640	12.00	11.72	-0.28	-2.30

STANDARD DEVIATION = 0.47

2-METHYL-3-BUTYN-2-OL C5H8O

CONNER A.Z.,ET AL.: IND.ENG.CHEM.42,106(1950).

A = 6.65747 B = 976.540 C = 154.080

B.P.(760) = 104.493

T	P EXPTL.	P CALCD.	DEV.	PERCENT
21.600	12.90	12.56	-0.34	-2.67
33.200	27.60	27.74	0.14	0.51
42.300	48.50	48.39	-0.11	-0.23
49.200	70.00	71.37	1.37	1.96
54.000	91.80	92.12	0.32	0.35
55.600	101.40	100.04	-1.36	-1.34
59.700	123.00	122.88	-0.12	-0.09
64.300	154.00	153.36	-0.64	-0.42
69.400	194.10	193.98	-0.12	-0.06
73.000	228.70	227.53	-1.17	-0.51
77.100	272.00	271.21	-0.79	-0.29
81.000	317.30	318.70	1.40	0.44
84.700	367.10	369.62	2.52	0.69
90.000	453.10	453.48	0.38	0.08
93.400	511.40	514.67	3.27	0.64
96.600	578.50	577.96	-0.54	-0.09
99.000	632.80	629.27	-3.53	-0.56
106.200	805.60	804.59	-1.01	-0.13

STANDARD DEVIATION = 1.62

METHYL METHACRYLATE C5H8O2

VONBROCKHAUS A.,JENCKEL E.: MAKROMOL.CHEM. 18/19,262(1956).

A = 8.40919 B = 2050.467 C = 274.369

B.P.(100) = 45.557

T	P EXPTL.	P CALCD.	DEV.	PERCENT
39.200	74.00	74.14	0.14	0.19
59.000	182.00	181.32	-0.68	-0.37
80.800	430.00	432.49	2.49	0.58
89.200	590.00	587.98	-2.02	-0.34

STANDARD DEVIATION = 3.28

ISOPROPYL CHLOROACETATE C5H9CLO2

NELSON O.A.: IND.ENG.CHEM.20,1380(1928).

A = 8.38175 B = 2328.297 C = 274.774

B.P.(760) = 148.480

T	P EXPTL.	P CALCD.	DEV.	PERCENT
35.000	7.50	7.34	-0.16	-2.14
40.000	10.00	9.66	-0.34	-3.39
45.000	12.00	12.61	0.61	5.08
50.000	16.40	16.32	-0.08	-0.47
55.150	20.90	21.12	0.22	1.05
60.200	26.90	26.98	0.08	0.31
65.250	34.30	34.22	-0.08	-0.23
70.300	44.30	43.10	-1.20	-2.70
75.300	52.20	53.81	1.61	3.09
80.300	67.20	66.76	-0.44	-0.65
85.300	83.60	82.34	-1.26	-1.51
90.200	104.60	100.56	-4.04	-3.87
95.200	121.50	122.64	1.14	0.94
100.000	147.40	147.65	0.25	0.17
110.000	215.70	214.13	-1.57	-0.73
120.000	305.90	304.76	-1.14	-0.37
125.000	350.20	361.18	10.98	3.14
129.900	424.40	424.86	0.46	0.11
134.900	497.70	499.42	1.72	0.35
139.700	577.30	581.14	3.84	0.67
144.800	674.50	680.08	5.58	0.83
147.700	730.00	742.43	12.43	1.70
148.700	758.20	765.02	6.82	0.90
149.600	785.70	785.83	0.13	0.02
150.600	818.10	809.52	-8.58	-1.05
151.600	847.00	833.80	-13.20	-1.56
152.600	877.80	858.70	-19.10	-2.18

STANDARD DEVIATION = 6.49

VALERONITRILE C5H9N

DREISBACH R.R.,SHRADER S.A.:
IND.ENG.CHEM. 41,2879(1949).

A = 7.10487 B = 1519.367 C = 218.435

B.P.(760) = 141.259

T	P EXPTL.	P CALCD.	DEV.	PERCENT
69.070	66.39	66.10	-0.29	-0.44
72.550	75.86	76.45	0.59	0.78
84.590	123.76	123.27	-0.49	-0.39
111.480	315.52	315.89	0.37	0.12
126.920	507.50	507.49	-0.01	0.00
141.250	760.00	759.81	-0.19	-0.03

STANDARD DEVIATION = 0.53

CYCLOPENTANE C_5H_{10}

WILLINGHAM C.J.,TAYLOR W.J.,PIGNOCCO J.M.,ROSSINI F.D.:
J.RESEARCH NATL.BUR.STANDARDS 35,219(1945).

A = 6.90325 B = 1132.868 C = 232.375

B.P.(760) = 49.262

T	P EXPTL.	P CALCD.	DEV.	PERCENT
15.707	217.19	217.15	-0.04	-0.02
20.196	261.71	261.77	0.06	0.02
25.598	324.94	324.96	0.02	0.01
31.172	402.45	402.45	0.00	0.00
37.119	500.74	500.69	-0.05	-0.01
43.574	627.97	627.92	-0.05	-0.01
48.131	732.12	732.14	0.02	0.00
48.621	744.10	744.11	0.01	0.00
49.073	755.30	755.29	-0.01	0.00
49.587	768.07	768.16	0.09	0.01
50.031	779.47	779.42	-0.05	-0.01

STANDARD DEVIATION = 0.05

CYCLOPENTANE C_5H_{10}

ASTON J.G.,FINK H.L.,SCHUMANN S.C.:
J.AM.CHEM.SOC. 65,341(1943).

A = 7.13275 B = 1240.113 C = 242.925

B.P.(100) = -1.317

T	P EXPTL.	P CALCD.	DEV.	PERCENT
-47.250	6.23	6.24	0.01	0.15
-38.360	11.78	11.76	-0.02	-0.14
-31.640	18.35	18.34	-0.01	-0.06
-24.710	28.15	28.17	0.02	0.07
-18.610	40.20	40.21	0.01	0.02
-12.910	55.13	55.12	-0.01	-0.02
-7.300	74.07	74.07	0.00	0.01
-1.940	96.99	96.99	0.00	0.00
4.460	131.78	131.78	0.00	0.00
9.320	164.63	164.61	-0.02	-0.01
14.240	204.39	204.41	0.02	0.01

STANDARD DEVIATION = 0.02

2-METHYL-1-BUTENE C_5H_{10}

SCOTT D.W., WADDINGTON G., SMITH J.C., HUFFMAN H.M.: J.AM.CHEM.SOC. 71,2767(1949).

A = 6.86364 B = 1048.876 C = 232.194

B.P.(760) = 31.156

T	P EXPTL.	P CALCD.	DEV.	PERCENT
1.155	233.74	233.75	0.01	0.01
6.054	289.13	289.19	0.06	0.02
10.993	355.21	355.30	0.09	0.03
15.973	433.54	433.65	0.12	0.03
20.996	526.86	526.01	-0.85	-0.16
26.062	634.06	634.24	0.18	0.03
31.162	759.96	760.16	0.20	0.03
36.308	906.02	906.23	0.21	0.02
41.500	1074.70	1074.83	0.13	0.01
46.728	1268.00	1268.15	0.15	0.01
52.005	1489.30	1489.35	0.05	0.00
57.320	1740.90	1740.81	-0.09	-0.01
62.675	2025.80	2025.55	-0.25	-0.01

STANDARD DEVIATION = 0.31

2-METHYL-2-BUTENE C5H10

SCOTT D.W.,WADDINGTON G.,SMITH J.C.,HUFFMAN H.M.: J.AM.CHEM.SOC. 71,2767(1949).

A = 6.92322 B = 1099.075 C = 233.317

B.P.(760) = 38.569

T	P EXPTL.	P CALCD.	DEV.	PERCENT
3.042	187.57	187.58	0.01	0.01
8.008	233.81	233.82	0.01	0.01
12.987	289.03	289.04	0.01	0.00
18.033	355.26	355.25	-0.01	0.00
23.103	433.52	433.50	-0.02	0.00
28.220	525.87	525.83	-0.04	-0.01
33.373	633.97	633.94	-0.04	-0.01
38.567	759.96	759.94	-0.03	0.00
43.806	906.15	906.15	0.00	0.00
49.078	1074.60	1074.58	-0.02	0.00
54.399	1268.20	1268.29	0.09	0.01
59.753	1489.20	1489.37	0.17	0.01
65.151	1740.80	1741.12	0.32	0.02
70.590	2026.90	2026.43	-0.47	-0.02

STANDARD DEVIATION = 0.18

2-METHYL-2-BUTENE C5H10

BOUBLIKOVA L. THESIS,UTZCHT,PRAGUE(1972).

A = 6.93402 B = 1104.007 C = 233.891

B.P.(760) = 38.488

T	P EXPTL.	P CALCD.	DEV.	PERCENT
-12.102	90.35	90.42	0.07	0.08
-8.421	109.13	109.03	-0.10	-0.09
-4.158	134.42	134.40	-0.02	-0.01
-0.487	159.87	159.95	0.08	0.05
3.643	193.37	193.30	-0.07	-0.04
7.001	224.44	224.40	-0.04	-0.02
10.974	266.22	266.30	0.08	0.03
14.692	311.05	311.04	-0.01	0.00
19.185	372.81	372.96	0.15	0.04
24.567	459.75	459.72	-0.03	-0.01
30.739	578.49	578.26	-0.23	-0.04
37.723	740.15	740.27	0.12	0.02

STANDARD DEVIATION = 0.12

3-METHYL-1-BUTENE C5H10

SCOTT D.W.,WADDINGTON G.:
J.AM.CHEM.SOC. 72,4310(1950).

A = 6.82643 B = 1013.605 C = 236.833

B.P.(760) = 20.061

T	P EXPTL.	P CALCD.	DEV.	PERCENT
0.218	355.25	355.25	0.00	0.00
5.112	433.53	433.54	0.01	0.00
10.053	525.86	525.86	0.00	0.00
15.033	633.94	633.94	0.00	0.00
20.061	760.00	759.99	-0.01	0.00
25.128	906.00	906.00	0.00	0.00
30.245	1074.60	1074.64	0.04	0.00
35.402	1268.10	1268.10	0.00	0.00
40.602	1489.20	1489.15	-0.05	0.00
45.847	1740.70	1740.72	0.02	0.00
51.139	2025.90	2025.93	0.03	0.00

STANDARD DEVIATION = 0.02

1-PENTENE C5H10

FORZIATI A.D.,CAMIN D.L.,ROSSINI F.D.:
J.RESEARCH NATL.BUR.STANDARDS 45,406(1950).

A = 6.84268 B = 1043.206 C = 233.344

B.P.(760) = 29.967

T	P EXPTL.	P CALCD.	DEV.	PERCENT
12.834	402.81	402.79	-0.02	-0.01
18.468	501.02	501.06	0.04	0.01
24.584	628.21	628.23	0.02	0.00
28.900	732.40	732.29	-0.11	-0.01
29.362	744.27	744.19	-0.08	-0.01
29.796	755.52	755.49	-0.03	0.00
30.289	768.46	768.51	0.05	0.01
30.723	779.98	780.10	0.12	0.02

STANDARD DEVIATION = 0.09

CIS-2-PENTENE C5H10

SCOTT D.W.,WADDINGTON G.: J.AM.CHEM.SOC. 72,4310(1950).

A = 6.87497 B = 1069.243 C = 230.759

B.P.(760) = 36.943

T	P EXPTL.	P CALCD.	DEV.	PERCENT
1.595	187.58	187.58	0.00	0.00
6.522	233.74	233.74	0.00	0.00
11.486	289.12	289.12	0.00	0.00
16.494	355.21	355.21	0.00	0.00
21.541	433.50	433.50	0.00	0.00
26.633	525.81	525.80	-0.01	0.00
31.766	633.94	633.94	0.00	0.00
36.944	760.03	760.03	0.00	0.00
42.161	906.11	906.10	-0.01	0.00
47.423	1074.70	1074.69	-0.01	0.00
52.724	1268.10	1268.12	0.02	0.00
58.070	1489.30	1489.27	-0.03	0.00
63.456	1740.80	1740.78	-0.02	0.00
68.882	2025.60	2025.60	0.00	0.00

STANDARD DEVIATION = 0.01

TRANS-2-PENTENE C5H10

SCOTT D.W.,WADDINGTON G.: J.AM.CHEM.SOC. 72,4310(1950).

A = 6.90546 B = 1083.828 C = 232.945

B.P.(760) = 36.352

	T	P EXPTL.	P CALCD.	DEV.	PERCENT
	1.026	187.56	187.56	0.00	0.00
	5.956	233.74	233.74	0.00	0.00
	10.922	289.15	289.15	0.00	0.00
	15.927	355.22	355.22	0.00	0.00
	20.969	433.49	433.49	0.00	0.00
	26.055	525.78	525.77	-0.01	0.00
	31.183	633.93	633.93	0.00	0.00
	36.354	760.03	760.03	0.00	0.00
	41.561	906.09	906.10	0.01	0.00
	46.813	1074.70	1074.72	0.02	0.00
*	52.100	1267.50	1268.10	0.60	0.05
	57.430	1489.20	1489.18	-0.02	0.00
	62.803	1740.80	1740.84	0.04	0.00
	68.211	2025.70	2025.66	-0.04	0.00

STANDARD DEVIATION = 0.02

DIETHYL KETONE C5H10O

COLLERSON R.R., ET AL.:

J.CHEM.SOC. 3697(1965)

A = 7.02529 B = 1310.281 C = 214.192

B.P.(760) = 101.959

T	P EXPTL.	P CALCD.	DEV.	PERCENT
56.544	153.32	153.31	-0.01	0.00
63.032	198.99	199.00	0.01	0.00
68.735	247.83	247.81	-0.02	-0.01
73.908	300.09	300.11	0.02	0.01
78.158	349.45	349.46	0.01	0.00
82.114	401.08	401.08	0.00	0.00
85.494	449.88	449.89	0.01	0.00
88.612	499.03	499.03	0.00	0.00
91.605	550.15	550.14	-0.01	0.00
94.314	599.94	599.93	-0.01	0.00
96.897	650.69	650.68	-0.01	0.00
99.177	698.25	698.25	0.00	0.00
101.566	751.03	751.02	-0.01	0.00
103.724	801.33	801.34	0.01	0.00
105.737	850.67	850.64	-0.03	0.00
107.682	900.53	900.53	0.00	0.00
109.486	948.79	948.82	0.03	0.00
111.303	999.49	999.50	0.01	0.00

STANDARD DEVIATION = 0.01

METHYL PROPYL KETONE C5H10O

COLLERSON R.R., ET AL.: J.CHEM.SOC. 3697(1965).

A = 7.02193 B = 1313.847 C = 215.009

B.P.(760) = 102.260

T	P EXPTL.	P CALCD.	DEV.	PERCENT
56.649	153.31	153.29	-0.02	-0.01
63.184	199.11	199.13	0.02	0.01
68.897	247.82	247.84	0.02	0.01
74.077	299.96	299.98	0.02	0.01
78.340	349.29	349.25	-0.04	-0.01
82.326	401.03	401.03	0.00	0.00
85.708	449.64	449.65	0.01	0.00
88.893	499.63	499.65	0.02	0.00
91.834	549.67	549.67	0.00	0.00
94.549	599.34	599.31	-0.03	0.00
97.179	650.75	650.74	-0.01	0.00
99.470	698.36	698.34	-0.02	0.00
101.845	750.53	750.56	0.03	0.00
104.031	801.30	801.30	0.00	0.00
106.114	852.16	852.14	-0.02	0.00
108.023	900.92	900.92	0.00	0.00
109.830	949.09	949.10	0.01	0.00
111.655	999.78	999.79	0.01	0.00

STANDARD DEVIATION = 0.02

3-PENTANONE C5H10O

DREISBACH R.R., SHRADER S.A.: IND.ENG.CHEM. 41,2879(1949).

A = 7.23064 B = 1477.021 C = 237.517

B.P.(760) = 102.042

T	P EXPTL.	P CALCD.	DEV.	PERCENT
36.360	75.86	68.81	-7.05	-9.30
51.240	123.76	130.48	6.72	5.43
75.040	315.52	319.91	4.39	1.39
88.910	507.50	507.95	0.45	0.09
101.700	760.00	752.36	-7.64	-1.00

STANDARD DEVIATION = 9.29

BUTYL FORMATE $C_5H_{10}O_2$

NELSON O.A.: IND.ENG.CHEM.20,1382(1928).

A = 7.69363 B = 1698.735 C = 247.413

B.P.(760) = 105.548

T	P EXPTL.	P CALCD.	DEV.	PERCENT
29.100	34.80	35.50	0.70	2.01
34.300	45.80	46.09	0.29	0.63
41.500	65.70	65.15	-0.56	-0.84
49.000	93.10	91.76	-1.34	-1.44
57.800	134.90	134.24	-0.66	-0.49
65.380	184.20	183.13	-1.07	-0.58
71.100	229.00	229.24	0.24	0.11
75.500	266.80	271.00	4.20	1.57
80.500	323.10	325.97	2.87	0.89
85.700	397.80	392.69	-5.11	-1.28
93.200	505.00	508.56	3.56	0.70
100.400	641.90	645.03	3.13	0.49
102.000	683.50	679.12	-4.38	-0.64
106.000	770.90	770.85	-0.05	-0.01
106.700	787.30	787.90	0.60	0.08
* 109.300	828.10	853.95	25.85	3.12
109.800	872.40	867.16	-5.24	-0.60
111.300	903.30	907.79	4.49	0.50
112.400	941.10	938.57	-2.53	-0.27

STANDARD DEVIATION = 3.19

SEC BUTYL FORMATE C5H10O2

NELSON O.A.: IND.ENG.CHEM.20,1382(1928).

A = 6.49273 B = 972.957 C = 175.946

B.P.(760) = 93.428

T	P EXPTL.	P CALCD.	DEV.	PERCENT
29.700	58.80	57.74	-1.06	-1.80
35.700	81.20	78.64	-2.56	-3.16
41.700	108.00	105.28	-2.72	-2.52
46.500	139.90	131.47	-8.43	-6.03
52.900	172.80	174.24	1.44	0.83
58.100	212.20	216.57	4.37	2.06
61.800	246.50	251.36	4.86	1.97
65.800	309.80	293.78	-16.02	-5.17
71.300	340.60	361.03	20.43	6.00
76.800	427.30	439.72	12.43	2.91
83.500	534.40	552.83	18.43	3.45
88.500	635.50	650.87	15.37	2.42
95.000	798.40	797.56	-0.84	-0.11
97.000	857.79	847.37	-10.42	-1.21
98.500	907.50	886.25	-21.25	-2.34
99.800	945.00	921.03	-23.98	-2.54

STANDARD DEVIATION =14.33

ETHYL PROPIONATE C5H10O2

MERTL I.,POLAK J.:
COLLECTION CZECHOSLOV.CHEM.COMM. 30,3526(1965).

A = 6.99492 B = 1260.621 C = 207.396

B.P.(760) = 99.018

T	P EXPTL.	P CALCD.	DEV.	PERCENT
33.785	58.79	58.62	-0.17	-0.29
37.035	69.02	68.79	-0.23	-0.33
39.710	78.12	78.23	0.11	0.14
42.545	89.03	89.38	0.35	0.39
44.685	98.67	98.64	-0.03	-0.03
48.155	115.28	115.34	0.06	0.05
52.050	136.84	136.78	-0.06	-0.04
56.335	164.03	164.05	0.02	0.01
59.815	189.33	189.33	0.00	0.00
64.825	231.35	231.23	-0.12	-0.05
69.965	281.40	281.75	0.35	0.12
75.055	340.12	340.23	0.11	0.03
81.100	423.30	421.97	-1.33	-0.31
87.330	521.10	521.98	0.88	0.17
93.845	645.55	645.89	0.34	0.05
98.310	743.80	743.49	-0.31	-0.04

STANDARD DEVIATION = 0.49

3-HYDROXY-3-METHYL-2-BUTANONE C5H10O2

CONNER A.Z.,ET AL.: IND.ENG.CHEM.42,106(1950).

A = 7.34085 B = 1653.599 C = 227.521

B.P.(760) = 143.238

T	P EXPTL.	P CALCD.	DEV.	PERCENT
44.700	18.80	18.47	-0.33	-1.78
54.700	30.60	30.31	-0.29	-0.94
63.400	44.40	45.38	0.98	2.20
70.800	62.70	62.78	0.08	0.13
79.900	93.10	91.60	-1.50	-1.61
89.200	130.00	131.78	1.78	1.37
97.000	176.80	175.93	-0.87	-0.49
103.800	225.10	223.83	-1.27	-0.56
111.400	289.30	289.63	0.33	0.11
121.500	398.30	400.89	2.59	0.65
123.400	426.30	425.28	-1.02	-0.24
126.600	468.90	469.09	0.19	0.04
131.800	549.20	548.07	-1.13	-0.21
134.200	585.10	587.99	2.89	0.49
137.400	644.70	644.85	0.15	0.02
144.000	776.60	776.17	-0.43	-0.06
145.700	815.80	813.27	-2.53	-0.31

STANDARD DEVIATION = 1.54

METHYL CELLOSOLVE ACETATE C5H10O3

DYKYJ J.,SEPRAKOVA M.,PAULECH J.: CHEM.ZVESTI 11,461(1957).

A = 7.12513 B = 1447.037 C = 196.103

B.P.(760) = 144.832

T	P EXPTL.	P CALCD.	DEV.	PERCENT
70.000	49.60	48.67	-0.93	-1.88
80.900	78.90	79.66	0.76	0.96
86.500	100.80	101.10	0.30	0.29
93.100	132.00	132.31	0.31	0.23
93.300	134.10	133.37	-0.73	-0.55
97.600	156.90	157.85	0.95	0.61
111.700	268.20	265.43	-2.77	-1.03
119.100	342.80	342.23	-0.57	-0.17
123.900	397.10	401.03	3.93	0.99
144.000	743.60	742.03	-1.57	-0.21

STANDARD DEVIATION = 2.03

BETA-METHYL-ALPHA-KETOBUTANOL C5H10O2

DREISBACH R.R.,SHRADER S.A.: IND.ENG.CHEM. 41,2879(1949).

A = 7.37171 B = 1691.491 C = 187.022

B.P.(760) = 189.626

T	P EXPTL.	P CALCD.	DEV.	PERCENT
109.500	47.16	46.48	-0.68	-1.44
114.000	57.04	56.57	-0.47	-0.83
118.300	66.39	67.87	1.48	2.23
121.080	75.86	76.15	0.29	0.38
133.160	123.76	122.69	-1.07	-0.86
160.200	315.52	316.39	0.87	0.27
189.600	760.00	759.44	-0.56	-0.07

STANDARD DEVIATION = 1.14

PROPYL ACETATE C5H10O2

MERTL I.,POLAK J.: COLLECTION CZECHOSLOV.CHEM.COMM. 30,3526(1965).

A = 7.01615 B = 1282.282 C = 208.600

B.P.(760) = 101.479

T	P EXPTL.	P CALCD.	DEV.	PERCENT
39.075	69.02	69.00	-0.02	-0.02
41.665	78.12	78.06	-0.06	-0.07
44.480	89.03	89.01	-0.02	-0.02
46.765	98.67	98.80	0.13	0.14
50.220	115.28	115.29	0.01	0.01
54.160	136.84	136.80	-0.04	-0.03
58.485	164.03	164.10	0.07	0.04
61.990	189.33	189.37	0.04	0.02
67.030	231.35	231.18	-0.17	-0.07
72.175	281.40	281.32	-0.08	-0.03
77.325	340.12	339.99	-0.13	-0.04
83.535	423.30	423.44	0.14	0.03
89.680	521.10	521.46	0.36	0.07
96.250	645.55	645.46	-0.09	-0.01
100.885	746.40	746.22	-0.18	-0.02

STANDARD DEVIATION = 0.15

VALERIC ACID

$C_5H_{10}O_2$

KAHLBAUM G.W.A.:

BER. 16,2476(1883).

A = 5.41163 B = 591.146 C = 59.867

B.P.(760) = 173.712

T	P EXPTL.	P CALCD.	DEV.	PERCENT
72.400	10.58	8.76	-1.82	-17.24
72.600	10.68	8.89	-1.79	-16.73
78.400	14.90	13.68	-1.22	-8.16
90.000	27.28	29.32	2.04	7.48
99.200	45.92	49.58	3.66	7.97
105.800	71.94	69.72	-2.22	-3.09
173.700	760.00	759.76	-0.24	-0.03

STANDARD DEVIATION = 2.76

ISOVALERIC ACID C5H10O2

KAHLBAUM G.W.A.:

Z.PHYS.CHEM. 13,14(1894).

A = 3.94655 B = 255.406 C = 11.293

B.P.(10) = 75.387

	T	P EXPTL.	P CALCD.	DEV.	PERCENT
	86.600	21.90	21.75	-0.15	-0.67
	87.000	22.50	22.29	-0.21	-0.93
	87.200	22.60	22.56	-0.04	-0.16
	88.000	22.90	23.68	0.78	3.39
	87.600	23.20	23.11	-0.09	-0.37
*	88.700	23.20	24.68	1.48	6.37
	87.900	23.40	23.53	0.13	0.57
	87.800	23.60	23.39	-0.21	-0.87
	87.800	23.70	23.39	-0.31	-1.29
	88.200	24.30	23.96	-0.34	-1.40
	88.600	24.90	24.53	-0.37	-1.48
	89.200	25.60	25.41	-0.19	-0.74
	90.000	26.40	26.61	0.21	0.80
	89.600	26.50	26.01	-0.49	-1.86
	90.200	27.40	26.92	-0.48	-1.76
	90.800	28.10	27.85	-0.25	-0.89
	91.400	29.00	28.80	-0.20	-0.67
	91.800	29.20	29.45	0.25	0.86
	92.800	30.80	31.11	0.31	1.01
	93.000	31.30	31.45	0.15	0.48
	93.200	31.80	31.79	-0.01	-0.03
	93.400	31.80	32.13	0.33	1.05
	93.800	32.40	32.83	0.43	1.32
	93.800	33.00	32.83	-0.17	-0.52
	94.200	33.10	33.53	0.43	1.31
	94.200	33.30	33.53	0.23	0.70
	94.400	33.50	33.89	0.39	1.16
	94.800	34.30	34.61	0.31	0.90
	95.000	34.80	34.97	0.17	0.49
	96.000	36.60	36.82	0.22	0.60
	96.600	37.80	37.96	0.16	0.42
	97.400	39.10	39.51	0.41	1.06
	97.800	40.50	40.31	-0.19	-0.48
	97.800	40.60	40.31	-0.29	-0.73
	98.400	41.40	41.51	0.11	0.27
	98.600	41.80	41.92	0.12	0.28
	98.800	42.20	42.33	0.13	0.30
	99.200	43.40	43.16	-0.25	-0.56
	99.600	43.40	43.99	0.59	1.36
	99.600	44.20	43.99	-0.21	-0.47
	100.100	45.10	45.05	-0.05	-0.11
	100.400	45.10	45.69	0.59	1.32
	100.600	46.10	46.13	0.03	0.06
	101.000	46.90	47.00	0.10	0.21
	101.400	48.40	47.88	-0.52	-1.07
	101.900	49.50	49.00	-0.50	-1.02
	102.400	50.60	50.13	-0.47	-0.93
	103.200	51.90	51.97	0.07	0.14
	103.600	53.40	52.91	-0.49	-0.91
*	104.200	54.60	54.34	-0.26	-0.48

STANDARD DEVIATION = 0.33

CYCLOPENTANETHIOL

$C_5H_{10}S$

BERG W.T.,ET AL.:

J.PHYS.CHEM. 65,1425(1961).

A = 6.91497 B = 1388.633 C = 212.053

B.P.(760) = 132.166

T	P EXPTL.	P CALCD.	DEV.	PERCENT
80.874	149.41	149.43	0.02	0.01
87.107	187.57	187.58	0.01	0.01
93.390	233.72	233.71	-0.01	0.00
99.729	289.13	289.14	0.01	0.00
106.113	355.22	355.20	-0.02	-0.01
112.548	433.56	433.51	-0.05	-0.01
119.037	525.86	525.83	-0.03	-0.01
125.577	633.99	633.99	0.00	0.00
132.165	760.00	759.98	-0.02	0.00
138.806	906.06	906.06	0.00	0.00
145.501	1074.60	1074.65	0.05	0.00
152.245	1268.00	1268.13	0.13	0.01
159.040	1489.10	1489.22	0.12	0.01
165.887	1740.80	1740.81	0.01	0.00
172.783	2026.00	2025.77	-0.23	-0.01

STANDARD DEVIATION = 0.09

2-METHYLTHIACYCLOPENTANE C5H10S

OSBORN A.N.,DOUSLIN D.R.: J.CHEM.ENG.DATA 11,502(1966).

A = 6.94094 B = 1407.366 C = 214.156

B.P.(760) = 132.475

T	P EXPTL.	P CALCD.	DEV.	PERCENT
62.633	71.87	71.83	-0.04	-0.05
65.683	81.64	81.61	-0.03	-0.04
68.744	92.52	92.50	-0.02	-0.02
71.813	104.63	104.60	-0.03	-0.03
74.898	118.06	118.05	-0.01	-0.01
77.997	132.95	132.96	0.01	0.01
81.114	149.41	149.47	0.06	0.04
87.359	187.57	187.62	0.05	0.03
93.661	233.72	233.80	0.08	0.03
100.004	289.13	289.17	0.04	0.02
106.398	355.22	355.24	0.02	0.00
112.840	433.56	433.52	-0.04	-0.01
119.336	525.86	525.83	-0.03	-0.01
125.877	633.99	633.91	-0.08	-0.01
132.471	760.00	759.91	-0.09	-0.01
139.114	906.06	905.96	-0.10	-0.01
145.809	1074.60	1074.49	-0.11	-0.01
152.555	1268.00	1268.02	0.02	0.00
159.352	1489.10	1489.24	0.14	0.01
166.197	1740.80	1740.91	0.11	0.01
173.090	2026.00	2026.00	0.00	0.00

STANDARD DEVIATION = 0.07

3-METHYLTHIACYCLOPENTANE C5H10S

OSBORN A.N.,DOUSLIN D.R.:

J.CHEM.ENG.DATA 11,502(1966).

A = 6.94912 B = 1431.804 C = 213.607

B.P.(760) = 138.333

T	P EXPTL.	P CALCD.	DEV.	PERCENT
67.540	71.87	71.85	-0.02	-0.03
70.633	81.64	81.62	-0.02	-0.02
73.733	92.52	92.51	-0.01	-0.02
76.849	104.63	104.62	-0.01	-0.01
79.973	118.06	118.05	-0.01	-0.01
83.112	132.95	132.94	-0.01	0.00
86.270	149.41	149.45	0.04	0.02
92.606	187.57	187.62	0.05	0.03
98.990	233.72	233.76	0.04	0.02
105.423	289.13	289.16	0.03	0.01
111.905	355.22	355.22	0.00	0.00
118.436	433.56	433.53	-0.03	-0.01
125.019	525.86	525.83	-0.03	-0.01
131.648	633.99	633.92	-0.07	-0.01
138.330	760.00	759.93	-0.07	-0.01
145.062	906.06	906.00	-0.06	-0.01
151.846	1074.60	1074.57	-0.03	-0.01
158.679	1268.00	1268.07	0.07	0.00
165.561	1489.10	1489.19	0.09	0.01
172.495	1740.80	1740.87	0.07	0.01
179.476	2026.00	2025.93	-0.07	0.00

STANDARD DEVIATION = 0.05

THIACYCLOHEXANE C5H10S

MCCULLOUGH J.P.,ET AL.:

J.AM.CHEM.SOC. 76,2661(1954).

A = 6.90130 B = 1419.904 C = 211.419

B.P.(760) = 141.748

T	P EXPTL.	P CALCD.	DEV.	PERCENT
78.280	100.00	100.00	0.00	0.00
95.750	190.00	190.00	0.00	0.00
117.150	380.00	380.03	0.03	0.01
131.100	570.00	569.93	-0.07	-0.01
141.750	760.00	760.04	0.04	0.01

STANDARD DEVIATION = 0.06

PIPERIDINE C5H11N

OSBORN A.G.,DOUSLIN D.R.:

J.CHEM.ENG.DATA13,534(1968).

A = 6.85569 B = 1238.792 C = 205.434

B.P.(760) = 106.222

T	P EXPTL.	P CALCD.	DEV.	PERCENT
42.361	71.87	71.85	-0.02	-0.03
45.141	81.64	81.64	0.00	0.00
47.924	92.52	92.51	-0.01	-0.01
50.725	104.63	104.63	0.00	0.00
53.529	118.06	118.04	-0.02	-0.02
56.355	132.95	132.94	-0.01	-0.01
59.198	149.41	149.45	0.04	0.03
64.904	187.57	187.63	0.06	0.03
70.655	233.72	233.75	0.03	0.01
76.459	289.13	289.16	0.03	0.01
82.311	355.22	355.23	0.01	0.00
88.209	433.56	433.49	-0.07	-0.02
94.164	525.86	525.82	-0.04	-0.01
100.167	633.99	633.96	-0.04	-0.01
106.219	760.00	759.94	-0.07	-0.01
112.324	906.06	906.03	-0.03	0.00
118.481	1074.60	1074.61	0.01	0.00
124.687	1268.00	1268.07	0.07	0.01
130.946	1489.10	1489.25	0.15	0.01
137.254	1740.80	1740.84	0.04	0.00
143.613	2026.00	2025.88	-0.12	-0.01

STANDARD DEVIATION = 0.06

N,N-DIETHYLFORMAMIDE C5H11NO

RAM G.,SHARAF A.R.:

J.IND.CHEM.SOC. 45,13(1968).

A = 6.39537 B = 1203.812 C = 165.635

B.P.(10) = 57.484

T	P EXPTL.	P CALCD.	DEV.	PERCENT
30.000	2.00	1.75	-0.25	-12.70
40.000	3.50	3.48	-0.02	-0.64
50.000	6.50	6.50	0.00	-0.04
60.000	11.50	11.49	-0.01	-0.12
70.000	19.00	19.35	0.35	1.82
80.000	31.50	31.23	-0.27	-0.86
90.000	48.50	48.56	0.06	0.12

STANDARD DEVIATION = 0.26

2,2-DIMETHYLPROPANE C5H12

WHITMORE F.C.,FLEMMING G.H.: J.AM.CHEM.SOC. 55,3803(1933).

A = 7.26335 B = 1276.110 C = 281.629

B.P.(760) = 9.552

T	P EXPTL.	P CALCD.	DEV.	PERCENT
-42.000	87.10	86.70	-0.40	-0.46
-38.000	104.60	106.03	1.43	1.37
-33.100	135.10	134.49	-0.61	-0.45
-30.500	155.60	152.01	-3.59	-2.31
-28.100	172.10	169.81	-2.29	-1.33
-24.000	206.10	204.21	-1.89	-0.92
-21.700	222.00	225.89	3.89	1.75
-19.400	247.00	249.43	2.43	0.99
-17.600	269.80	269.23	-0.57	-0.21
-17.200	276.00	273.81	-2.20	-0.80
-15.800	286.90	290.31	3.41	1.19
-13.400	316.50	320.49	3.99	1.26
-8.000	394.00	397.83	3.83	0.97
-1.000	523.70	520.03	-3.67	-0.70
3.100	605.70	604.66	-1.04	-0.17
9.400	760.00	756.01	-3.99	-0.52

STANDARD DEVIATION = 3.08

2-METHYLBUTANE C5H12

WILLINGHAM C.J.,TAYLOR W.J.,PIGNOCCO J.M.,ROSSINI F.D.: J.RESEARCH NATL.BUR.STANDARDS 35,219(1945).

A = 6.73457 B = 992.019 C = 229.564

B.P.(760) = 27.852

T	P EXPTL.	P CALCD.	DEV.	PERCENT
16.291	500.74	500.72	-0.02	0.00
22.435	627.97	628.02	0.05	0.01
26.773	732.12	732.13	0.01	0.00
27.240	744.11	744.09	-0.02	0.00
27.673	755.31	755.32	0.01	0.00
28.160	768.08	768.10	0.02	0.00
28.587	779.48	779.44	-0.04	-0.01

STANDARD DEVIATION = 0.04

2-METHYLBUTANE C5H12

SCHUMANN S.C.,ASTON J.G.,SAGENKAHN M.:
J.AM.CHEM.SOC. 64,1039(1942).

A = 6.92405 B = 1081.654 C = 239.805

B.P.(100) = -20.138

T	P EXPTL.	P CALCD.	DEV.	PERCENT
-55.954	10.95	10.98	0.03	0.31
-41.739	29.08	29.04	-0.04	-0.14
-34.349	45.67	45.65	-0.02	-0.05
-27.828	66.28	66.28	0.00	0.00
-22.766	87.14	87.17	0.03	0.04
-17.399	114.89	114.98	0.09	0.08
-9.556	168.56	168.39	-0.18	-0.10
-2.563	231.49	231.62	0.13	0.06
3.780	304.43	304.44	0.01	0.00
8.471	369.37	369.32	-0.05	-0.01
13.827	456.51	456.45	-0.06	-0.01
19.048	556.43	556.43	0.00	0.00
22.036	620.95	621.00	0.06	0.01

STANDARD DEVIATION = 0.08

PENTANE C5H12

WILLINGHAM C.J.,TAYLOR W.J.,PIGNOCCO J.M.,ROSSINI F.D.:
J.RESEARCH NATL.BUR.STANDARDS 35,219(1945).

A = 6.84471 B = 1060.793 C = 231.541

B.P.(760) = 36.073

T	P EXPTL.	P CALCD.	DEV.	PERCENT
13.282	324.94	324.94	0.00	0.00
18.647	402.46	402.45	-0.01	0.00
24.374	500.71	500.73	0.02	0.00
30.592	627.95	627.95	0.00	0.00
34.981	732.09	732.09	0.00	0.00
35.453	744.07	744.05	-0.02	0.00
35.890	755.27	755.25	-0.01	0.00
36.379	768.01	767.95	-0.06	-0.01
36.818	779.40	779.49	0.09	0.01

STANDARD DEVIATION = 0.05

PENTANE C_5H_{12}

MESSERLY G.H., KENNEDY R.H.: J.AM.CHEM.SOC. 62,2988(1940).

A = 6.99680 B = 1130.719 C = 238.884

B.P.(100) = -12.595

T	P EXPTL.	P CALCD.	DEV.	PERCENT
-65.178	3.08	3.07	-0.01	-0.26
-47.811	12.03	12.00	-0.03	-0.28
-39.537	21.14	21.12	-0.02	-0.10
-27.420	44.64	44.64	0.00	-0.01
-17.476	77.56	77.60	0.04	0.05
-9.793	115.07	115.11	0.04	0.04
-2.359	164.51	164.53	0.02	0.01
3.618	215.77	215.81	0.04	0.02
9.621	279.80	279.71	-0.09	-0.03
14.653	344.63	344.36	-0.27	-0.08
18.613	403.33	403.28	-0.05	-0.01
21.679	454.08	454.23	0.15	0.03
24.828	511.64	511.79	0.15	0.03

STANDARD DEVIATION = 0.12

ETHYL PROPYL ETHER $C_5H_{12}O$

CIDLINSKY J., POLAK J.: COLL.CZECH.CHEM.COMM. 34,1317(1969).

A = 6.98512 B = 1188.477 C = 226.435

B.P.(760) = 63.133

T	P EXPTL.	P CALCD.	DEV.	PERCENT
20.380	147.85	147.86	0.01	0.01
23.120	167.02	167.00	-0.02	-0.01
26.210	191.03	190.97	-0.06	-0.03
28.900	214.15	214.06	-0.09	-0.04
31.980	243.10	243.23	0.13	0.05
35.090	275.85	275.87	0.02	0.01
38.480	315.43	315.39	-0.04	-0.01
40.930	346.65	346.71	0.06	0.02
47.600	444.36	444.80	0.44	0.10
50.270	489.74	489.79	0.05	0.01
52.720	534.28	534.20	-0.08	-0.02
55.140	581.72	581.16	-0.56	-0.10
57.660	633.82	633.48	-0.34	-0.05
59.380	671.31	671.29	-0.02	0.00
60.720	702.17	701.96	-0.21	-0.03
62.510	743.94	744.65	0.71	0.10

STANDARD DEVIATION = 0.30

2-METHYL-1-BUTANOL C5H12O

BUTLER J.A.V., RAMCHANDANI C.N., THOMPSON D.W.: J.CHEM.SOC.138,280(1935).

A = 7.27396 B = 1317.406 C = 169.138

B.P.(100) = 80.656

T	P EXPTL.	P CALCD.	DEV.	PERCENT
24.970	3.11	3.07	-0.04	-1.32
29.990	4.61	4.55	-0.06	-1.28
35.020	6.75	6.62	-0.13	-1.87
50.070	18.09	18.37	0.28	1.55
60.110	33.78	33.68	-0.10	-0.31
70.290	59.27	59.11	-0.16	-0.27
80.580	99.52	99.63	0.11	0.11
90.830	160.60	160.84	0.24	0.15
101.160	251.50	251.22	-0.28	-0.11
110.190	361.20	361.09	-0.11	-0.03
120.550	532.30	532.46	0.16	0.03

STANDARD DEVIATION = 0.20

2-METHYL-2-BUTANOL C5H12O

BUTLER J.A.V., RAMCHANDANI C.N., THOMPSON D.W.: J.CHEM.SOC.138,280(1935).

A = 7.32218 B = 1252.216 C = 180.301

B.P.(100) = 54.982

T	P EXPTL.	P CALCD.	DEV.	PERCENT
24.970	16.72	16.67	-0.05	-0.31
29.990	23.30	23.31	0.01	0.03
35.020	32.00	32.11	0.11	0.33
50.070	77.14	77.01	-0.13	-0.17
60.110	129.80	129.88	0.08	0.06
70.290	211.40	211.41	0.01	0.00
80.580	332.80	332.83	0.03	0.01
90.830	505.50	505.45	-0.05	-0.01

STANDARD DEVIATION = 0.09

METHYL BUTYL ETHER C5H12O

CIDLINSKY J.,POLAK J.:
COLL.CZECH.CHEM.COMM. 34,1317(1969).

A = 6.88707 B = 1162.102 C = 219.930

B.P.(760) = 70.141

T	P EXPTL.	P CALCD.	DEV.	PERCENT
22.960	126.59	126.65	0.06	0.05
25.860	144.14	144.23	0.09	0.06
28.680	163.24	163.19	-0.05	-0.03
30.560	176.90	176.92	0.02	0.01
33.280	198.52	198.43	-0.09	-0.05
35.840	220.89	220.56	-0.33	-0.15
36.990	231.15	231.14	-0.01	-0.01
41.050	271.49	271.79	0.30	0.11
42.130	283.91	283.52	-0.39	-0.14
48.160	356.54	356.72	0.18	0.05
50.870	393.91	394.19	0.28	0.07
53.030	426.04	426.25	0.21	0.05
56.190	476.46	476.86	0.40	0.08
58.060	508.43	508.98	0.55	0.11
62.800	598.91	598.12	-0.79	-0.13
64.610	636.51	635.23	-1.28	-0.20
67.920	708.17	707.78	-0.39	-0.06
68.950	730.41	731.63	1.22	0.17

STANDARD DEVIATION = 0.57

1-PENTANOL C5H12O

AMBROSE D.,SPRAKE C.H.S.: J.CHEM.THERMODYNAMICS 2,631(1970).

A = 7.18246 B = 1287.625 C = 161.330

B.P.(760) = 138.002

T	P EXPTL.	P CALCD.	DEV.	PERCENT
74.763	53.82	53.53	-0.29	-0.54
83.542	84.04	83.97	-0.07	-0.09
87.571	102.12	102.15	0.03	0.03
90.768	118.70	118.80	0.10	0.09
95.514	147.46	147.64	0.18	0.12
101.629	192.87	193.10	0.23	0.12
106.362	235.47	235.71	0.24	0.10
110.658	280.62	280.76	0.14	0.05
115.564	340.52	340.58	0.06	0.02
119.681	398.50	398.43	-0.07	-0.02
124.099	469.36	469.11	-0.25	-0.05
128.665	552.84	552.46	-0.38	-0.07
133.309	649.63	649.07	-0.56	-0.09
136.688	727.97	727.51	-0.46	-0.06
138.514	773.39	772.94	-0.45	-0.06
142.435	878.36	878.18	-0.18	-0.02
146.598	1001.71	1002.04	0.33	0.03
151.332	1157.94	1159.31	1.37	0.12
155.977	1328.78	1331.95	3.17	0.24

STANDARD DEVIATION = 0.47

2-PENTANOL C5H12O

BUTLER J.A.V.,RAMCHANDANI C.N.,THOMPSON D.W.: J.CHEM.SOC.138,280(1935).

A = 7.22556 B = 1254.854 C = 169.609

B.P.(100) = 70.529

T	P EXPTL.	P CALCD.	DEV.	PERCENT
24.970	6.03	5.98	-0.05	-0.88
29.990	8.79	8.68	-0.11	-1.22
35.020	12.44	12.39	-0.05	-0.37
50.070	32.40	32.61	0.21	0.64
60.110	57.92	57.94	0.02	0.04
70.290	99.12	98.81	-0.31	-0.32
80.580	161.80	162.15	0.35	0.22
90.830	255.50	255.46	-0.04	-0.02
101.160	390.40	390.07	-0.33	-0.08
110.190	550.20	550.44	0.24	0.04

STANDARD DEVIATION =

DIETHOXYMETHANE $C_5H_{12}O_2$

NICOLINI E.,LAFFITTE P.:

COMPTES RENDUS 229,757(1949).

A = 6.90841 B = 1229.522 C = 217.012

B.P.(100) = 33.481

T	P EXPTL.	P CALCD.	DEV.	PERCENT
0.000	17.70	17.49	-0.21	-1.20
5.000	23.50	23.46	-0.04	-0.17
10.000	31.00	31.07	0.07	0.22
15.000	40.60	40.65	0.05	0.11
20.000	52.70	52.58	-0.12	-0.23
25.000	67.10	67.30	0.20	0.29
30.000	85.10	85.28	0.18	0.21
35.000	106.80	107.05	0.25	0.23
40.000	133.50	133.20	-0.30	-0.23
45.000	164.70	164.36	-0.34	-0.21
50.000	201.10	201.21	0.11	0.06
55.000	244.60	244.51	-0.09	-0.04
60.000	295.00	295.04	0.04	0.02
65.000	353.70	353.66	-0.04	-0.01
70.000	420.80	421.24	0.44	0.11
75.000	499.00	498.75	-0.25	-0.05

STANDARD DEVIATION = 0.23

PROPYL CELLOSOLVE $C_5H_{12}O_2$

DYKYJ J.,SEPRAKOVA M.,PAULECH J.:

CHEM.ZVESTI 11,461(1957).

A = 7.14643 B = 1440.603 C = 187.655

B.P.(760) = 150.069

T	P EXPTL.	P CALCD.	DEV.	PERCENT
77.100	51.20	50.72	-0.48	-0.94
88.900	86.10	86.56	0.46	0.54
103.600	158.30	158.58	0.28	0.18
114.500	238.90	239.15	0.25	0.10
123.800	334.10	331.92	-2.18	-0.65
128.300	385.30	386.29	0.99	0.26
138.700	538.50	539.77	1.27	0.24
148.700	730.90	730.21	-0.69	-0.10

STANDARD DEVIATION = 1.30

ISOPROPYL CELLOSOLVE $C_5H_{12}O_2$

DYKYJ J.,SEPRAKOVA M.,PAULECH J.: CHEM.ZVESTI 11,461(1957).

A = 7.49964 B = 1639.201 C = 213.250

B.P.(760) = 141.645

T	P EXPTL.	P CALCD.	DEV.	PERCENT
67.600	46.30	46.03	-0.27	-0.58
84.100	96.70	97.04	0.34	0.35
99.400	179.70	180.60	0.90	0.50
99.500	182.30	181.30	-1.00	-0.55
116.300	335.20	335.42	0.22	0.06
126.500	473.60	473.06	-0.54	-0.11
140.600	736.20	736.49	0.29	0.04

STANDARD DEVIATION = 0.78

METHYLCARBITOL $C_5H_{12}O_3$

DYKYJ J.,SEPRAKOVA M.,PAULECH J.: CHEM.ZVESTI 11,461(1957).

A = 7.42422 B = 1751.142 C = 191.770

B.P.(760) = 193.655

T	P EXPTL.	P CALCD.	DEV.	PERCENT
112.300	47.00	46.26	-0.74	-1.58
119.600	63.70	63.13	-0.57	-0.90
120.400	65.30	65.26	-0.04	-0.06
137.800	129.10	129.06	-0.04	-0.03
158.700	262.70	267.71	5.01	1.91
172.200	411.60	410.19	-1.41	-0.34
181.200	543.40	535.90	-7.50	-1.38
193.000	742.10	746.58	4.48	0.60

STANDARD DEVIATION = 4.57

3,3-DIMETHYL-2-THIABUTANE C5H12S

SCOTT D.W.,ET AL.:

J.CHEM.PHYS. 36,410(1962).

A = 6.84636 B = 1259.192 C = 218.637

B.P.(760) = 98.896

T	P EXPTL.	P CALCD.	DEV.	PERCENT
33.713	71.87	71.86	-0.01	-0.01
36.543	81.64	81.63	-0.01	-0.02
39.388	92.52	92.52	0.00	0.00
42.242	104.63	104.62	-0.01	-0.01
45.107	118.06	118.05	-0.01	-0.01
47.990	132.95	132.95	0.00	0.00
50.890	149.41	149.45	0.04	0.03
56.709	187.57	187.60	0.03	0.01
62.586	233.72	233.77	0.05	0.02
68.507	289.13	289.15	0.02	0.01
74.478	355.22	355.19	-0.03	-0.01
80.507	433.56	433.54	-0.02	0.00
86.585	525.86	525.84	-0.02	0.00
92.706	633.99	633.82	-0.17	-0.03
98.892	760.00	759.90	-0.10	-0.01
105.128	906.06	906.02	-0.04	0.00
111.420	1074.60	1074.68	0.08	0.01
117.761	1268.00	1268.21	0.21	0.02
124.156	1489.10	1489.44	0.34	0.02
130.595	1740.80	1740.81	0.01	0.00
137.088	2026.00	2025.64	-0.36	-0.02

STANDARD DEVIATION = 0.14

2-METHYL-1-BUTANETHIOL $C_5H_{12}S$

OSBORN A.N., DOUSLIN D.R.: J.CHEM.ENG.DATA 11,502(1966).

A = 6.91515 B = 1348.146 C = 215.169

B.P.(760) = 118.999

T	P EXPTL.	P CALCD.	DEV.	PERCENT
51.339	71.87	71.88	0.01	0.01
54.284	81.64	81.63	-0.01	-0.01
57.243	92.52	92.52	0.00	0.00
60.219	104.63	104.64	0.01	0.01
63.194	118.06	118.04	-0.02	-0.02
66.193	132.95	132.94	-0.01	-0.01
69.207	149.41	149.43	0.02	0.01
75.263	187.57	187.62	0.05	0.03
81.361	233.72	233.74	0.02	0.01
87.510	289.13	289.13	0.00	0.00
93.708	355.22	355.20	-0.02	-0.01
99.955	433.56	433.51	-0.05	-0.01
106.253	525.86	525.81	-0.05	-0.01
112.600	633.99	633.94	-0.06	-0.01
118.999	760.00	759.99	-0.01	0.00
125.446	906.06	906.09	0.03	0.00
131.944	1074.60	1074.64	0.04	0.00
138.492	1268.00	1268.15	0.15	0.01
145.089	1489.10	1489.28	0.18	0.01
151.733	1740.80	1740.77	-0.03	0.00
158.428	2026.00	2025.76	-0.24	-0.01

STANDARD DEVIATION = 0.08

2-METHYL-2-BUTANETHIOL C5H12S

SCOTT D.W.,ET AL.:

J.PHYS.CHEM.66,1334(1962).

A = 6.82885 B = 1255.164 C = 218.791

B.P.(760) = 99.130

T	P EXPTL.	P CALCD.	DEV.	PERCENT
50.888	149.41	149.47	0.06	0.04
56.725	187.57	187.57	0.00	0.00
62.625	233.72	233.71	-0.01	-0.01
68.578	289.13	289.11	-0.02	-0.01
74.579	355.22	355.15	-0.07	-0.02
80.638	433.56	433.50	-0.06	-0.01
86.749	525.86	525.81	-0.05	-0.01
92.914	633.99	633.98	-0.01	0.00
99.132	760.00	760.04	0.04	0.00
105.401	906.06	906.10	0.04	0.00
111.728	1074.60	1074.71	0.11	0.01
118.106	1268.00	1268.19	0.19	0.01
124.537	1489.10	1489.26	0.16	0.01
131.021	1740.80	1740.74	-0.06	0.00
137.559	2026.00	2025.66	-0.34	-0.02

STANDARD DEVIATION = 0.13

3-METHYL-2-BUTANETHIOL C5H12S

OSBORN A.N.,DOUSLIN D.R.:
J.CHEM.ENG.DATA 11,502(1966).

A = 6.87585 B = 1307.562 C = 217.535

B.P.(760) = 109.761

T	P EXPTL.	P CALCD.	DEV.	PERCENT
42.969	71.87	71.86	-0.01	-0.01
45.876	81.64	81.64	0.00	0.00
48.791	92.52	92.52	0.00	0.00
51.720	104.63	104.62	-0.01	-0.01
54.658	118.06	118.05	-0.01	-0.01
57.613	132.95	132.94	-0.01	-0.01
60.592	149.41	149.47	0.06	0.04
66.556	187.57	187.60	0.03	0.02
72.575	233.72	233.74	0.02	0.01
78.645	289.13	289.13	0.00	0.00
84.765	355.22	355.20	-0.02	-0.01
90.936	433.56	433.51	-0.05	-0.01
97.161	525.86	525.84	-0.02	0.00
103.431	633.99	633.90	-0.09	-0.02
109.760	760.00	759.97	-0.03	0.00
116.139	906.06	906.08	0.02	0.00
122.571	1074.60	1074.68	0.08	0.01
129.051	1268.00	1268.12	0.12	0.01
135.585	1489.10	1489.25	0.15	0.01
142.170	1740.80	1740.82	0.02	0.00
148.805	2026.00	2025.78	-0.22	-0.01

STANDARD DEVIATION = 0.08

1-PENTANETHIOL C5H12S

FINKE H.L.,ET AL.:
J.AM.CHEM.SOC. 74,2804(1952).

A = 6.93426 B = 1370.252 C = 211.408

B.P.(760) = 126.638

T	P EXPTL.	P CALCD.	DEV.	PERCENT
76.470	149.41	149.43	0.02	0.01
82.569	187.57	187.58	0.01	0.00
88.721	233.72	233.73	0.01	0.01
94.918	289.13	289.12	-0.01	0.00
101.167	355.22	355.23	0.01	0.00
107.457	433.56	433.49	-0.07	-0.02
113.802	525.86	525.81	-0.05	-0.01
120.193	633.99	633.91	-0.08	-0.01
126.638	760.00	760.00	0.00	0.00
133.131	906.06	906.15	0.09	0.01
139.671	1074.60	1074.70	0.10	0.01
146.255	1268.00	1268.05	0.05	0.00
152.896	1489.10	1489.28	0.18	0.01
159.580	1740.80	1740.76	-0.04	0.00
166.314	2026.00	2025.76	-0.24	-0.01

STANDARD DEVIATION = 0.10

2-THIAHEXANE C5H12S

MCCULLOUGH J.P.,ET AL.:
J.PHYS.CHEM. 65,784(1961).

A = 6.94620 B = 1364.053 C = 212.104

B.P.(760) = 123.425

T	P EXPTL.	P CALCD.	DEV.	PERCENT
73.752	149.41	149.41	0.00	0.00
79.798	187.57	187.59	0.02	0.01
85.888	233.72	233.73	0.01	0.00
92.025	289.13	289.12	-0.01	0.00
98.211	355.22	355.21	-0.01	0.00
104.442	433.56	433.53	-0.03	-0.01
110.725	525.86	525.88	0.02	0.00
117.048	633.99	633.94	-0.05	-0.01
123.423	760.00	759.96	-0.04	-0.01
129.847	906.06	906.07	0.01	0.00
136.317	1074.60	1074.57	-0.03	0.00
142.839	1268.00	1268.17	0.17	0.01
149.403	1489.10	1489.21	0.11	0.01
156.019	1740.80	1740.88	0.08	0.00
162.676	2026.00	2025.75	-0.25	-0.01

STANDARD DEVIATION = 0.10

CHLOROPENTAFLUOROBENZENE C6CLF5

AMBROSE D.:

J.CHEM.SOC.(A) 1968,1381.

A = 7.06883 B = 1389.187 C = 213.752

B.P.(760) = 117.953

T	P EXPTL.	P CALCD.	DEV.	PERCENT
35.576	31.40	31.41	0.01	0.04
44.790	49.61	49.62	0.01	0.02
55.927	82.72	82.71	-0.01	-0.01
61.832	106.63	106.65	0.02	0.01
65.865	126.09	126.08	-0.01	-0.01
70.085	149.48	149.46	-0.02	-0.02
75.343	183.46	183.46	0.00	0.00
81.691	232.70	232.69	-0.01	0.00
86.763	279.35	279.35	0.00	0.00
92.135	336.78	336.76	-0.02	-0.01
97.254	400.02	400.02	0.00	0.00
102.357	472.26	472.26	0.00	0.00
107.155	549.43	549.40	-0.03	-0.01
112.764	652.01	652.01	0.00	0.00
116.436	727.01	727.06	0.05	0.01
117.448	748.80	748.90	0.10	0.01
118.893	780.95	780.99	0.04	0.01
123.580	892.55	892.63	0.08	0.01
127.992	1008.96	1008.88	-0.08	-0.01
133.467	1169.39	1169.32	-0.07	-0.01
138.689	1340.40	1340.33	-0.07	-0.01
144.131	1539.05	1538.69	-0.36	-0.02

STANDARD DEVIATION = 0.05

PERDEUTEROBENZENE C6D6

DAVIS R.T.,SCHIESSLER R.W.:

J.PHYS.CHEM. 57,966(1953)

A = 6.89235 B = 1198.394 C = 219.432

B.P.(760) = 79.305

T	P EXPTL.	P CALCD.	DEV.	PERCENT
10.000	46.66	46.67	0.01	0.02
20.000	77.10	77.12	0.02	0.03
30.000	122.43	122.42	-0.01	-0.01
40.000	187.57	187.52	-0.05	-0.03
50.000	278.35	278.28	-0.07	-0.02
60.000	401.39	401.48	0.09	0.02
65.000	477.59	477.59	0.00	0.00
70.000	564.65	564.73	0.08	0.01
75.000	663.81	663.98	0.17	0.03
80.000	776.51	776.46	-0.05	-0.01
82.000	825.61	825.42	-0.19	-0.02

STANDARD DEVIATION = 0.11

PERDEUTEROCYCLOHEXANE C6D12

DAVIS R.T.,SCHIESSLER R.W.:

J.PHYS.CHEM. 57,966(1953)

A = 6.83786 B = 1190.384 C = 222.401

B.P.(760) = 78.426

T	P EXPTL.	P CALCD.	DEV.	PERCENT
10.000	51.92	51.97	0.05	0.09
20.000	84.52	84.54	0.02	0.02
30.000	132.41	132.31	-0.10	-0.07
40.000	200.14	200.14	0.00	0.00
50.000	293.80	293.68	-0.12	-0.04
60.000	419.33	419.39	0.06	0.02
65.000	496.32	496.54	0.22	0.04
70.000	584.18	584.50	0.32	0.05
75.000	684.61	684.27	-0.34	-0.05
80.000	797.03	796.90	-0.13	-0.02

STANDARD DEVIATION = 0.21

HEXAFLUOROBENZENE C6F6

DOUSLIN D.R.,OSBORN A.: J.SCI.INSTRUM. 42,369(1965).

A = 7.03295 B = 1227.984 C = 215.491

B.P.(760) = 80.256

	T	P EXPTL.	P CALCD.	DEV.	PERCENT
	5.200	29.42	29.42	0.00	0.01
*	5.329	29.61	29.64	0.03	0.12
	21.738	71.87	71.88	0.01	0.01
	24.297	81.64	81.63	-0.01	-0.02
	26.870	92.52	92.51	-0.01	-0.01
	29.453	104.63	104.62	-0.01	0.00
	32.042	118.06	118.05	-0.01	-0.01
	34.643	132.95	132.94	-0.01	-0.01
	37.259	149.41	149.44	0.03	0.02
	42.503	187.57	187.59	0.02	0.01
	47.786	233.72	233.74	0.02	0.01
	53.105	289.13	289.13	0.00	0.00
	58.461	355.22	355.20	-0.02	0.00
	63.854	433.56	433.53	-0.03	-0.01
	69.284	525.86	525.82	-0.04	-0.01
	74.752	633.99	633.98	-0.01	0.00
	80.255	760.00	759.97	-0.03	0.00
	85.797	906.06	906.09	0.03	0.00
	91.375	1074.60	1074.64	0.04	0.00
	96.989	1268.00	1268.11	0.11	0.01
	102.639	1489.10	1489.19	0.09	0.01
	108.323	1740.80	1740.63	-0.17	-0.01
*	114.042	2026.00	2025.44	-0.56	-0.03

STANDARD DEVIATION = 0.06

HEXAFLUOROBENZENE C6F6

FINDLAY T.J.V.:

J.CHEM.ENG.DATA 14,229(1969).

A = 6.81199 B = 1129.458 C = 206.302

B.P.(100) = 28.415

T	P EXPTL.	P CALCD.	DEV.	PERCENT
4.740	28.87	28.85	-0.02	-0.06
5.700	30.49	30.51	0.02	0.06
7.070	33.20	33.01	-0.19	-0.58
7.140	33.22	33.14	-0.08	-0.24
9.770	38.39	38.44	0.05	0.13
17.230	57.35	57.44	0.09	0.16
17.230	57.38	57.44	0.06	0.10
19.880	65.68	65.83	0.15	0.23
24.280	81.87	81.98	0.11	0.13
24.400	82.42	82.46	0.04	0.05
25.100	85.52	85.32	-0.20	-0.23
27.080	93.84	93.86	0.02	0.02
29.920	107.23	107.31	0.08	0.08
29.920	107.26	107.31	0.05	0.05
29.920	107.28	107.31	0.03	0.03
35.920	141.18	140.96	-0.22	-0.16
41.810	182.18	181.88	-0.30	-0.17
47.320	228.44	228.39	-0.05	-0.02
47.850	233.00	233.33	0.33	0.14

STANDARD DEVIATION = 0.16

PERFLUOROCYCLOHEXANE C6F12

ROWLINSON J.S.,THACKER R.:

TRANS.FARADAY SOC. 53, 1(1957).

A = 6.04045 B = 596.749 C = 136.318

B.P.(760) = 52.549

T	P EXPTL.	P CALCD.	DEV.	PERCENT
19.810	169.32	165.30	-4.02	-2.38
20.610	176.30	172.88	-3.42	-1.94
25.250	221.74	222.31	0.57	0.26
30.000	278.88	283.43	4.55	1.63
40.000	440.04	452.83	12.79	2.91
50.000	675.81	688.00	12.20	1.80
60.000	1014.00	1001.70	-12.30	-1.21
63.160	1139.70	1119.16	-20.54	-1.80
65.000	1207.00	1191.88	-15.12	-1.25
70.000	1405.80	1406.27	0.47	0.03
74.000	1579.50	1596.17	16.67	1.06
109.000	4025.00	4054.07	29.07	0.72
121.300	5320.00	5297.05	-22.95	-0.43

STANDARD DEVIATION =16.77

PERFLUORO-1,2-DIMETHYLCYCLOBUTANE C6F12

CROWDER G.A.,TAYLOR Z.L.,REED T.M.,YOUNG J.A.:
J.CHEM.ENG.DATA 12,481(1967).

A = 6.80907 B = 1026.889 C = 217.033

B.P.(760) = 44.378

T	P EXPTL.	P CALCD.	DEV.	PERCENT
-30.540	19.19	20.08	0.89	4.64
-25.000	29.32	28.95	-0.37	-1.27
-16.010	50.50	50.21	-0.29	-0.58
-8.510	76.56	76.65	0.09	0.11
-5.290	91.61	91.07	-0.54	-0.59
0.210	120.44	120.82	0.38	0.32
3.830	145.20	144.42	-0.78	-0.54
10.160	193.66	194.61	0.95	0.49
16.360	256.71	256.59	-0.12	-0.05
26.700	394.31	394.36	0.05	0.01
31.110	466.61	468.57	1.96	0.42
35.910	561.48	561.44	-0.04	-0.01
40.780	672.33	669.87	-2.46	-0.37
44.860	772.40	772.76	0.36	0.05

STANDARD DEVIATION = 1.09

PERFLUOROHEXANE C6F14

DUNLAP R.D.,MURPHY C.J.,JR.,BEDFORD R.G.:
J.AM.CHEM.SOC. 80,83(1958).

A = 6.87520 B = 1080.779 C = 213.423

B.P.(760) = 57.151

T	P EXPTL.	P CALCD.	DEV.	PERCENT
30.060	273.48	273.14	-0.34	-0.13
34.890	332.70	333.21	0.51	0.15
39.900	405.99	406.26	0.27	0.07
44.850	490.61	490.42	-0.19	-0.04
47.430	539.81	539.46	-0.35	-0.07
49.760	587.18	586.99	-0.19	-0.03
52.310	642.75	642.75	0.00	0.00
54.280	688.47	688.61	0.14	0.02
57.400	766.30	766.45	0.15	0.02

STANDARD DEVIATION = 0.34

PERFLUORO-2,3-DIMETHYLBUTANE C_6F_{14}

CROWDER G.A., TAYLOR Z.L., REED T.M., YOUNG J.A.: J.CHEM.ENG.DATA 12,481(1967).

A = 7.21136 B = 1274.155 C = 234.502

B.P.(760) = 59.723

T	P EXPTL.	P CALCD.	DEV.	PERCENT
-10.820	32.75	32.74	-0.01	-0.03
-2.810	51.87	51.53	-0.35	-0.67
1.520	64.83	65.00	0.17	0.26
5.010	77.61	77.91	0.30	0.38
10.570	102.65	102.86	0.21	0.21
15.750	131.92	131.79	-0.13	-0.10
20.260	162.07	162.19	0.12	0.07
23.760	191.09	189.58	-1.51	-0.79
27.110	219.11	219.26	0.15	0.07
30.820	256.50	256.49	-0.01	0.00
35.630	310.84	312.31	1.47	0.47
40.600	379.10	380.01	0.91	0.24
45.460	456.62	457.29	0.67	0.15
50.150	546.11	543.48	-2.63	-0.48
55.230	650.89	651.12	0.23	0.04
59.480	753.45	753.77	0.32	0.04

STANDARD DEVIATION = 1.01

PERFLUORO-2-METHYLPENTANE C_6F_{14}

STILES V.E., CADY G.H.: J.AM.CHEM.SOC.74,3771(1952).

A = 6.89709 B = 1100.946 C = 216.388

B.P.(760) = 57.733

T	P EXPTL.	P CALCD.	DEV.	PERCENT
4.380	81.40	81.32	-0.08	-0.10
* 17.250	151.90	153.07	1.17	0.77
23.820	206.00	205.96	-0.04	-0.02
29.200	259.30	259.53	0.23	0.09
33.370	308.30	308.34	0.04	0.01
37.570	364.40	364.70	0.30	0.08
40.480	408.40	408.36	-0.04	-0.01
43.570	458.90	459.19	0.29	0.06
46.480	512.00	511.53	-0.47	-0.09
49.920	579.20	579.39	0.19	0.03
53.200	650.70	650.54	-0.16	-0.03
55.400	703.10	701.99	-1.11	-0.16
60.680	838.20	838.54	0.34	0.04
63.650	924.50	923.98	-0.52	-0.06
66.700	1018.30	1018.64	0.34	0.03
68.320	1071.20	1071.89	0.69	0.06

STANDARD DEVIATION = 0.48

PERFLUORO-3-METHYLPENTANE C6F14

CROWDER G.A.,TAYLOR Z.L.,REED T.M.,YOUNG J.A.:
J.CHEM.ENG.DATA 12,481(1967).

A = 6.89969 B = 1114.155 C = 218.972

B.P.(760) = 58.258

T	P EXPTL.	P CALCD.	DEV.	PERCENT
-18.000	22.45	22.69	0.24	1.08
-10.820	35.38	35.24	-0.14	-0.38
-9.140	39.20	38.90	-0.30	-0.77
-4.290	51.61	51.27	-0.34	-0.65
4.330	80.65	81.33	0.68	0.84
9.640	106.26	106.20	-0.06	-0.06
15.400	139.79	139.93	0.14	0.10
21.120	181.83	181.62	-0.21	-0.12
27.630	241.02	240.80	-0.22	-0.09
35.520	331.42	332.45	1.04	0.31
40.510	404.00	403.57	-0.43	-0.11
46.130	499.01	497.67	-1.34	-0.27
50.720	586.84	586.78	-0.06	-0.01
56.230	708.45	709.89	1.44	0.20
59.070	781.30	780.80	-0.50	-0.06

STANDARD DEVIATION = 0.72

PENTAFLUOROBENZENE C6HF5

AMBROSE D.:

J.CHEM.SOC.(A) 1968,1381.

A = 7.03665 B = 1254.070 C = 216.016

B.P.(760) = 85.745

T	P EXPTL.	P CALCD.	DEV.	PERCENT
48.729	199.40	199.41	0.01	0.01
54.355	250.22	250.22	0.00	0.00
59.041	300.20	300.15	-0.05	-0.02
63.033	348.77	348.79	0.02	0.01
66.698	398.87	398.87	0.00	0.00
69.910	447.35	447.36	0.01	0.00
72.828	495.42	495.41	-0.01	0.00
75.604	544.87	544.87	0.00	0.00
78.158	593.77	593.78	0.01	0.00
80.643	644.70	644.67	-0.03	0.00
82.998	696.00	696.04	0.04	0.01
85.123	745.06	745.12	0.06	0.01
85.496	753.97	754.02	0.05	0.01
87.212	795.99	796.01	0.02	0.00
89.093	844.25	844.14	-0.11	-0.01
91.014	895.70	895.63	-0.07	-0.01
92.509	937.50	937.40	-0.10	-0.01
94.296	989.17	989.30	0.13	0.01

STANDARD DEVIATION = 0.06

PENTAFLUOROPHENOL C6HF5O

AMBROSE D.:

J.CHEM.SOC.(A) 1968,1381.

A = 7.06603 B = 1379.153 C = 183.906

B.P.(760) = 145.624

T	P EXPTL.	P CALCD.	DEV.	PERCENT
105.448	199.35	199.39	0.04	0.02
111.328	248.09	248.11	0.02	0.01
116.192	295.47	295.36	-0.11	-0.04
120.720	345.68	345.67	-0.01	0.00
124.827	397.08	397.09	0.01	0.00
128.292	445.29	445.11	-0.18	-0.04
131.556	494.32	494.51	0.19	0.04
134.647	545.07	545.25	0.18	0.03
137.240	591.08	590.95	-0.13	-0.02
140.302	648.72	648.80	0.08	0.01
142.507	693.15	693.18	0.03	0.01
145.002	746.36	746.27	-0.09	-0.01
147.069	792.70	792.66	-0.04	-0.01
149.090	840.14	840.19	0.05	0.01
151.190	891.95	891.93	-0.02	0.00
153.041	939.60	939.59	-0.01	0.00
154.963	991.23	991.19	-0.04	0.00

STANDARD DEVIATION = 0.10

CHLORO-2,4,6-TRINITROBENZENE C6H2CLN3O6

MAKSIMOV YU.YA.:
ZH.FIZ.KHIM. 42,2921(1968).

A = 3.08091 B = 184.926 C = -117.888

B.P.(10) = 206.756

T	P EXPTL.	P CALCD.	DEV.	PERCENT
200.000	9.50	6.74	-2.76	-29.03
210.000	14.00	11.84	-2.16	-15.44
220.000	18.50	18.62	0.12	0.63
230.000	24.00	27.00	3.00	12.52
230.000	28.00	27.00	-1.00	-3.56
240.000	36.00	36.86	0.86	2.38
240.000	38.00	36.86	-1.14	-3.01
250.000	45.00	47.99	2.99	6.64
250.000	47.00	47.99	0.99	2.10
250.000	46.00	47.99	1.99	4.32
250.000	47.00	47.99	0.99	2.10
250.000	48.00	47.99	-0.01	-0.02
250.000	50.00	47.99	-2.01	-4.02
260.000	58.00	60.21	2.21	3.80
260.000	61.00	60.21	-0.79	-1.30
270.000	75.00	73.31	-1.69	-2.25
270.000	75.00	73.31	-1.69	-2.25

STANDARD DEVIATION = 1.97

1,2,3,4-TETRAFLUOROBENZENE C6H2F4

FINDLAY T.J.V.: J.CHEM.ENG.DATA 14,229(1969).

A = 7.08456 B = 1339.226 C = 223.485

B.P.(100) = 39.906

T	P EXPTL.	P CALCD.	DEV.	PERCENT
6.050	17.89	17.78	-0.11	-0.59
6.060	17.83	17.79	-0.04	-0.20
10.360	22.81	22.78	-0.03	-0.13
15.210	29.73	29.78	0.05	0.17
15.210	29.78	29.78	0.00	0.00
20.640	39.58	39.69	0.11	0.29
20.640	39.61	39.69	0.08	0.21
24.820	49.09	49.10	0.01	0.02
24.820	49.11	49.10	-0.01	-0.02
29.980	63.25	63.22	-0.03	-0.04
29.990	63.25	63.25	0.00	0.00
35.050	80.42	80.26	-0.16	-0.20
35.060	80.42	80.29	-0.13	-0.16
40.070	100.79	100.73	-0.06	-0.06
40.070	100.80	100.73	-0.07	-0.07
43.990	119.29	119.57	0.28	0.23
44.010	119.40	119.67	0.27	0.23
46.910	135.19	135.42	0.23	0.17
46.910	135.19	135.42	0.23	0.17
48.820	147.16	146.70	-0.46	-0.31
48.830	147.17	146.76	-0.41	-0.28
49.840	152.86	153.03	0.17	0.11
49.870	153.22	153.22	0.00	0.00

STANDARD DEVIATION = 0.19

1,2,3,5-TETRAFLUOROBENZENE C6H2F4

FINDLAY T.J.V.:

J.CHEM.ENG.DATA 14,229(1969).

A = 6.98617 B = 1245.204 C = 218.347

B.P.(100) = 31.385

T	P EXPTL.	P CALCD.	DEV.	PERCENT
6.080	27.54	27.40	-0.14	-0.50
6.120	27.49	27.46	-0.03	-0.09
10.290	34.64	34.67	0.03	0.09
10.290	34.69	34.67	-0.02	-0.06
15.210	45.12	45.15	0.03	0.07
15.240	45.20	45.22	0.02	0.05
19.770	57.08	57.12	0.04	0.07
19.770	57.04	57.12	0.08	0.14
24.810	73.23	73.31	0.08	0.11
24.810	73.28	73.31	0.03	0.04
29.780	92.78	92.84	0.06	0.07
29.780	92.85	92.84	-0.01	-0.01
35.010	117.98	117.85	-0.13	-0.11
35.030	118.07	117.96	-0.11	-0.10
39.890	146.04	145.95	-0.09	-0.06
39.890	146.08	145.95	-0.13	-0.09
44.810	179.45	179.62	0.17	0.10
44.810	179.51	179.62	0.11	0.06
49.830	220.24	220.26	0.02	0.01
49.830	220.31	220.26	-0.05	-0.02

STANDARD DEVIATION = 0.09

1,3,5-TRIFLUOROBENZENE C6H3F3

FINDLAY T.J.V.:
J.CHEM.ENG.DATA 14,229(1969).

A = 6.91979 B = 1197.129 C = 219.118

B.P.(100) = 24.211

T	P EXPTL.	P CALCD.	DEV.	PERCENT
6.180	40.39	40.39	0.00	-0.01
6.190	40.42	40.41	-0.01	-0.02
10.030	49.62	49.61	-0.01	-0.03
15.010	63.98	64.07	0.09	0.14
15.010	64.01	64.07	0.06	0.09
19.990	81.87	81.87	0.00	0.00
19.990	81.84	81.87	0.03	0.04
20.490	84.02	83.87	-0.15	-0.18
20.530	84.08	84.03	-0.05	-0.06
22.790	93.52	93.56	0.04	0.04
25.210	104.77	104.74	-0.03	-0.03
25.240	104.90	104.88	-0.02	-0.02
29.980	129.99	130.00	0.01	0.01
29.980	130.05	130.00	-0.05	-0.04
35.150	162.61	162.80	0.19	0.12
35.250	163.56	163.49	-0.07	-0.04
35.290	163.86	163.77	-0.09	-0.05
40.050	199.92	199.83	-0.09	-0.04
40.110	200.30	200.32	0.02	0.01
43.850	232.66	233.03	0.37	0.16
43.940	233.93	233.87	-0.06	-0.03
44.200	235.99	236.30	0.31	0.13
46.970	263.69	263.51	-0.18	-0.07
46.970	263.78	263.51	-0.27	-0.10
49.980	295.92	295.88	-0.04	-0.01
49.990	295.99	295.99	0.00	0.00
49.990	295.99	295.99	0.00	0.00

STANDARD DEVIATION = 0.14

1,2,4-TRINITROBENZENE C6H3N3O6

MAKSIMOV YU.YA.: ZH.FIZ.KHIM. 42,2921(1968).

A = 3.19402 B = 86.677 C = -199.236

B.P.(100) = 271.828

T	P EXPTL.	P CALCD.	DEV.	PERCENT
250.000	44.50	30.66	-13.84	-31.10
260.000	62.00	58.56	-3.44	-5.56
270.000	88.00	93.14	5.14	5.84
280.000	120.00	132.07	12.07	10.06
280.000	127.00	132.07	5.07	3.99
290.000	162.00	173.40	11.40	7.04
290.000	171.00	173.40	2.40	1.40
300.000	233.00	215.69	-17.31	-7.43

STANDARD DEVIATION =12.94

1,3,5-TRINITROBENZENE C6H3N3O6

MAKSIMOV YU.YA.: ZH.FIZ.KHIM. 42,2921(1968).

A = 5.53457 B = 993.582 C = 11.198

B.P.(100) = 269.907

T	P EXPTL.	P CALCD.	DEV.	PERCENT
202.500	11.00	7.68	-3.32	-30.23
210.500	14.00	11.29	-2.71	-19.33
223.500	22.00	20.00	-2.00	-9.07
234.500	31.00	30.95	-0.05	-0.17
245.500	46.00	46.12	0.12	0.27
250.000	56.00	53.78	-2.22	-3.97
250.000	52.00	53.78	1.78	3.42
256.000	64.50	65.47	0.97	1.50
267.000	89.00	91.85	2.85	3.20
271.000	100.00	103.20	3.20	3.20
271.000	101.00	103.20	2.20	2.18
271.000	102.00	103.20	1.20	1.18
288.000	165.00	163.58	-1.42	-0.86
288.000	166.00	163.58	-2.42	-1.46
292.000	184.00	180.95	-3.05	-1.66
312.000	287.00	288.63	1.63	0.57
312.000	285.00	288.63	3.63	1.27
312.000	293.00	288.63	-4.37	-1.49

STANDARD DEVIATION = 2.69

2,4,6-TRINITROBENZENE C6H3N3O6

MAKSIMOV YU.YA.:

ZH.FIZ.KHIM. 42,2921(1968).

A = 9.62112 B = 4987.855 C = 329.932

B.P.(100) = 324.546

T	P EXPTL.	P CALCD.	DEV.	PERCENT
249.000	10.00	10.13	0.13	1.28
279.000	27.00	26.91	-0.09	-0.32
284.000	32.50	31.38	-1.12	-3.44
293.000	41.00	41.12	0.12	0.29
302.000	53.00	53.47	0.47	0.88
310.500	67.50	68.05	0.55	0.82
342.500	160.00	159.77	-0.23	-0.15

STANDARD DEVIATION = 0.68

1,2-DICHLOROBENZENE C6H4CL2

MCDONALD R.A.,SHRADER S.A.,STULL D.R.:

J.CHEM.ENG.DATA 4,311(1959).

A = 7.14378 B = 1704.488 C = 219.416

B.P.(760) = 180.421

T	P EXPTL.	P CALCD.	DEV.	PERCENT
130.820	189.27	189.27	0.00	0.00
154.830	388.44	388.43	-0.01	0.00
178.660	727.55	727.70	0.15	0.02
179.730	747.17	747.20	0.03	0.00
180.540	762.48	762.22	-0.26	-0.03
181.620	782.56	782.64	0.08	0.01

STANDARD DEVIATION = 0.18

1,3-DICHLOROBENZENE C6H4CL2

DREISBACH R.R.,SHRADER S.A.: IND.ENG.CHEM. 41, 2879(1949).

A = 7.04012 B = 1607.051 C = 213.378

B.P.(760) = 172.997

T	P EXPTL.	P CALCD.	DEV.	PERCENT
90.720	57.04	56.95	-0.09	-0.16
98.050	75.86	75.83	-0.03	-0.04
111.490	123.76	123.97	0.21	0.17
140.490	315.52	315.30	-0.22	-0.07
157.370	507.50	507.57	0.07	0.01
173.000	760.00	760.05	0.05	0.01

STANDARD DEVIATION = 0.19

1,4-DICHLOROBENZENE C6H4CL2

MCDONALD R.A.,SHRADER S.A.,STULL D.R.: J.CHEM.ENG.DATA 4,311(1959).

A = 7.02075 B = 1590.883 C = 210.226

B.P.(760) = 174.050

T	P EXPTL.	P CALCD.	DEV.	PERCENT
94.800	63.86	63.85	-0.01	-0.01
98.550	73.84	73.88	0.04	0.06
104.810	93.34	93.52	0.18	0.19
116.280	141.07	140.70	-0.37	-0.26
143.520	333.52	333.82	0.30	0.09
159.790	526.37	526.33	-0.04	-0.01
171.870	719.65	719.76	0.11	0.01
174.040	760.00	759.80	-0.20	-0.03

STANDARD DEVIATION = 0.25

1,3-DINITROBENZENE C6H4N2O4

MAKSIMOV YU.YA.:

ZH.FIZ.KHIM. 42,2921(1968).

A = 4.33737 B = 229.248 C = -137.030

B.P.(760) = 294.420

T	P EXPTL.	P CALCD.	DEV.	PERCENT
252.000	220.00	220.49	0.49	0.22
253.000	230.00	229.39	-0.61	-0.26
290.000	687.00	689.80	2.80	0.41
291.500	716.00	713.31	-2.69	-0.38

STANDARD DEVIATION = 3.96

BROMOBENZENE C6H5BR

DREYER R.,MARTIN W.,VON WEBER U.:

J.PRAKT.CHEM. 273,324(1954/55).

A = 6.86064 B = 1438.817 C = 205.441

B.P.(760) = 156.086

T	P EXPTL.	P CALCD.	DEV.	PERCENT
56.070	23.00	22.84	-0.16	-0.69
63.520	33.00	32.44	-0.56	-1.69
69.700	42.80	42.78	-0.02	-0.04
74.880	53.40	53.44	0.04	0.08
78.250	61.60	61.50	-0.10	-0.16
81.900	71.60	71.33	-0.27	-0.37
85.550	82.40	82.43	0.03	0.04
88.530	92.50	92.52	0.02	0.02
90.820	100.70	100.94	0.24	0.24
93.490	111.30	111.54	0.24	0.22
97.090	127.00	127.26	0.26	0.21
99.410	138.50	138.33	-0.17	-0.13
101.260	147.80	147.70	-0.10	-0.07
105.000	168.20	168.23	0.03	0.01
109.500	195.90	195.94	0.04	0.02
115.590	239.10	239.21	0.11	0.05
122.790	299.80	299.98	0.18	0.06
129.610	368.70	368.40	-0.30	-0.08
135.480	437.10	436.78	-0.32	-0.07
139.940	496.00	495.17	-0.83	-0.17
144.730	566.50	564.60	-1.90	-0.34
149.510	640.80	641.32	0.52	0.08
151.950	682.90	683.52	0.62	0.09
154.240	725.20	725.07	-0.13	-0.02

STANDARD DEVIATION = 0.32

CHLOROBENZENE C6H5CL

BROWN I.:

AUSTR.J.SCI.RESEARCH 5,530(1952).

A = 6.97808 B = 1431.053 C = 217.550

B.P.(760) = 131.719

T	P EXPTL.	P CALCD.	DEV.	PERCENT
62.040	72.43	72.39	-0.04	-0.05
62.060	72.48	72.45	-0.03	-0.04
66.380	86.69	86.68	-0.01	-0.01
74.130	117.91	117.99	0.08	0.07
81.600	156.48	156.44	-0.04	-0.02
89.060	204.36	204.53	0.17	0.08
94.040	242.84	242.85	0.01	0.01
94.470	246.45	246.42	-0.03	-0.01
98.790	284.73	284.65	-0.08	-0.03
103.180	328.29	328.26	-0.03	-0.01
106.250	361.75	361.85	0.10	0.03
110.350	410.99	410.95	-0.04	-0.01
114.480	465.85	465.66	-0.19	-0.04
115.890	485.69	485.62	-0.07	-0.01
117.940	515.85	515.85	0.00	0.00
121.100	565.34	565.36	0.02	0.00
123.390	606.63	603.54	-3.09	-0.51
126.180	652.69	652.80	0.11	0.02
128.450	695.11	695.17	0.06	0.01
130.370	732.69	732.68	-0.01	0.00
131.400	753.43	753.45	0.02	0.00
131.700	759.54	759.59	0.05	0.01
131.730	760.28	760.21	-0.07	-0.01

STANDARD DEVIATION = 0.08

PHENYLTRICHLOROSILANE C6H5CL3SI

JENKINS A.C.,CHAMBERS G.F.: IND.ENG.CHEM. 46,2367(1954).

A = 6.89071 B = 1575.652 C = 191.853

B.P.(760) = 201.088

T	P EXPTL.	P CALCD.	DEV.	PERCENT
101.700	33.80	33.36	-0.44	-1.31
108.800	44.40	44.66	0.26	0.59
153.800	214.70	214.89	0.19	0.09
175.600	400.30	400.56	0.26	0.07
194.800	657.90	654.03	-3.87	-0.59
197.400	692.90	696.34	3.44	0.50

STANDARD DEVIATION = 3.01

FLUOROBENZENE C_6H_5F

YOUNG S.: J.CHEM.SOC. 55,486(1889).

A = 7.18703 B = 1381.828 C = 235.563

B.P.(760) = 85.328

T	P EXPTL.	P CALCD.	DEV.	PERCENT
-17.850	6.95	6.92	-0.03	-0.45
* -15.100	8.15	8.30	0.15	1.87
-13.000	9.70	9.51	-0.19	-1.93
* -11.350	10.85	10.57	-0.28	-2.59
-9.450	12.00	11.91	-0.09	-0.78
-8.050	13.00	12.98	-0.02	-0.12
-5.950	14.85	14.76	-0.09	-0.64
-4.550	16.05	16.05	0.00	-0.01
-3.000	17.80	17.59	-0.21	-1.17
-0.950	19.95	19.82	-0.13	-0.63
0.150	21.05	21.12	0.07	0.33
1.800	23.10	23.20	0.10	0.42
2.600	24.10	24.27	0.17	0.69
4.100	26.10	26.38	0.28	1.08
4.700	27.10	27.27	0.17	0.63
6.050	29.40	29.37	-0.03	-0.11
6.650	30.20	30.34	0.14	0.46
7.300	31.30	31.43	0.13	0.40
9.500	35.20	35.35	0.15	0.42
10.650	37.50	37.56	0.06	0.15
13.850	44.45	44.33	-0.12	-0.27
16.750	51.55	51.33	-0.22	-0.42
19.500	59.00	58.81	-0.19	-0.33
22.000	66.25	66.38	0.13	0.19
24.900	76.85	76.16	-0.69	-0.89
27.500	86.45	85.94	-0.51	-0.59
28.250	89.05	88.94	-0.11	-0.12
30.350	97.50	97.83	0.33	0.34
30.600	99.00	98.94	-0.06	-0.06
33.300	111.30	111.56	0.26	0.23
36.050	125.70	125.76	0.06	0.05
39.150	143.40	143.53	0.13	0.09
42.050	161.70	161.99	0.29	0.18
45.550	187.50	186.84	-0.66	-0.35
49.600	219.90	219.42	-0.48	-0.22
49.900	222.80	222.01	-0.79	-0.35
46.900	197.20	197.22	0.02	0.01
47.600	201.60	202.79	1.19	0.59
48.100	206.25	206.85	0.60	0.29
52.600	246.75	246.45	-0.30	-0.12
52.950	248.90	249.77	0.87	0.35
56.200	282.10	282.42	0.32	0.11
56.400	283.80	284.54	0.74	0.26
59.600	320.20	320.22	0.02	0.01
62.750	359.20	358.83	-0.37	-0.10
63.250	364.00	365.29	1.29	0.35
65.650	398.70	397.64	-1.07	-0.27
68.250	437.00	435.25	-1.75	-0.40
71.000	480.00	478.13	-1.88	-0.39
73.700	522.50	523.47	0.97	0.19

(CONTINUED)

	T	P EXPTL.	P CALCD.	DEV.	PERCENT
	75.900	563.05	562.93	-0.12	-0.02
	77.300	588.40	589.26	0.86	0.15
	78.600	613.10	614.59	1.49	0.24
	78.750	618.60	617.57	-1.04	-0.17
	79.150	624.10	625.57	1.47	0.23
	79.550	633.20	633.64	0.44	0.07
	80.700	659.80	657.34	-2.46	-0.37
	82.450	694.20	694.76	0.56	0.08
*	83.850	730.40	725.91	-4.49	-0.62

STANDARD DEVIATION = 0.74

IODOBENZENE C6H5I

YOUNG S.:

J.CHEM.SOC. 55,486(1889).

A = 7.01187 B = 1640.124 C = 208.765

B.P.(760) = 188.258

	T	P EXPTL.	P CALCD.	DEV.	PERCENT
*	29.300	1.60	1.33	-0.27	-17.14
	33.000	1.80	1.69	-0.11	-6.11
	40.100	2.75	2.64	-0.11	-4.04
	46.300	3.85	3.82	-0.03	-0.88
	52.500	5.65	5.42	-0.23	-4.02
	57.000	7.00	6.93	-0.07	-1.05
	61.450	8.90	8.75	-0.15	-1.65
	64.500	10.40	10.23	-0.17	-1.63
	67.800	12.20	12.06	-0.14	-1.11
	70.900	14.45	14.04	-0.41	-2.86
	74.450	17.00	16.63	-0.37	-2.20
	78.400	20.25	19.97	-0.28	-1.37
	81.900	23.75	23.40	-0.35	-1.48
	85.500	27.55	27.43	-0.12	-0.43
	87.450	30.20	29.85	-0.35	-1.16
	88.750	32.00	31.56	-0.44	-1.38
	91.550	36.15	35.52	-0.63	-1.73
	91.600	36.15	35.60	-0.55	-1.53
	94.750	41.00	40.56	-0.44	-1.07
	100.350	50.15	50.82	0.67	1.33
	100.500	50.70	51.12	0.42	0.82
	101.400	51.90	52.96	1.06	2.04
	105.650	62.40	62.44	0.04	0.06
	109.500	71.75	72.20	0.45	0.63
	113.400	83.10	83.35	0.25	0.30
	116.950	94.80	94.71	-0.09	-0.09
	117.350	96.30	96.07	-0.23	-0.24
	120.650	107.90	107.88	-0.02	-0.01
	120.900	108.70	108.83	0.13	0.12
	123.900	121.15	120.67	-0.48	-0.40
	124.650	123.50	123.79	0.29	0.24
	126.850	134.00	133.33	-0.67	-0.50
*	130.500	154.20	150.49	-3.71	-2.40
	131.200	154.20	153.98	-0.22	-0.14
	135.500	177.95	176.90	-1.05	-0.59
	135.950	176.45	179.45	3.00	1.70
	136.700	182.05	183.77	1.72	0.95
	139.700	202.40	201.91	-0.49	-0.24
	142.000	214.20	216.78	2.58	1.20
	144.000	231.85	230.42	-1.43	-0.62
	147.650	256.30	257.12	0.82	0.32
	148.150	261.75	260.96	-0.79	-0.30
	148.250	260.95	261.74	0.79	0.30
	153.100	302.30	301.61	-0.69	-0.23
	156.800	336.30	335.21	-1.09	-0.32
	160.200	370.55	368.69	-1.86	-0.50
	163.500	404.20	403.71	-0.49	-0.12
	166.800	442.60	441.35	-1.25	-0.28
	170.650	489.60	488.75	-0.85	-0.17
	173.800	532.90	530.50	-2.40	-0.45

(CONTINUED)

T	P EXPTL.	P CALCD.	DEV.	PERCENT
177.350	581.40	580.90	-0.50	-0.09
181.200	639.40	639.79	0.39	0.06
184.000	683.20	685.52	2.32	0.34
188.200	756.40	758.93	2.53	0.33

STANDARD DEVIATION = 1.05

NITROBENZENE C6H5NO2

BROWN I.:

AUSTR.J.SCI.RESEARCH 5,530(1952).

A = 7.11562 B = 1746.586 C = 201.783

B.P.(760) = 210.653

T	P EXPTL.	P CALCD.	DEV.	PERCENT
134.100	82.23	82.34	0.11	0.14
139.750	100.30	100.38	0.08	0.08
145.170	120.64	120.65	0.01	0.01
149.730	140.20	140.23	0.03	0.02
154.610	164.01	164.02	0.01	0.00
159.770	192.80	192.67	-0.13	-0.07
164.450	222.33	222.11	-0.22	-0.10
168.720	252.39	252.07	-0.32	-0.13
172.960	285.04	285.01	-0.03	-0.01
178.480	333.44	333.05	-0.39	-0.12
182.070	367.64	367.68	0.04	0.01
185.700	405.18	405.60	0.42	0.10
188.900	441.33	441.59	0.26	0.06
192.980	490.74	491.16	0.42	0.09
196.630	538.87	539.21	0.34	0.06
200.410	592.53	592.87	0.34	0.06
203.880	645.62	645.81	0.19	0.03
206.620	690.23	690.23	0.00	0.00
209.490	739.54	739.32	-0.22	-0.03
210.626	759.98	759.51	-0.47	-0.06
210.629	760.04	759.56	-0.48	-0.06

STANDARD DEVIATION = 0.29

BENZENE C6H6

WILLINGHAM C.J., TAYLOR W.J., PIGNOCCO J.M., ROSSINI F.D.:
J.RESEARCH NATL.BUR.STANDARDS 35,219(1945).

A = 6.89272 B = 1203.531 C = 219.888

B.P.(760) = 80.102

T	P EXPTL.	P CALCD.	DEV.	PERCENT
14.548	57.41	57.41	0.00	0.00
17.720	67.22	67.22	0.00	0.00
20.594	77.28	77.28	0.00	0.00
23.270	87.75	87.73	-0.02	-0.03
26.886	103.64	103.67	0.03	0.03
31.004	124.67	124.65	-0.02	-0.01
35.191	149.43	149.43	0.00	0.00
39.078	175.89	175.90	0.01	0.00
44.284	217.16	217.20	0.04	0.02
49.066	261.75	261.73	-0.02	-0.01
54.832	324.93	324.92	-0.01	0.00
60.784	402.42	402.42	0.00	0.00
67.135	500.69	500.68	-0.01	0.00
74.028	627.93	627.91	-0.02	0.00
78.891	732.07	732.06	-0.01	0.00
79.413	744.04	744.00	-0.04	-0.01
79.898	755.23	755.23	0.00	0.00
80.442	767.94	767.98	0.04	0.01
80.922	779.34	779.37	0.03	0.00

STANDARD DEVIATION = 0.02

M-CHLORANILINE C6H6CLN

ZALIKIN A.A.,KOCHETKOV V.L.,STREPIKHEEV YU.A.:
KHIM.PROM. 41,338(1965).

A = 3.03767 B = 171.347 C = -14.988

B.P.(10) = 99.078

T	P EXPTL.	P CALCD.	DEV.	PERCENT
61.000	0.50	0.21	-0.29	-58.82
70.000	1.50	0.84	-0.66	-44.17
84.000	4.00	3.59	-0.41	-10.31
91.000	6.00	6.07	0.07	1.23
101.000	10.50	11.11	0.61	5.77
103.000	12.00	12.33	0.33	2.71
113.000	20.00	19.47	-0.53	-2.63
114.500	21.00	20.69	-0.31	-1.47
119.500	23.50	25.01	1.51	6.44
125.000	31.00	30.21	-0.79	-2.55

STANDARD DEVIATION = 0.80

PHENOL C6H6O

DREISBACH R.R.,SHRADER S.A.:
IND.ENG.CHEM. 41,2879(1949).

A = 7.13301 B = 1516.790 C = 174.954

B.P.(760) = 181.754

T	P EXPTL.	P CALCD.	DEV.	PERCENT
107.150	57.04	57.06	0.02	0.03
113.810	75.86	75.91	0.05	0.07
125.950	123.76	123.66	-0.10	-0.08
152.370	315.52	315.57	0.05	0.01
167.630	507.50	507.58	0.08	0.02
181.750	760.00	759.92	-0.08	-0.01

STANDARD DEVIATION = 0.10

HYDROCHINONE C6H6O2

VONTERRES E.,ET AL.:
BRENNSTOFF CHEM. 36,272(1955).

A = 8.13697 B = 2461.045 C = 182.821

B.P.(760) = 285.400

T	P EXPTL.	P CALCD.	DEV.	PERCENT
159.100	10.00	8.70	-1.31	-13.05
181.000	25.00	23.58	-1.42	-5.68
199.100	50.00	49.33	-0.67	-1.34
210.000	75.00	74.46	-0.54	-0.72
218.500	100.00	101.07	1.07	1.07
230.100	150.00	150.27	0.27	0.18
239.200	200.00	202.02	2.02	1.01
246.100	250.00	250.73	0.73	0.29
252.000	300.00	299.96	-0.04	-0.01
254.800	325.00	326.04	1.04	0.32
257.200	350.00	349.90	-0.10	-0.03
259.000	400.00	368.75	-31.25	-7.81
266.100	450.00	451.68	1.68	0.37
269.500	500.00	496.64	-3.36	-0.67
273.000	550.00	546.79	-3.21	-0.58
276.700	600.00	604.35	4.35	0.73
278.800	650.00	639.22	-10.78	-1.66
282.000	700.00	695.60	-4.40	-0.63
286.000	760.00	771.85	11.85	1.56

STANDARD DEVIATION = 4.70

PYROCATECHOL C6H6O2

VONTERRES E.,ET AL.:
BRENNSTOFF CHEM. 36,272(1955).

A = 7.57729 B = 2024.422 C = 186.533

B.P.(760) = 244.518

T	P EXPTL.	P CALCD.	DEV.	PERCENT
118.500	10.00	8.72	-1.28	-12.79
139.300	25.00	23.13	-1.87	-7.47
157.000	50.00	48.34	-1.66	-3.31
168.000	75.00	73.65	-1.35	-1.80
176.200	100.00	99.14	-0.86	-0.86
188.200	150.00	149.62	-0.38	-0.25
198.000	200.00	205.43	5.43	2.72
204.800	250.00	253.60	3.60	1.44
210.900	300.00	304.47	4.47	1.49
213.300	325.00	326.68	1.68	0.52
215.800	350.00	351.22	1.22	0.35
220.200	400.00	398.12	-1.88	-0.47
224.600	450.00	450.07	0.07	0.02
228.200	500.00	496.62	-3.38	-0.68
231.400	550.00	541.25	-8.75	-1.59
235.000	600.00	595.34	-4.66	-0.78
237.800	650.00	640.41	-9.59	-1.48
241.000	700.00	695.29	-4.71	-0.67
245.500	760.00	778.91	18.91	2.49

STANDARD DEVIATION = 6.43

RESORCINOL C6H6O2

VONTERRES E.,ET AL.:
BRENNSTOFF CHEM. 36,272(1955).

A = 7.88906 B = 2231.138 C = 169.288

B.P.(760) = 276.205

	T	P EXPTL.	P CALCD.	DEV.	PERCENT
	151.500	10.00	8.59	-1.41	-14.12
	173.000	25.00	23.48	-1.52	-6.07
	190.300	50.00	48.34	-1.66	-3.31
	201.700	75.00	74.99	-0.01	-0.01
	210.000	100.00	101.54	1.54	1.54
	221.700	150.00	152.28	2.28	1.52
	230.100	200.00	200.75	0.75	0.38
	237.000	250.00	249.77	-0.23	-0.09
*	240.200	300.00	275.71	-24.29	-8.10
	246.000	325.00	328.51	3.51	1.08
	248.000	350.00	348.58	-1.42	-0.41
	252.000	400.00	391.80	-8.20	-2.05
	257.000	450.00	452.05	2.05	0.45
	261.000	500.00	505.64	5.64	1.13
	264.000	550.00	549.21	-0.79	-0.14
	267.000	600.00	595.86	-4.14	-0.69
	270.000	650.00	645.76	-4.24	-0.65
	273.000	700.00	699.07	-0.93	-0.13
	276.500	760.00	765.82	5.82	0.77

STANDARD DEVIATION = 3.68

PYROGALLOL C6H6O3

VONTERRES E.,ET AL.:
BRENNSTOFF CHEM. 36,272(1955).

A = 6.09158 B = 1030.542 C = 12.289

B.P.(760) = 308.675

T	P EXPTL.	P CALCD.	DEV.	PERCENT
177.300	10.00	4.53	-5.47	-54.72
200.200	25.00	17.45	-7.55	-30.21
218.000	50.00	41.36	-8.64	-17.28
230.000	75.00	68.90	-6.10	-8.13
238.600	100.00	96.39	-3.61	-3.61
251.000	150.00	150.48	0.48	0.32
261.500	200.00	212.61	12.61	6.30
268.100	250.00	260.72	10.72	4.29
274.000	300.00	310.40	10.40	3.47
277.000	325.00	338.26	13.26	4.08
279.000	350.00	357.86	7.86	2.25
284.000	400.00	410.60	10.60	2.65
288.000	450.00	456.82	6.82	1.52
292.000	500.00	506.83	6.83	1.37
295.700	550.00	556.60	6.60	1.20
296.200	600.00	563.60	-36.40	-6.07
299.000	650.00	603.97	-46.03	-7.08
305.000	700.00	697.62	-2.38	-0.34
309.000	760.00	765.70	5.70	0.75

STANDARD DEVIATION =16.89

BENZENETHIOL C6H6S

SCOTT D.W.,ET AL.:
J.AM.CHEM.SOC. 78,5457(1956).

A = 6.99054 B = 1529.712 C = 203.079

B.P.(760) = 169.138

T	P EXPTL.	P CALCD.	DEV.	PERCENT
114.543	149.41	149.42	0.01	0.01
121.191	187.57	187.56	-0.01	-0.01
127.897	233.72	233.73	0.01	0.01
134.649	289.13	289.15	0.02	0.01
141.447	355.22	355.22	0.00	0.00
148.294	433.56	433.52	-0.04	-0.01
155.194	525.86	525.84	-0.02	0.00
162.140	633.99	633.96	-0.03	-0.01
169.137	760.00	759.97	-0.03	0.00
176.188	906.06	906.16	0.10	0.01
183.278	1074.60	1074.53	-0.07	-0.01
190.426	1268.00	1268.06	0.06	0.00
197.623	1489.10	1489.23	0.13	0.01
204.867	1740.80	1740.81	0.01	0.00
212.160	2026.00	2025.86	-0.14	-0.01

STANDARD DEVIATION = 0.07

ANILINE C6H7N

MCDONALD R.A.,SHRADER S.A.,STULL D.R.:
J.CHEM.ENG.DATA 4,311(1959).

A = 7.32010 B = 1731.515 C = 206.049

B.P.(760) = 183.994

T	P EXPTL.	P CALCD.	DEV.	PERCENT
102.590	51.06	51.28	0.22	0.43
117.220	92.22	92.01	-0.21	-0.23
137.500	190.73	190.55	-0.18	-0.09
160.080	389.38	389.81	0.43	0.11
182.400	728.33	728.77	0.44	0.06
184.240	764.40	764.90	0.50	0.07
185.150	784.48	783.29	-1.19	-0.15

STANDARD DEVIATION = 0.74

2-METHYLPYRIDINE C6H7N

SCOTT D.W.,ET AL.:

J.PHYS.CHEM. 67,680(1963).

A = 7.03237 B = 1415.725 C = 211.625

B.P.(760) = 129.386

T	P EXPTL.	P CALCD.	DEV.	PERCENT
79.794	149.41	149.39	-0.02	-0.01
85.853	187.57	187.61	0.04	0.02
91.942	233.72	233.74	0.02	0.01
98.074	289.13	289.11	-0.02	-0.01
104.252	355.22	355.20	-0.02	-0.01
110.472	433.56	433.53	-0.03	-0.01
116.736	525.86	525.86	0.00	0.00
123.038	633.99	633.96	-0.03	-0.01
129.387	760.00	760.01	0.01	0.00
135.773	906.06	906.02	-0.04	0.00
142.207	1074.60	1074.58	-0.02	0.00
148.683	1268.00	1268.10	0.10	0.01
155.201	1489.10	1489.26	0.16	0.01
161.761	1740.80	1740.90	0.10	0.01
168.356	2026.00	2025.73	-0.27	-0.01

STANDARD DEVIATION = 0.10

3-METHYLPYRIDINE C_6H_7N

SCOTT D.W.,ET AL.:

J.PHYS.CHEM. 67,685(1963).

A = 7.05021 B = 1481.775 C = 211.253

B.P.(760) = 144.140

T	P EXPTL.	P CALCD.	DEV.	PERCENT
74.036	71.87	71.82	-0.05	-0.07
77.115	81.64	81.61	-0.04	-0.04
80.202	92.52	92.50	-0.02	-0.02
83.303	104.63	104.63	0.00	0.00
86.403	118.06	118.05	-0.01	-0.01
89.524	132.95	132.96	0.01	0.01
92.658	149.41	149.46	0.05	0.03
98.946	187.57	187.65	0.08	0.04
105.270	233.72	233.77	0.05	0.02
111.640	289.13	289.17	0.04	0.01
118.052	355.22	355.22	0.00	0.00
124.508	433.56	433.53	-0.03	-0.01
131.008	525.86	525.82	-0.04	-0.01
137.551	633.99	633.94	-0.05	-0.01
144.135	760.00	759.89	-0.11	-0.02
150.767	906.06	906.01	-0.05	-0.01
157.441	1074.60	1074.54	-0.06	-0.01
164.156	1268.00	1267.98	-0.02	0.00
170.918	1489.10	1489.19	0.09	0.01
177.721	1740.80	1740.85	0.05	0.00
184.568	2026.00	2026.09	0.09	0.00

STANDARD DEVIATION = 0.06

4-METHYLPYRIDINE C6H7N

OSBORN A.G.,DOUSLIN D.R.:
J.CHEM.ENG.DATA13,534(1968).

A = 7.04177 B = 1480.684 C = 210.495

B.P.(760) = 145.357

T	P EXPTL.	P CALCD.	DEV.	PERCENT
75.052	71.87	71.84	-0.03	-0.05
78.135	81.64	81.61	-0.03	-0.04
81.229	92.52	92.50	-0.02	-0.02
84.333	104.63	104.61	-0.02	-0.02
87.448	118.06	118.05	-0.01	-0.01
90.576	132.95	132.96	0.01	0.01
93.719	149.41	149.46	0.05	0.03
100.022	187.57	187.64	0.07	0.04
106.371	233.72	233.81	0.09	0.04
112.754	289.13	289.16	0.03	0.01
119.184	355.22	355.21	-0.01	0.00
125.659	433.56	433.50	-0.06	-0.01
132.178	525.86	525.76	-0.10	-0.02
138.747	633.99	633.96	-0.03	0.00
145.354	760.00	759.94	-0.06	-0.01
152.007	906.06	906.04	-0.02	0.00
158.701	1074.60	1074.50	-0.10	-0.01
165.445	1268.00	1268.09	0.09	0.01
172.223	1489.10	1489.04	-0.06	0.00
179.060	1740.80	1741.03	0.23	0.01
185.925	2026.00	2025.96	-0.04	0.00

STANDARD DEVIATION = 0.08

ALPHA-PICOLINE C6H7N

HERINGTON E.F.G.,MARTIN J.F.:
TRANS.FARADAY SOC.49,154(1953).

A = 7.03897 B = 1420.461 C = 212.200

B.P.(760) = 129.409

T	P EXPTL.	P CALCD.	DEV.	PERCENT
64.363	79.96	79.95	-0.01	-0.01
69.916	100.95	100.91	-0.04	-0.04
76.836	133.20	133.19	-0.01	0.00
82.362	164.67	164.70	0.03	0.02
88.566	207.02	207.09	0.07	0.03
93.617	247.85	247.83	-0.02	-0.01
101.283	321.82	321.91	0.09	0.03
108.594	408.29	408.33	0.04	0.01
114.552	491.90	491.75	-0.15	-0.03
117.647	540.20	540.18	-0.02	0.00
122.132	617.28	617.04	-0.24	-0.04
125.664	683.61	683.48	-0.13	-0.02
126.992	710.03	709.88	-0.15	-0.02
127.828	726.70	726.91	0.21	0.03
128.591	742.72	742.74	0.02	0.00
129.290	757.45	757.47	0.02	0.00
129.608	764.16	764.25	0.09	0.01
130.037	773.28	773.47	0.19	0.03

STANDARD DEVIATION = 0.12

BETA-PICOLINE

C6H7N

HERINGTON E.F.G.,MARTIN J.F.:

TRANS.FARADAY SOC.49,154(1953).

A = 7.03577 B = 1472.074 C = 210.152

B.P.(760) = 144.142

T	P EXPTL.	P CALCD.	DEV.	PERCENT
81.282	96.54	96.52	-0.02	-0.02
85.275	112.97	112.95	-0.02	-0.02
92.059	146.10	146.13	0.03	0.02
97.519	178.31	178.31	0.00	0.00
103.922	223.12	223.22	0.10	0.04
109.006	265.14	265.09	-0.05	-0.02
115.583	328.50	328.49	-0.01	0.00
121.932	400.87	400.79	-0.08	-0.02
129.368	501.21	501.19	-0.02	0.00
132.165	543.58	543.79	0.21	0.04
137.714	637.09	636.84	-0.25	-0.04
140.871	695.26	695.17	-0.09	-0.01
142.132	719.67	719.62	-0.05	-0.01
142.639	729.55	729.63	0.08	0.01
143.293	742.69	742.72	0.03	0.00
143.577	748.43	748.46	0.04	0.00
143.993	757.02	756.93	-0.09	-0.01
144.320	763.50	763.65	0.15	0.02
144.659	770.50	770.65	0.15	0.02
145.101	779.98	779.87	-0.11	-0.01

STANDARD DEVIATION = 0.11

GAMA-PICOLINE C6H7N

HERINGTON E.F.G.,MARTIN J.F.:
TRANS.FARADAY SOC.49,154(1953).

A = 7.04557 B = 1483.232 C = 210.783

B.P.(760) = 145.356

T	P EXPTL.	P CALCD.	DEV.	PERCENT
76.906	77.62	77.61	-0.01	-0.02
79.600	86.65	86.64	-0.01	-0.01
82.830	98.60	98.61	0.01	0.01
88.166	121.35	121.36	0.01	0.01
94.654	154.69	154.69	0.00	0.00
98.323	176.65	176.65	0.00	0.00
104.738	221.17	221.14	-0.03	-0.01
109.660	261.10	261.14	0.04	0.02
116.506	326.35	326.36	0.01	0.00
123.021	400.06	400.08	0.02	0.01
129.837	490.98	490.98	0.00	0.00
135.281	574.92	574.86	-0.06	-0.01
137.174	606.70	606.57	-0.13	-0.02
140.126	658.81	658.78	-0.03	0.00
142.127	696.22	696.16	-0.06	-0.01
143.604	724.87	724.81	-0.06	-0.01
144.246	737.41	737.55	0.14	0.02
144.745	747.41	747.58	0.17	0.02
145.034	753.54	753.44	-0.10	-0.01
145.378	760.59	760.46	-0.13	-0.02
145.462	762.00	762.18	0.18	0.02

STANDARD DEVIATION = 0.09

P-AMINOPHENOL C6H7NO

DUNN S.A.:
J.AM.CHEM.SOC. 76,6191(1954).

A = -3.35750 B = 699.157 C = -331.343

B.P.(10) = 170.894

T	P EXPTL.	P CALCD.	DEV.	PERCENT
130.200	0.30	1.31	1.01	337.81
143.000	2.20	2.26	0.06	2.85
145.000	2.20	2.48	0.28	12.73
150.000	3.00	3.15	0.15	4.91
151.100	3.20	3.32	0.12	3.83
157.500	4.70	4.62	-0.08	-1.79
159.600	5.10	5.17	0.07	1.36
167.000	8.00	7.88	-0.12	-1.45
171.100	9.90	10.13	0.23	2.32
176.500	14.90	14.38	-0.52	-3.49
185.300	26.70	26.90	0.20	0.77

STANDARD DEVIATION = 0.44

TRIVINYLARSINE C6H9AS

MAIER L.,SEYFERTH D.,STONE F.G.A.,ROCHOV E.:
J.AM.CHEM.SOC. 79,5884(1957).

A = 7.89409 B = 2115.593 C = 293.863

B.P.(100) = 65.071

T	P EXPTL.	P CALCD.	DEV.	PERCENT
22.700	16.40	16.26	-0.14	-0.86
25.000	18.30	18.17	-0.13	-0.72
31.500	24.50	24.65	0.15	0.62
39.000	34.50	34.54	0.04	0.12
43.900	42.70	42.71	0.01	0.03
49.900	54.60	54.94	0.34	0.62
53.600	64.20	63.89	-0.31	-0.49
63.000	92.50	92.43	-0.08	-0.08
65.900	103.10	103.18	0.08	0.07

STANDARD DEVIATION = 0.22

TRIVINYLBISMUTH C6H9BI

MAIER L.,SEYFERTH D.,STONE F.G.A.,ROCHOV E.:
J.AM.CHEM.SOC. 79,5884(1957).

A = 7.23717 B = 1667.037 C = 215.121

B.P.(10) = 52.154

T	P EXPTL.	P CALCD.	DEV.	PERCENT
19.900	1.40	1.39	-0.01	-0.48
39.400	4.90	4.87	-0.03	-0.63
53.000	10.40	10.46	0.06	0.61
56.500	12.60	12.58	-0.02	-0.13
59.900	15.10	14.99	-0.11	-0.76
63.000	17.50	17.51	0.01	0.05
66.000	20.20	20.29	0.09	0.43
70.000	24.50	24.57	0.07	0.28
73.500	29.00	28.93	-0.07	-0.25

STANDARD DEVIATION = 0.08

1-BUTYL TRIFLUOROACETATE C6H9F3O2

SHEEHAN R.J.,LANGER S.H.:
J.CHEM.ENG.DATA 14,248(1969).

A = 8.56794 B = 2305.221 C = 301.057

B.P.(760) = 104.283

T	P EXPTL.	P CALCD.	DEV.	PERCENT
70.900	235.30	234.63	-0.67	-0.28
85.500	399.70	402.22	2.52	0.63
94.300	548.20	546.00	-2.20	-0.40
103.500	740.70	740.96	0.26	0.04

STANDARD DEVIATION = 3.42

2,5-DIMETHYLPYRROLE C6H9N

OSBORN A.G.,DOUSLIN D.R.: J.CHEM.ENG.DATA13,534(1968).

A = 7.20306 B = 1509.600 C = 181.764

B.P.(760) = 167.499

T	P EXPTL.	P CALCD.	DEV.	PERCENT
100.560	71.87	71.78	-0.09	-0.12
103.523	81.64	81.57	-0.07	-0.08
106.491	92.52	92.48	-0.04	-0.05
109.464	104.63	104.59	-0.04	-0.04
112.450	118.06	118.06	0.00	0.00
115.443	132.95	132.98	0.03	0.02
118.448	149.41	149.49	0.08	0.05
124.468	187.57	187.70	0.13	0.07
130.516	233.72	233.85	0.13	0.06
136.598	289.13	289.26	0.13	0.04
142.707	355.22	355.27	0.05	0.01
148.853	433.56	433.56	0.00	0.00
155.030	525.86	525.77	-0.09	-0.02
161.241	633.99	633.80	-0.19	-0.03
167.489	760.00	759.78	-0.22	-0.03
173.768	906.06	905.79	-0.27	-0.03
180.084	1074.60	1074.33	-0.27	-0.02
186.435	1268.00	1267.95	-0.05	0.00
192.820	1489.10	1489.32	0.22	0.01
199.239	1740.80	1741.34	0.54	0.03

STANDARD DEVIATION = 0.20

TRIVINYLPHOSPHINE C6H9P

MAIER L.,SEYFERTH D.,STONE F.G.A.,ROCHOV E.: J.AM.CHEM.SOC. 79,5884(1957).

A = 7.92838 B = 2102.007 C = 301.282

B.P.(100) = 53.285

T	P EXPTL.	P CALCD.	DEV.	PERCENT
16.200	20.40	20.30	-0.10	-0.49
21.200	25.70	25.71	0.01	0.05
26.500	32.70	32.78	0.08	0.23
35.000	47.90	47.60	-0.30	-0.62
40.100	58.90	59.02	0.12	0.21
44.900	71.40	71.85	0.45	0.63
49.900	88.00	87.67	-0.33	-0.37
54.900	106.30	106.38	0.08	0.08
60.900	133.30	133.24	-0.06	-0.04

STANDARD DEVIATION = 0.27

TRIVINYLSTIBINE C6H9SB

MAIER L.,SEYFERTH D.,STONE F.G.A.,ROCHOV E.: J.AM.CHEM.SOC. 79,5884(1957).

A = 8.32212 B = 2446.308 C = 303.798

B.P.(10) = 30.301

T	P EXPTL.	P CALCD.	DEV.	PERCENT
20.200	6.10	5.91	-0.19	-3.08
23.900	7.40	7.19	-0.21	-2.78
30.000	9.80	9.85	0.05	0.50
35.000	12.70	12.63	-0.07	-0.51
39.900	15.90	16.01	0.11	0.72
45.100	20.20	20.44	0.24	1.21
50.500	26.20	26.15	-0.05	-0.20
55.000	31.70	31.92	0.22	0.69
59.800	39.40	39.27	-0.13	-0.33
65.000	49.10	48.86	-0.24	-0.50
70.000	59.80	59.93	0.13	0.22

STANDARD DEVIATION = 0.19

2,3-DIMETHYL-1,3-BUTADIENE C6H10

CUMMINGS G.A.M.,MCLAUGHLIN E.: J.CHEM.SOC. 1955,1391.

A = 7.11967 B = 1299.692 C = 238.092

B.P.(760) = 68.523

T	P EXPTL.	P CALCD.	DEV.	PERCENT
0.040	46.00	45.90	-0.10	-0.22
12.000	83.70	83.71	0.01	0.02
13.460	89.60	89.73	0.13	0.15
15.420	98.40	98.38	-0.02	-0.02
16.960	105.90	105.65	-0.25	-0.24
17.240	106.80	107.02	0.22	0.20
19.860	120.40	120.55	0.15	0.12
25.740	156.20	156.11	-0.09	-0.06
31.350	197.80	197.70	-0.10	-0.05
68.400	757.00	757.03	0.03	0.00

STANDARD DEVIATION = 0.16

3-HEXYNE C6H10

RONDEAU R.E.,HARRAN L.A.:
J.CHEM.ENG.DATA 10,84(1965).

A = 5.89462 B = 863.264 C = 193.745

B.P.(10) = -17.375

T	P EXPTL.	P CALCD.	DEV.	PERCENT
-19.950	10.00	8.46	-1.54	-15.38
-11.050	16.00	14.77	-1.23	-7.67
-5.250	21.00	20.65	-0.35	-1.69
0.250	28.00	27.84	-0.16	-0.57
1.050	30.00	29.04	-0.96	-3.21
5.050	35.00	35.66	0.66	1.87
8.550	41.00	42.39	1.39	3.39
9.950	44.00	45.35	1.35	3.07
12.250	50.00	50.57	0.57	1.14
14.850	57.00	57.03	0.03	0.06
16.050	60.00	60.23	0.23	0.38
16.850	61.00	62.44	1.44	2.36
19.050	69.00	68.84	-0.16	-0.24
19.450	70.00	70.05	0.05	0.08
21.550	77.00	76.72	-0.28	-0.36
22.850	82.00	81.10	-0.90	-1.10
23.950	85.00	84.94	-0.06	-0.07
24.450	88.00	86.74	-1.26	-1.43

STANDARD DEVIATION = 0.97

1,5-HEXADIENE C6H10

CUMMINGS G.A.M.,MCLAUGHLIN E.:
J.CHEM.SOC. 1955,1391.

A = 6.57412 B = 1013.484 C = 214.814

B.P.(760) = 59.597

T	P EXPTL.	P CALCD.	DEV.	PERCENT
0.040	71.90	71.95	0.05	0.07
12.960	133.30	133.23	-0.07	-0.05
13.900	138.90	138.96	0.06	0.04
14.830	144.80	144.82	0.02	0.01
15.640	149.90	150.09	0.19	0.13
16.520	156.50	155.98	-0.52	-0.33
17.830	164.90	165.10	0.20	0.12
31.000	282.60	282.58	-0.02	-0.01
38.220	370.30	370.50	0.20	0.05
59.200	750.80	750.69	-0.11	-0.02

STANDARD DEVIATION = 0.24

MESITYL OXIDE C6H10O

FUGE E.T.J.,BOWDEN S.T.,JONES W.J.:
J.PHYS.CHEM. 56,1013(1952).

A = 6.63583 B = 1186.058 C = 186.039

B.P.(760) = 129.820

T	P EXPTL.	P CALCD.	DEV.	PERCENT
14.000	5.70	5.09	-0.61	-10.71
23.200	10.00	9.28	-0.72	-7.23
31.800	16.00	15.53	-0.47	-2.94
44.000	30.20	30.19	-0.01	-0.02
56.500	55.50	55.67	0.17	0.31
65.800	84.50	84.38	-0.12	-0.14
79.600	145.50	148.22	2.72	1.87
94.500	257.50	255.88	-1.62	-0.63
103.100	344.00	341.82	-2.18	-0.63
129.900	760.00	761.65	1.65	0.22

STANDARD DEVIATION = 1.63

PROPIONIC ANHYDRIDE C6H10O3

KAHLBAUM G.W.A.:
BER. 16,2476(1883).

A = 5.81952 B = 810.310 C = 108.735

B.P.(760) = 167.002

T	P EXPTL.	P CALCD.	DEV.	PERCENT
67.500	17.94	16.66	-1.28	-7.15
72.000	21.44	21.68	0.24	1.13
77.200	28.06	28.94	0.88	3.13
80.000	33.42	33.58	0.16	0.49
82.700	38.06	38.61	0.55	1.44
85.000	44.02	43.35	-0.67	-1.53
167.000	760.00	759.96	-0.04	-0.01

STANDARD DEVIATION = 0.90

GLYCOL DIACETATE C6H10O4

TAYLOR C.A.,RINKENBACH W.R.:
J.AM.CHEM.SOC.48,1305(1926).

A = 6.40731 B = 1092.015 C = 119.298

B.P.(760) = 190.362

T	P EXPTL.	P CALCD.	DEV.	PERCENT
100.000	26.00	26.77	0.77	2.98
110.000	44.00	44.14	0.14	0.33
120.000	70.00	69.81	-0.19	-0.28
130.000	107.00	106.40	-0.60	-0.56
140.000	158.00	156.99	-1.01	-0.64
150.000	225.00	225.04	0.04	0.02
160.000	313.00	314.37	1.37	0.44
170.000	428.00	429.14	1.14	0.27
180.000	573.00	573.74	0.74	0.13
190.000	755.00	752.80	-2.20	-0.29

STANDARD DEVIATION = 1.23

DIALLYL SULFIDE C6H10S

BAUER H.,BURSCHKIES K.:
BER. 68,1243(1935).

A = 4.82930 B = 643.178 C = 142.342

B.P.(10) = 25.621

T	P EXPTL.	P CALCD.	DEV.	PERCENT
10.000	4.40	4.05	-0.35	-7.98
12.000	4.80	4.59	-0.21	-4.32
14.000	5.40	5.19	-0.21	-3.84
16.000	6.00	5.85	-0.15	-2.46
18.000	6.60	6.58	-0.02	-0.36
20.000	7.30	7.37	0.07	0.95
22.000	8.10	8.23	0.13	1.66
24.000	9.10	9.18	0.08	0.84
26.000	10.10	10.20	0.10	1.00
28.000	11.20	11.31	0.11	0.99
30.000	12.30	12.51	0.21	1.72
32.000	13.50	13.81	0.31	2.28
34.000	15.00	15.20	0.20	1.36
37.000	17.50	17.50	0.00	-0.02
40.000	20.50	20.04	-0.46	-2.23

STANDARD DEVIATION = 0.24

SEC BUTYL CHLOROACETATE C6H11CLO2

NELSON O.A.: IND.ENG.CHEM.20,1380(1928).

A = 7.93338 B = 2103.296 C = 249.287

B.P.(760) = 166.995

	T	P EXPTL.	P CALCD.	DEV.	PERCENT
	30.000	2.50	2.53	0.03	1.04
	40.000	4.50	4.60	0.10	2.23
*	50.100	7.00	8.09	1.09	15.60
	55.100	10.50	10.55	0.05	0.52
	60.200	14.00	13.72	-0.28	-2.01
	70.300	21.90	22.50	0.60	2.72
	80.300	34.40	35.63	1.23	3.57
	85.300	43.30	44.38	1.08	2.48
	90.200	54.30	54.69	0.39	0.71
	95.000	66.80	66.72	-0.08	-0.12
	100.000	83.20	81.60	-1.60	-1.92
	105.000	104.10	99.24	-4.86	-4.67
	109.400	117.50	117.36	-0.14	-0.12
	119.200	167.40	168.06	0.66	0.39
	125.200	208.70	207.45	-1.25	-0.60
	135.200	288.80	290.40	1.60	0.55
	145.100	395.90	398.39	2.49	0.63
	150.000	455.20	463.19	7.99	1.75
	154.800	528.40	534.97	6.57	1.24
	159.500	612.10	614.01	1.91	0.31
	164.500	705.80	708.51	2.71	0.38
	166.500	748.10	749.54	1.44	0.19
	167.700	771.00	775.09	4.09	0.53
	168.500	792.00	792.52	0.52	0.07
	169.300	825.90	810.27	-15.63	-1.89
	170.500	849.80	837.53	-12.28	-1.44
*	171.500	895.70	860.81	-34.89	-3.90

STANDARD DEVIATION = 5.10

CAPRONITRILE C6H11N

DREISBACH R.R.,SHRADER S.A.: IND.ENG.CHEM. 41,2879(1949).

A = 7.12312 B = 1597.173 C = 212.799

B.P.(760) = 163.688

T	P EXPTL.	P CALCD.	DEV.	PERCENT
91.980	75.86	76.33	0.47	0.62
104.540	123.76	123.06	-0.70	-0.57
132.620	315.52	315.68	0.16	0.05
148.800	507.50	508.33	0.83	0.16
163.650	760.00	759.25	-0.75	-0.10

STANDARD DEVIATION = 1.00

CYCLOHEXANE C6H12

WILLINGHAM C.J.,TAYLOR W.J.,PIGNOCCO J.M.,ROSSINI F.D.: J.RESEARCH NATL.BUR.STANDARDS 35,219(1945).

A = 6.84941 B = 1206.001 C = 223.148

B.P.(760) = 80.737

T	P EXPTL.	P CALCD.	DEV.	PERCENT
19.915	77.28	77.22	-0.06	-0.08
22.657	87.72	87.72	0.00	0.00
26.347	103.67	103.67	0.00	0.00
30.556	124.65	124.69	0.04	0.03
34.821	149.39	149.43	0.04	0.03
38.798	175.91	175.96	0.05	0.03
44.108	217.22	217.21	-0.01	0.00
48.991	261.73	261.73	0.00	0.00
54.884	324.96	324.92	-0.04	-0.01
60.969	402.48	402.42	-0.06	-0.01
67.467	500.76	500.72	-0.04	-0.01
74.520	627.98	627.94	-0.04	-0.01
79.502	732.13	732.17	0.04	0.00
80.037	744.13	744.12	-0.01	0.00
80.534	755.32	755.35	0.03	0.00
81.093	768.09	768.15	0.06	0.01
81.582	779.49	779.49	0.00	0.00

STANDARD DEVIATION = 0.04

CYCLOHEXANE C_6H_{12}

ASTON J.G.,SZASZ G.J.,FINK H.L.: J.AM.CHEM.SOC. 65,1135(1943).

A = 7.57364 B = 1601.198 C = 261.885

B.P.(10) = -18.306

T	P EXPTL.	P CALCD.	DEV.	PERCENT
6.320	40.11	40.14	0.03	0.08
9.880	48.16	48.06	-0.10	-0.21
13.620	57.73	57.78	0.05	0.08
17.640	69.96	70.04	0.08	0.12
20.960	81.83	81.77	-0.06	-0.07

STANDARD DEVIATION = 0.11

3,3-DIMETHYL-1-BUTENE C_6H_{12}

BAGHDOYAN A.,MALIK J.,FRIED V.: J.CHEM.ENG.DATA 16,96(1971).

A = 6.77402 B = 1060.326 C = 231.088

B.P.(760) = 41.265

T	P EXPTL.	P CALCD.	DEV.	PERCENT
-9.460	97.72	97.67	-0.05	-0.05
-4.140	126.43	126.45	0.02	0.01
0.480	156.65	156.72	0.07	0.04
4.390	186.63	186.70	0.07	0.04
9.710	234.81	234.76	-0.05	-0.02
13.620	276.16	276.05	-0.11	-0.04
17.640	324.29	324.35	0.06	0.02
21.190	372.50	372.40	-0.10	-0.03
24.180	416.94	417.10	0.16	0.04
26.880	461.08	461.01	-0.07	-0.02
29.770	511.97	511.97	0.00	0.00
32.660	567.25	567.27	0.02	0.00
35.120	617.99	617.93	-0.06	-0.01
38.450	692.04	692.07	0.03	0.00
41.460	764.81	764.87	0.06	0.01
41.510	766.17	766.13	-0.04	0.00

STANDARD DEVIATION = 0.08

2,3-DIMETHYL-2-BUTENE C6H12

SCOTT D.W.,AT AL.:
J.AM.CHEM.SOC. 77,4993(1955).

A = 6.93690 B = 1208.041 C = 224.627

B.P.(760) = 73.207

T	P EXPTL.	P CALCD.	DEV.	PERCENT
29.026	149.41	149.39	-0.02	-0.01
34.404	187.57	187.59	0.02	0.01
39.820	233.72	233.73	0.01	0.01
45.278	289.13	289.14	0.01	0.00
50.778	355.22	355.21	-0.01	0.00
56.321	433.56	433.54	-0.02	-0.01
61.909	525.86	525.87	0.01	0.00
67.535	633.99	633.97	-0.02	0.00
73.208	760.00	760.02	0.02	0.00

STANDARD DEVIATION = 0.02

1-HEXENE C6H12

FORZIATI A.F.,CAMIN D.L.,ROSSINI F.D.:
J.RESEARCH NATL.BUR.STANDARDS 45,406(1950).

A = 6.86880 B = 1154.646 C = 226.046

B.P.(760) = 63.485

T	P EXPTL.	P CALCD.	DEV.	PERCENT
15.890	124.85	124.82	-0.03	-0.03
19.950	149.61	149.64	0.03	0.02
23.720	176.15	176.15	0.00	0.00
28.762	217.44	217.45	0.01	0.01
33.399	262.04	262.03	-0.01	0.00
38.993	325.27	325.30	0.03	0.01
44.763	402.82	402.82	0.00	0.00
50.914	501.03	500.96	-0.07	-0.01
62.323	732.42	732.39	-0.03	0.00
62.827	744.29	744.27	-0.02	0.00
63.299	755.54	755.52	-0.02	0.00
63.837	768.49	768.52	0.03	0.00
64.311	780.03	780.11	0.08	0.01

STANDARD DEVIATION = 0.04

METHYLCYCLOPENTANE C6H12

WILLINGHAM C.J.,TAYLOR W.J.,PIGNOCCO J.M.,ROSSINI F.D.: J.RESEARCH NATL.BUR.STANDARDS 35,219(1945).

A = 6.86640 B = 1188.050 C = 226.276

B.P.(760) = 71.811

T	P EXPTL.	P CALCD.	DEV.	PERCENT
15.035	87.73	87.71	-0.02	-0.02
18.642	103.66	103.65	-0.01	-0.01
22.757	124.65	124.66	0.01	0.01
26.935	149.40	149.43	0.03	0.02
30.816	175.91	175.91	0.00	0.00
36.013	217.20	217.19	-0.01	0.00
40.791	261.74	261.74	0.00	0.00
46.552	324.95	324.95	0.00	0.00
52.499	402.46	402.44	-0.02	0.00
58.847	500.73	500.71	-0.02	0.00
65.739	627.96	627.96	0.00	0.00
70.604	732.11	732.15	0.04	0.01
71.128	744.10	744.14	0.04	0.01
71.612	755.29	755.34	0.05	0.01
72.150	768.05	767.95	-0.10	-0.01
72.634	779.44	779.43	-0.01	0.00

STANDARD DEVIATION = 0.04

1,2-BIS-DIFLUOROAMINO-4-METHYLPENTANE C6H12F4N2

GOOD W.D.,ET AL.: J.PHYS.CHEM. 67,1312(1963).

A = 8.00911 B = 1944.924 C = 245.439

B.P.(10) = 32.046

T	P EXPTL.	P CALCD.	DEV.	PERCENT
-20.000	0.24	0.24	0.00	2.08
-15.000	0.36	0.37	0.01	1.84
-10.000	0.56	0.56	0.00	0.02
-5.000	0.84	0.83	-0.01	-1.32
0.000	1.24	1.22	-0.02	-1.56
5.000	1.74	1.75	0.01	0.40
10.000	2.47	2.48	0.02	0.71
15.000	3.48	3.48	0.00	-0.02
20.000	4.81	4.81	0.00	-0.10

STANDARD DEVIATION = 0.01

BUTYL ACETATE C6H12O2

KLIMENT V.,FRIED V.,PICK J.: COLL.CZECH.CHEM.COMMUN. 29,2008(1964).

A = 7.12712 B = 1430.418 C = 210.745

B.P.(760) = 126.116

T	P EXPTL.	P CALCD.	DEV.	PERCENT
59.740	69.08	68.99	-0.09	-0.13
62.530	78.20	78.12	-0.08	-0.10
65.520	89.08	89.01	-0.07	-0.08
67.960	98.70	98.80	0.10	0.10
71.700	115.30	115.53	0.23	0.20
75.880	136.90	136.95	0.05	0.04
80.490	164.20	164.27	0.07	0.04
84.210	189.50	189.45	-0.05	-0.02
89.580	231.50	231.32	-0.18	-0.08
95.060	281.60	281.56	-0.04	-0.02
100.530	340.20	340.23	0.03	0.01
107.070	423.40	422.99	-0.41	-0.10
113.620	521.20	521.45	0.25	0.05
120.620	645.45	646.21	0.76	0.12
126.090	760.00	759.41	-0.59	-0.08

STANDARD DEVIATION = 0.33

CAPROIC ACID C6H12O2

ROSE A.,ACCIARI J.A.,JOHNSON R.C.,SANDERS N.N.: IND.ENG.CHEM.49,104(1957).

A = 6.92485 B = 1340.830 C = 126.625

B.P.(100) = 145.633

T	P EXPTL.	P CALCD.	DEV.	PERCENT
98.100	9.40	9.08	-0.32	-3.36
108.800	17.50	16.96	-0.54	-3.07
117.000	26.60	26.37	-0.23	-0.85
123.300	36.30	36.30	0.00	0.00
126.200	41.40	41.83	0.43	1.03
129.800	49.40	49.65	0.25	0.50
136.600	67.30	67.76	0.46	0.69
141.100	82.10	82.53	0.43	0.52
144.000	94.30	93.38	-0.92	-0.97
145.500	99.80	99.45	-0.35	-0.36
154.000	141.80	140.23	-1.57	-1.11
160.800	180.30	181.92	1.62	0.90
163.500	202.60	201.04	-1.56	-0.77
166.700	222.40	225.79	3.39	1.52
170.300	255.90	256.52	0.62	0.24
174.700	298.20	298.58	0.38	0.13
179.100	348.80	346.02	-2.78	-0.80

STANDARD DEVIATION = 1.45

ISOCAPROIC ACID $C_6H_{12}O_2$

KAHLBAUM G.W.A.:

Z.PHYS.CHEM. 13,14(1894).

A = 6.25830 B = 1038.647 C = 102.711

B.P.(10) = 94.814

T	P EXPTL.	P CALCD.	DEV.	PERCENT
95.900	11.00	10.68	-0.32	-2.87
97.100	11.40	11.49	0.09	0.75
96.400	11.50	11.01	-0.49	-4.24
97.000	11.60	11.42	-0.18	-1.58
97.000	11.60	11.42	-0.18	-1.58
97.500	11.90	11.76	-0.14	-1.14
98.100	12.20	12.19	-0.01	-0.07
99.200	13.40	13.01	-0.39	-2.92
99.300	13.50	13.09	-0.42	-3.07
99.600	13.50	13.32	-0.18	-1.36
99.900	13.80	13.55	-0.25	-1.80
100.800	14.50	14.28	-0.22	-1.53
101.300	15.00	14.70	-0.30	-2.03
101.900	15.20	15.21	0.01	0.06
102.700	16.20	15.92	-0.28	-1.74
104.600	17.80	17.71	-0.09	-0.51
105.400	18.80	18.51	-0.29	-1.53
105.600	18.80	18.72	-0.08	-0.43
105.600	19.20	18.72	-0.48	-2.51
106.200	19.40	19.35	-0.05	-0.28
106.200	19.50	19.35	-0.15	-0.79
107.000	19.70	20.21	0.51	2.59
107.900	21.10	21.22	0.12	0.56
109.100	22.80	22.63	-0.17	-0.75
110.200	23.80	23.99	0.19	0.79
111.000	25.20	25.02	-0.18	-0.72
111.800	26.20	26.08	-0.12	-0.44
112.400	26.20	26.91	0.71	2.70
113.200	27.30	28.04	0.74	2.71
114.300	29.50	29.66	0.16	0.54
115.200	30.40	31.04	0.64	2.11
116.200	32.20	32.64	0.44	1.36
116.700	34.10	33.46	-0.64	-1.88
117.500	34.90	34.81	-0.09	-0.26
118.700	37.30	36.92	-0.38	-1.02
120.400	39.50	40.09	0.59	1.49
121.100	40.90	41.46	0.56	1.36
121.400	41.50	42.05	0.55	1.33
122.400	43.70	44.09	0.39	0.90
123.000	45.10	45.36	0.26	0.57
123.900	46.90	47.31	0.41	0.87
125.200	50.10	50.24	0.14	0.28
126.300	52.10	52.84	0.74	1.42
127.300	54.90	55.29	0.39	0.72
128.100	56.80	57.32	0.52	0.92
128.600	58.70	58.62	-0.08	-0.14
129.300	61.10	60.48	-0.62	-1.02
129.600	62.40	61.29	-1.11	-1.78
130.000	62.70	62.38	-0.32	-0.51
133.200	72.70	71.71	-0.99	-1.36

STANDARD DEVIATION = 0.44

CYCLOHEXANOL C6H12O

NOVAK J.,MATOUS J.,PICK J.:
COLLECTION CZECH.CHEM.COMMUN.25,583(1960).

A = 6.25530 B = 912.866 C = 109.126

B.P.(760) = 161.394

T	P EXPTL.	P CALCD.	DEV.	PERCENT
93.730	53.83	56.92	3.09	5.73
97.020	69.08	67.15	-1.93	-2.79
99.710	78.20	76.57	-1.63	-2.08
102.520	89.08	87.52	-1.56	-1.75
104.830	98.70	97.43	-1.27	-1.29
108.820	115.30	116.63	1.33	1.15
112.700	136.90	138.06	1.16	0.85
116.860	164.20	164.37	0.17	0.10
120.350	189.50	189.34	-0.16	-0.08
125.540	231.50	231.86	0.36	0.16
130.950	281.60	283.72	2.12	0.75
136.220	340.20	342.43	2.23	0.65
142.410	423.40	422.80	-0.60	-0.14
148.720	521.20	518.73	-2.47	-0.47
155.710	645.40	643.26	-2.14	-0.33
160.700	744.00	744.96	0.96	0.13

STANDARD DEVIATION = 1.85

DIACETONE ALCOHOL C6H12O2

FUGE E.T.J.,BOWDEN S.T.,JONES W.J.:
J.PHYS.CHEM. 56,1013(1952).

A = 8.50242 B = 2400.556 C = 263.792

B.P.(100) = 105.387

T	P EXPTL.	P CALCD.	DEV.	PERCENT
28.100	2.20	1.90	-0.30	-13.73
41.400	4.50	4.33	-0.17	-3.73
51.200	7.90	7.61	-0.29	-3.66
61.700	13.00	13.40	0.40	3.11
71.900	22.50	22.46	-0.04	-0.19
81.700	35.70	35.83	0.13	0.35
91.100	55.50	54.73	-0.77	-1.39
102.000	86.00	87.05	1.05	1.23
114.700	145.00	144.54	-0.46	-0.32

STANDARD DEVIATION = 0.62

ETHYL PROPYL KETONE C6H12O

COLLERSON R.R., ET AL.:
J.CHEM.SOC. 3697(1965).

A = 7.00082 B = 1365.792 C = 208.006

B.P.(760) = 123.496

T	P EXPTL.	P CALCD.	DEV.	PERCENT
75.613	153.19	153.19	0.00	0.00
82.486	199.14	199.15	0.01	0.00
88.481	247.87	247.88	0.01	0.00
93.872	299.58	299.57	-0.01	0.00
98.385	349.27	349.25	-0.01	0.00
102.558	400.90	400.90	0.00	0.00
106.128	449.80	449.80	0.00	0.00
109.419	498.99	498.99	0.00	0.00
112.558	549.82	549.83	0.01	0.00
115.416	599.62	599.62	0.00	0.00
118.125	650.07	650.06	-0.01	0.00
120.532	697.61	697.65	0.04	0.01
123.080	750.99	750.98	-0.01	0.00
125.345	801.07	801.05	-0.02	0.00
127.509	851.28	851.31	0.03	0.00
129.522	900.24	900.25	0.01	0.00
131.435	948.81	948.79	-0.02	0.00
133.365	999.80	999.80	0.00	0.00

STANDARD DEVIATION = 0.02

METHYL ISOBUTYL KETONE C6H12O

FUGE E.T.J.,BOWDEN S.T.,JONES W.J.:
J.PHYS.CHEM. 56,1013(1952).

A = 6.67272 B = 1168.408 C = 191.944

B.P.(760) = 116.188

T	P EXPTL.	P CALCD.	DEV.	PERCENT
21.700	16.50	15.99	-0.51	-3.11
32.700	29.50	29.62	0.12	0.40
41.500	47.00	46.52	-0.48	-1.02
50.200	69.50	70.38	0.88	1.27
60.800	112.50	112.16	-0.34	-0.31
70.000	162.00	163.00	1.00	0.62
80.100	240.00	238.67	-1.33	-0.56
90.900	348.00	348.17	0.17	0.05
116.200	760.00	760.25	0.25	0.03

STANDARD DEVIATION = 0.84

CYCLOPENTYL-1-THIAETHANE $C_6H_{12}S$

OSBORN A.N.,DOUSLIN D.R.:
J.CHEM.ENG.DATA 11,502(1966).

A = 6.94083 B = 1480.703 C = 208.468

B.P.(760) = 156.235

T	P EXPTL.	P CALCD.	DEV.	PERCENT
82.745	71.87	71.82	-0.05	-0.07
85.956	81.64	81.60	-0.04	-0.05
89.177	92.52	92.49	-0.03	-0.03
92.416	104.63	104.63	0.00	0.00
95.654	118.06	118.05	-0.01	-0.01
98.917	132.95	132.96	0.01	0.01
102.197	149.41	149.48	0.07	0.05
108.769	187.57	187.64	0.07	0.04
115.396	233.72	233.80	0.08	0.03
122.068	289.13	289.15	0.02	0.01
128.797	355.22	355.23	0.01	0.00
135.575	433.56	433.51	-0.05	-0.01
142.411	525.86	525.83	-0.03	-0.01
149.293	633.99	633.91	-0.08	-0.01
156.230	760.00	759.89	-0.11	-0.01
163.222	906.06	905.99	-0.07	-0.01
170.265	1074.60	1074.50	-0.10	-0.01
177.364	1268.00	1268.07	0.07	0.01
184.514	1489.10	1489.24	0.14	0.01
191.714	1740.80	1740.83	0.03	0.00
198.970	2026.00	2026.05	0.05	0.00

STANDARD DEVIATION = 0.07

CYCLOHEXANETHIOL C6H12S

OSBORN A.N.,DOUSLIN D.R.: J.CHEM.ENG.DATA 11,502(1966).

A = 6.88673 B = 1476.704 C = 209.828

B.P.(760) = 158.803

	T	P EXPTL.	P CALCD.	DEV.	PERCENT
	83.740	71.87	71.87	0.00	0.00
	87.006	81.64	81.64	-0.01	-0.01
*	90.189	92.52	92.18	-0.34	-0.36
	93.576	104.63	104.62	-0.01	-0.01
	96.881	118.06	118.05	-0.01	-0.01
	100.201	132.95	132.93	-0.02	-0.02
	103.549	149.41	149.45	0.04	0.03
	110.259	187.57	187.62	0.05	0.03
	117.023	233.72	233.75	0.03	0.01
	123.843	289.13	289.13	0.00	0.00
	130.719	355.22	355.19	-0.03	-0.01
	137.654	433.56	433.51	-0.05	-0.01
	144.647	525.86	525.82	-0.04	-0.01
	151.695	633.99	633.94	-0.05	-0.01
	158.803	760.00	760.00	0.00	0.00
	165.968	906.06	906.13	0.07	0.01
	173.186	1074.60	1074.59	-0.01	0.00
	180.464	1268.00	1268.05	0.05	0.00
	187.801	1489.10	1489.20	0.10	0.01
	195.196	1740.80	1740.84	0.04	0.00
	202.645	2026.00	2025.83	-0.17	-0.01

STANDARD DEVIATION = 0.06

1-CHLOROHEXANE C6H13CL

KEMME R.H.,KREPS S.I.:
J.CHEM.ENG.DATA 14,98(1969).

A = 7.05136 B = 1461.716 C = 215.566

B.P.(760) = 134.920

T	P EXPTL.	P CALCD.	DEV.	PERCENT
15.000	5.00	5.15	0.15	2.96
26.700	10.40	10.42	0.02	0.18
33.100	15.00	14.90	-0.10	-0.68
38.600	20.10	19.97	-0.13	-0.66
46.700	30.10	30.06	-0.04	-0.15
52.600	39.90	39.86	-0.04	-0.10
59.700	55.00	55.10	0.10	0.18
68.400	79.80	80.14	0.34	0.42
79.500	125.60	125.17	-0.43	-0.35
92.600	203.20	203.27	0.07	0.03
104.600	305.70	306.09	0.39	0.13
120.300	500.90	500.34	-0.56	-0.11
135.600	774.00	774.25	0.25	0.03

STANDARD DEVIATION = 0.30

CYCLOHEXYLAMINE C6H13N

NOVAK J.,MATOUS J.,PICK J.:
COLLECTION CZECH.CHEM.COMMUN.25, 2405(1960).

A = 6.68954 B = 1229.418 C = 188.802

B.P.(100) = 73.360

T	P EXPTL.	P CALCD.	DEV.	PERCENT
60.720	58.83	57.87	-0.96	-1.63
64.390	69.08	68.21	-0.87	-1.26
67.480	78.20	78.06	-0.14	-0.18
70.570	89.08	89.03	-0.05	-0.05
73.020	98.70	98.61	-0.09	-0.09
76.910	115.30	115.52	0.22	0.19
81.290	136.90	137.30	0.40	0.30
86.100	164.20	164.94	0.74	0.45
90.000	189.50	190.50	1.00	0.53
95.620	231.50	232.82	1.32	0.57
101.310	281.60	283.01	1.41	0.50
107.100	340.20	342.55	2.35	0.69
113.900	432.40	424.67	-7.73	-1.79
120.450	521.20	517.70	-3.50	-0.67
128.370	645.50	650.65	5.15	0.80

STANDARD DEVIATION = 3.04

2-METHYLPIPERIDINE C6H13N

OSBORN A.G.,DOUSLIN D.R.: J.CHEM.ENG.DATA13,534(1968).

A = 6.81859 B = 1274.612 C = 205.404

B.P.(760) = 118.284

T	P EXPTL.	P CALCD.	DEV.	PERCENT
51.480	71.87	71.91	0.04	0.05
54.374	81.64	81.67	0.03	0.03
57.280	92.52	92.54	0.02	0.02
60.197	104.63	104.62	-0.01	-0.01
63.132	118.06	118.05	-0.01	-0.01
66.078	132.95	132.91	-0.04	-0.03
69.048	149.41	149.41	0.00	0.00
75.011	187.57	187.56	-0.01	-0.01
81.035	233.72	233.74	0.02	0.01
87.103	289.13	289.09	-0.04	-0.01
93.229	355.22	355.16	-0.06	-0.02
99.412	433.56	433.52	-0.04	-0.01
105.650	525.86	525.85	-0.01	0.00
111.941	633.99	634.01	0.02	0.00
118.286	760.00	760.03	0.03	0.00
124.689	906.06	906.17	0.11	0.01
131.146	1074.60	1074.72	0.12	0.01
137.655	1268.00	1268.10	0.10	0.01
144.222	1489.10	1489.16	0.06	0.00
150.843	1740.80	1740.59	-0.21	-0.01
157.529	2026.00	2025.86	-0.14	-0.01

STANDARD DEVIATION = 0.08

2,2-DIMETHYBUTANE C6H14

WILLINGHAM C.J.,TAYLOR W.J.,PIGNOCCO J.M.,ROSSINI F.D.:
J.RESEARCH NATL.BUR.STANDARDS 35,219(1945).

A = 6.76189 B = 1084.935 C = 229.804

B.P.(760) = 49.740

T	P EXPTL.	P CALCD.	DEV.	PERCENT
15.376	217.18	217.19	0.01	0.00
19.946	261.75	261.71	-0.04	-0.02
25.472	324.93	324.97	0.04	0.01
31.175	402.44	402.46	0.02	0.00
37.269	500.71	500.70	-0.01	0.00
43.893	627.95	627.90	-0.05	-0.01
48.575	732.09	732.09	0.00	0.00
49.078	744.07	744.03	-0.04	0.00
49.544	755.26	755.24	-0.02	0.00
50.074	768.00	768.14	0.14	0.02
50.529	779.38	779.34	-0.04	-0.01

STANDARD DEVIATION = 0.06

2,2-DIMETHYLBUTANE C6H14

KILPATRICK J.E.,PITZER K.S.:
J.AM.CHEM.SOC. 68,1066(1946).

A = 6.76526 B = 1087.261 C = 230.172

B.P.(100) = -2.008

T	P EXPTL.	P CALCD.	DEV.	PERCENT
-62.160	1.91	1.97	0.06	3.01
-53.410	4.05	4.11	0.06	1.58
-43.250	8.90	8.88	-0.02	-0.18
-32.460	18.52	18.45	-0.07	-0.37
-22.870	33.11	33.15	0.04	0.11
-14.840	52.05	52.00	-0.05	-0.09
-6.180	81.45	81.52	0.07	0.08
0.000	110.12	110.04	-0.08	-0.07
2.140	121.60	121.64	0.04	0.03
9.710	170.85	170.91	0.06	0.04
16.230	225.33	225.27	-0.06	-0.03

STANDARD DEVIATION = 0.07

2,3-DIMETHYLBUTANE C6H14

WILLINGHAM C.B.,TAYLOR W.J.,PIGNOCCO J.M.,ROSSINI F.D.: J.RESEARCH NATL.BUR.STANDARDS 35,219(1945).

A = 6.81832 B = 1131.833 C = 229.462

B.P.(760) = 57.987

T	P EXPTL.	P CALCD.	DEV.	PERCENT
14.256	149.42	149.38	-0.04	-0.03
18.044	175.90	175.94	0.04	0.02
23.099	217.18	217.21	0.03	0.02
27.746	261.75	261.73	-0.02	-0.01
33.357	324.93	324.94	0.01	0.00
39.150	402.44	402.42	-0.02	-0.01
45.339	500.71	500.70	-0.01	0.00
52.060	627.94	627.92	-0.02	0.00
56.806	732.09	732.08	-0.01	0.00
57.317	744.07	744.06	-0.01	0.00
57.790	755.26	755.28	0.02	0.00
58.320	767.99	768.00	0.01	0.00
58.789	779.37	779.40	0.03	0.00

STANDARD DEVIATION = 0.03

HEXANE C6H14

WILLINGHAM C.J.,TAYLOR W.J.,PIGNOCCO J.M.,ROSSINI F.D.: J.RESEARCH NATL.BUR.STANDARDS 35,219(1945).

A = 6.88555 B = 1175.817 C = 224.867

B.P.(760) = 68.740

T	P EXPTL.	P CALCD.	DEV.	PERCENT
13.033	87.74	87.71	-0.03	-0.03
16.576	103.65	103.65	0.00	0.00
20.618	124.66	124.67	0.01	0.01
24.717	149.41	149.43	0.02	0.01
28.528	175.90	175.91	0.01	0.01
33.631	217.19	217.22	0.03	0.01
38.311	261.74	261.69	-0.05	-0.02
43.967	324.94	324.92	-0.02	-0.01
49.803	402.45	402.45	0.00	0.00
56.030	500.72	500.74	0.02	0.00
62.785	627.95	627.93	-0.02	0.00
67.554	732.10	732.10	0.00	0.00
68.067	744.09	744.07	-0.02	0.00
68.540	755.27	755.24	-0.03	0.00
69.081	768.02	768.18	0.16	0.02
69.541	779.41	779.31	-0.10	-0.01

STANDARD DEVIATION = 0.06

2-METHYLPENTANE C6H14

WILLINGHAM C.J.,TAYLOR W.J.,PIGNOCCO J.M.,ROSSINI F.D.: J.RESEARCH NATL.BUR.STANDARDS 35,219(1945).

A = 6.86830 B = 1151.401 C = 228.477

B.P.(760) = 60.270

T	P EXPTL.	P CALCD.	DEV.	PERCENT
12.758	124.66	124.58	-0.08	-0.07
16.820	149.42	149.44	0.02	0.02
20.584	175.90	175.96	0.06	0.03
25.617	217.18	217.26	0.08	0.04
30.237	261.75	261.76	0.01	0.00
35.810	324.93	324.90	-0.03	-0.01
41.567	402.44	402.37	-0.07	-0.02
47.714	500.70	500.64	-0.06	-0.01
54.388	627.94	627.89	-0.05	-0.01
59.099	732.09	732.10	0.01	0.00
59.607	744.07	744.09	0.02	0.00
60.074	755.26	755.25	0.00	0.00
60.602	767.99	768.04	0.05	0.01
61.066	779.37	779.41	0.04	0.01

STANDARD DEVIATION = 0.06

3-METHYLPENTANE C6H14

WILLINGHAM C.J.,TAYLOR W.J.,PIGNOCCO J.M.,ROSSINI F.D.: J.RESEARCH NATL.BUR.STANDARDS 35,219(1945).

A = 6.86532 B = 1161.429 C = 228.205

B.P.(760) = 63.281

T	P EXPTL.	P CALCD.	DEV.	PERCENT
15.290	124.66	124.59	-0.07	-0.05
19.393	149.42	149.46	0.04	0.03
23.189	175.90	175.94	0.04	0.02
28.270	217.18	217.21	0.03	0.02
32.941	261.75	261.75	0.00	0.00
38.574	324.94	324.93	-0.01	0.00
44.389	402.45	402.40	-0.05	-0.01
50.598	500.70	500.67	-0.03	-0.01
57.340	627.94	627.92	-0.02	0.00
62.098	732.10	732.10	0.00	0.00
62.610	744.07	744.07	0.00	0.00
63.084	755.26	755.29	0.03	0.00
63.617	768.00	768.05	0.05	0.01
64.083	779.38	779.36	-0.02	0.00

STANDARD DEVIATION = 0.04

DIPROPYL ETHER $C_6H_{14}O$

CIDLINSKY J.,POLAK J.:
COLL.CZECH.CHEM.COMM. 34,1317(1969).

A = 6.94763 B = 1256.499 C = 219.013

B.P.(760) = 89.950

T	P EXPTL.	P CALCD.	DEV.	PERCENT
26.590	67.81	67.87	0.06	0.08
31.420	85.09	85.18	0.09	0.10
36.480	107.04	107.08	0.04	0.03
40.830	129.40	129.43	0.03	0.02
45.080	154.98	154.82	-0.16	-0.10
46.000	160.99	160.83	-0.16	-0.10
50.470	192.95	192.75	-0.20	-0.10
52.940	212.55	212.50	-0.05	-0.03
55.890	238.22	238.19	-0.03	-0.01
58.820	266.16	266.16	0.00	0.00
60.700	285.29	285.45	0.16	0.06
65.980	345.74	345.74	0.00	0.00
68.760	381.20	381.37	0.17	0.04
71.440	418.14	418.44	0.30	0.07
73.770	452.70	452.96	0.26	0.06
76.270	491.98	492.49	0.51	0.10
80.790	571.19	570.89	-0.30	-0.05
82.130	596.00	595.93	-0.07	-0.01
83.570	624.07	623.81	-0.26	-0.04
85.770	669.11	668.39	-0.72	-0.11
87.340	702.42	701.70	-0.72	-0.10
88.650	729.44	730.50	1.06	0.15

STANDARD DEVIATION = 0.39

DIISOPROPYL ETHER C6H14O

CIDLINSKY J.,POLAK J.: COLL.CZECH.CHEM.COMM. 34,1317(1969).

A = 6.84953 B = 1139.340 C = 218.742

B.P.(760) = 68.338

T	P EXPTL.	P CALCD.	DEV.	PERCENT
23.500	139.92	140.03	0.11	0.08
25.290	151.47	151.60	0.13	0.09
27.420	166.34	166.38	0.04	0.03
32.000	202.24	202.14	-0.10	-0.05
34.220	221.95	221.58	-0.37	-0.17
36.930	247.37	247.33	-0.04	-0.02
41.170	292.56	292.39	-0.17	-0.06
44.080	327.32	326.96	-0.36	-0.11
46.700	360.67	360.82	0.15	0.04
48.570	386.39	386.65	0.26	0.07
50.960	421.57	421.78	0.21	0.05
54.520	478.76	478.76	0.00	0.00
56.120	505.33	506.28	0.95	0.19
58.170	543.25	543.35	0.10	0.02
60.680	591.73	591.61	-0.12	-0.02
63.210	643.83	643.61	-0.22	-0.03
64.660	675.02	674.99	-0.03	0.00
65.750	699.93	699.36	-0.57	-0.08
67.210	733.07	733.08	0.01	0.00

STANDARD DEVIATION = 0.33

ETHYL BUTYL ETHER C6H14O

CIDLINSKY J.POLAK J.: COLL.CZECH.CHEM.COMM. 34,1317(1969).

A = 6.94443 B = 1256.361 C = 216.907

B.P.(760) = 92.266

T	P EXPTL.	P CALCD.	DEV.	PERCENT
38.180	104.35	104.52	0.17	0.16
42.310	125.21	125.22	0.01	0.01
44.000	134.64	134.61	-0.03	-0.02
49.040	166.11	166.08	-0.03	-0.02
52.270	189.46	189.24	-0.22	-0.12
55.100	211.82	211.63	-0.19	-0.09
58.730	243.40	243.44	0.04	0.02
61.250	267.74	267.73	-0.01	0.00
63.220	288.02	288.04	0.02	0.01
65.850	317.06	317.08	0.02	0.01
68.510	348.91	348.81	-0.10	-0.03
71.660	389.47	389.61	0.14	0.04
73.740	418.59	418.59	0.00	0.00
77.120	468.89	469.33	0.44	0.09
79.640	509.79	510.26	0.47	0.09
82.250	555.65	555.59	-0.06	-0.01
85.080	608.10	608.29	0.19	0.03
86.710	640.86	640.39	-0.47	-0.07
89.370	696.47	695.64	-0.83	-0.12
91.380	739.42	739.83	0.41	0.05

STANDARD DEVIATION = 0.31

1-HEXANOL

$C_6H_{14}O$

KEMME R.H.,KREPS S.I.:

J.CHEM.ENG.DATA 14,98(1969).

A = 7.05401 B = 1293.831 C = 152.631

B.P.(760) = 157.402

T	P EXPTL.	P CALCD.	DEV.	PERCENT
52.200	5.70	5.46	-0.24	-4.16
60.400	9.70	9.56	-0.14	-1.42
67.400	15.30	14.92	-0.38	-2.48
72.600	20.50	20.40	-0.10	-0.51
79.200	29.80	29.72	-0.08	-0.26
85.000	40.70	40.67	-0.03	-0.07
91.100	55.30	55.66	0.36	0.66
98.700	80.00	80.55	0.55	0.69
107.600	120.80	120.82	0.02	0.02
119.700	201.00	200.93	-0.07	-0.03
130.100	301.80	300.48	-1.32	-0.44
144.500	500.00	500.72	0.72	0.14
157.300	757.30	757.58	0.28	0.04

STANDARD DEVIATION = 0.55

1-HEXANOL

$C_6H_{14}O$

ROSE A.,SUPINA W.R.:

J.CHEM.ENG.DATA6,173(1961).

A = 8.11170 B = 1872.743 C = 202.666

B.P.(100) = 103.753

T	P EXPTL.	P CALCD.	DEV.	PERCENT
60.900	10.30	10.15	-0.15	-1.50
64.700	12.90	12.80	-0.10	-0.76
69.200	16.70	16.72	0.02	0.11
71.900	19.40	19.54	0.14	0.73
73.700	21.60	21.65	0.05	0.21
80.800	32.00	32.00	0.00	-0.01
85.600	41.10	41.22	0.12	0.29
87.200	44.80	44.77	-0.03	-0.07
91.600	56.10	55.92	-0.18	-0.32
93.500	61.30	61.43	0.13	0.22
95.900	69.10	69.06	-0.04	-0.06
97.900	76.00	76.03	0.03	0.04
101.000	88.00	88.02	0.02	0.02
102.800	95.90	95.70	-0.20	-0.21
106.300	112.10	112.30	0.20	0.18
108.500	124.00	123.94	-0.06	-0.05

STANDARD DEVIATION = 0.13

2-HEXANOL C6H14O

HOVORKA F.,LANKELMA H.P.,STANFORD S.C.:
J.AM.CHEM.SOC. 60,820(1938).

A = 7.26101 B = 1371.734 C = 173.216

B.P.(760) = 139.952

T	P EXPTL.	P CALCD.	DEV.	PERCENT
25.000	3.30	2.19	-1.11	-33.61
35.000	5.40	4.71	-0.69	-12.79
45.000	10.00	9.44	-0.56	-5.62
55.000	18.80	17.80	-1.00	-5.34
65.000	33.00	31.82	-1.18	-3.59
75.000	53.00	54.28	1.28	2.41
85.000	86.00	88.85	2.85	3.31
95.000	140.00	140.19	0.19	0.13
105.000	216.30	214.06	-2.24	-1.04
115.000	320.50	317.40	-3.10	-0.97
125.000	455.00	458.35	3.35	0.74
135.000	647.20	646.31	-0.89	-0.14
140.000	760.50	761.18	0.68	0.09
142.000	811.50	811.47	-0.03	0.00

STANDARD DEVIATION = 1.94

3-HEXANOL C6H14O

HOVORKA F.,LANKELMA H.P.,STANFORD S.C.:
J.AM.CHEM.SOC. 60,820(1938).

A = 7.68850 B = 1669.898 C = 211.810

B.P.(760) = 135.530

T	P EXPTL.	P CALCD.	DEV.	PERCENT
25.000	7.20	4.33	-2.87	-39.81
35.000	10.00	8.37	-1.63	-16.33
45.000	16.10	15.35	-0.75	-4.68
55.000	26.80	26.90	0.10	0.37
65.000	44.00	45.27	1.27	2.89
75.000	71.50	73.48	1.98	2.77
85.000	115.00	115.44	0.44	0.38
95.000	175.60	176.08	0.48	0.27
105.000	264.00	261.53	-2.48	-0.94
115.000	380.70	379.14	-1.56	-0.41
125.000	537.80	537.66	-0.14	-0.03
135.000	746.50	747.25	0.75	0.10
138.000	820.50	821.79	1.29	0.16

STANDARD DEVIATION = 1.68

2-METHYL-1-PENTANOL $C_6H_{14}O$

HOVORKA F.,LANKELMA H.P.,STANFORD S.C.: J.AM.CHEM.SOC. 60,820(1938).

A = 7.52011 B = 1564.687 C = 189.183

B.P.(760) = 148.035

T	P EXPTL.	P CALCD.	DEV.	PERCENT
25.000	2.60	1.64	-0.96	-36.94
35.000	4.20	3.47	-0.73	-17.33
45.000	7.80	6.90	-0.90	-11.58
55.000	13.50	12.95	-0.55	-4.08
65.000	22.60	23.14	0.54	2.39
75.000	38.80	39.57	0.77	1.99
85.000	64.30	65.07	0.77	1.20
95.000	103.50	103.32	-0.18	-0.17
105.000	160.00	158.98	-1.02	-0.64
115.000	237.00	237.80	0.80	0.34
125.000	347.50	346.68	-0.82	-0.24
135.000	494.20	493.80	-0.40	-0.08
145.000	688.50	688.62	0.12	0.02
150.000	806.40	807.24	0.84	0.10

STANDARD DEVIATION = 0.81

2-METHYL-4-PENTANOL $C_6H_{14}O$

HOVORKA F.,LANKELMA H.P.,STANFORD S.C.: J.AM.CHEM.SOC. 60,820(1938).

A = 8.46706 B = 2174.869 C = 257.780

B.P.(760) = 131.545

T	P EXPTL.	P CALCD.	DEV.	PERCENT
25.000	8.20	5.97	-2.23	-27.18
35.000	12.00	10.93	-1.07	-8.89
45.000	21.00	19.23	-1.77	-8.41
55.000	32.60	32.64	0.04	0.12
65.000	52.20	53.60	1.40	2.68
75.000	83.50	85.43	1.93	2.31
85.000	131.10	132.52	1.42	1.08
95.000	201.20	200.50	-0.70	-0.35
105.000	299.00	296.52	-2.48	-0.83
115.000	432.50	429.42	-3.08	-0.71
125.000	609.00	609.95	0.95	0.16
130.000	719.80	722.03	2.23	0.31
133.000	796.70	797.28	0.58	0.07

STANDARD DEVIATION = 1.98

BUTYL CELLOSOLVE $C_6H_{14}O_2$

DYKYJ J.,SEPRAKOVA M.,PAULECH J.: CHEM.ZVESTI 11,461(1957).

A = 6.95659 B = 1399.903 C = 172.154

B.P.(760) = 171.315

T	P EXPTL.	P CALCD.	DEV.	PERCENT
92.600	47.50	46.67	-0.83	-1.75
107.900	90.30	90.76	0.46	0.51
120.900	150.70	151.23	0.53	0.35
130.900	217.00	217.41	0.41	0.19
134.000	242.70	242.13	-0.57	-0.24
146.400	369.80	364.78	-5.02	-1.36
147.600	376.30	378.90	2.60	0.69
147.900	380.00	382.50	2.50	0.66
170.200	737.60	737.12	-0.48	-0.06

STANDARD DEVIATION = 2.59

ISOBUTYL CELLOSOLVE $C_6H_{14}O_2$

DYKYJ J.,SEPRAKOVA M.,PAULECH J.: CHEM.ZVESTI 11,461(1957).

A = 7.69483 B = 1825.987 C = 219.585

B.P.(760) = 159.721

T	P EXPTL.	P CALCD.	DEV.	PERCENT
71.300	26.00	26.15	0.15	0.58
80.500	40.60	40.73	0.13	0.32
99.600	94.60	94.20	-0.40	-0.42
117.400	189.00	188.90	-0.10	-0.05
132.600	322.90	323.66	0.76	0.24
142.900	454.30	454.38	0.08	0.02
151.900	602.40	601.81	-0.59	-0.10
159.000	744.10	744.12	0.02	0.00

STANDARD DEVIATION = 0.48

1,1-DIETHOXYETHANE C6H14O2

NICOLINI E.,LAFFITTE P.:

COMPTES RENDUS 229,757(1949).

A = 6.75763 B = 1191.598 C = 203.115

B.P.(100) = 47.345

T	P EXPTL.	P CALCD.	DEV.	PERCENT
0.000	7.90	7.78	-0.12	-1.51
5.000	10.90	10.76	-0.14	-1.25
10.000	14.70	14.67	-0.03	-0.23
15.000	19.70	19.70	0.00	0.00
20.000	26.10	26.12	0.02	0.06
25.000	34.10	34.20	0.10	0.28
30.000	44.00	44.26	0.26	0.59
35.000	56.50	56.67	0.17	0.30
40.000	71.90	71.82	-0.08	-0.11
45.000	90.50	90.16	-0.34	-0.37
50.000	112.60	112.18	-0.42	-0.38
55.000	138.20	138.39	0.19	0.13
60.000	168.90	169.36	0.46	0.27
65.000	205.60	205.72	0.12	0.06
70.000	248.40	248.11	-0.29	-0.12

STANDARD DEVIATION = 0.26

CARBITOL C6H14O3

GARDNER G.S.,BREWER J.E.: IND.ENG.CHEM.29,179(1937).

A = 7.64081 B = 1801.305 C = 183.968

B.P.(100) = 135.366

T	P EXPTL.	P CALCD.	DEV.	PERCENT
40.400	0.80	0.41	-0.39	-48.79
45.500	0.90	0.62	-0.28	-31.35
50.400	1.30	0.90	-0.40	-30.65
55.000	1.50	1.27	-0.23	-15.50
61.000	1.80	1.94	0.14	7.72
65.600	2.60	2.65	0.05	1.89
70.800	3.70	3.72	0.02	0.52
76.500	5.50	5.31	-0.19	-3.44
79.700	6.40	6.44	0.04	0.67
89.900	11.70	11.58	-0.12	-1.06
100.900	20.60	20.77	0.17	0.84
111.700	34.70	35.36	0.66	1.90
119.800	51.70	51.40	-0.30	-0.59
135.200	99.60	99.32	-0.28	-0.28
151.000	183.20	183.34	0.14	0.08

STANDARD DEVIATION = 0.31

2,3-DIMETHYL-2-BUTANETHIOL C6H14S

OSBORN A.N.,DOUSLIN D.R.:
J.CHEM.ENG.DATA 11,502(1966).

A = 6.83956 B = 1354.240 C = 215.957

B.P.(760) = 126.130

T	P EXPTL.	P CALCD.	DEV.	PERCENT
55.814	71.87	71.87	0.00	0.00
58.867	81.64	81.64	0.00	0.00
61.931	92.52	92.52	0.00	0.00
65.011	104.63	104.63	0.00	0.00
68.099	118.06	118.05	-0.01	-0.01
71.208	132.95	132.94	-0.01	0.00
74.334	149.41	149.44	0.03	0.02
80.613	187.57	187.60	0.03	0.01
86.949	233.72	233.74	0.02	0.01
93.338	289.13	289.13	0.00	0.00
99.783	355.22	355.19	-0.03	-0.01
106.283	433.56	433.49	-0.07	-0.02
112.843	525.86	525.81	-0.05	-0.01
119.458	633.99	633.96	-0.04	-0.01
126.129	760.00	759.97	-0.03	0.00
132.858	906.06	906.07	0.01	0.00
139.644	1074.60	1074.61	0.01	0.00
146.492	1268.00	1268.24	0.24	0.02
153.391	1489.10	1489.35	0.25	0.02
160.344	1740.80	1740.78	-0.02	0.00
167.355	2026.00	2025.66	-0.34	-0.02

STANDARD DEVIATION = 0.12

2,3-DIMETHYL-2-BUTANETHIOL C6H14S

OSBORN A.N.,DOUSLIN D.R.:
J.CHEM.ENG.DATA 11,502(1966).

A = 8.76937 B = 2428.617 C = 296.809

B.P.(10) = 15.779

T	P EXPTL.	P CALCD.	DEV.	PERCENT
12.493	8.27	8.27	0.00	-0.01
15.000	9.56	9.56	0.00	0.03
17.500	11.03	11.03	0.00	-0.01
19.999	12.69	12.69	0.00	0.00

STANDARD DEVIATION = 0.00

2,4-DIMETHYL-3-THIAPENTANE C6H14S

OSBORN A.N.,DOUSLIN D.R.:
J.CHEM.ENG.DATA 11,502(1966).

A = 6.87147 B = 1326.832 C = 212.478

B.P.(760) = 120.007

T	P EXPTL.	P CALCD.	DEV.	PERCENT
52.101	71.87	71.88	0.01	0.01
55.052	81.64	81.64	0.00	0.00
58.010	92.52	92.50	-0.02	-0.02
60.990	104.63	104.62	-0.01	-0.01
63.980	118.06	118.05	-0.01	-0.01
66.983	132.95	132.94	-0.01	-0.01
70.011	149.41	149.46	0.05	0.04
76.075	187.57	187.60	0.03	0.02
82.193	233.72	233.73	0.01	0.00
88.366	289.13	289.14	0.01	0.00
94.588	355.22	355.20	-0.02	-0.01
100.863	433.56	433.51	-0.05	-0.01
107.192	525.86	525.82	-0.04	-0.01
113.574	633.99	633.99	0.00	0.00
120.007	760.00	760.00	0.00	0.00
126.493	906.06	906.09	0.03	0.00
133.032	1074.60	1074.62	0.02	0.00
139.623	1268.00	1268.05	0.05	0.00
146.269	1489.10	1489.18	0.08	0.01
152.965	1740.80	1740.67	-0.13	-0.01
* 159.712	2026.00	2025.51	-0.49	-0.02

STANDARD DEVIATION = 0.05

1-HEXANETHIOL C6H14S

OSBORN A.N.,DOUSLIN D.R.:
J.CHEM.ENG.DATA 11,502(1966).

A = 6.94505 B = 1452.908 C = 204.825

B.P.(760) = 152.661

T	P EXPTL.	P CALCD.	DEV.	PERCENT
80.694	71.87	71.85	-0.03	-0.03
83.837	81.64	81.62	-0.02	-0.02
86.991	92.52	92.51	-0.01	-0.01
90.157	104.63	104.63	0.00	0.00
93.334	118.06	118.07	0.01	0.01
96.530	132.95	132.99	0.04	0.03
99.733	149.41	149.45	0.04	0.03
106.168	187.57	187.59	0.02	0.01
112.658	233.72	233.73	0.01	0.01
119.198	289.13	289.13	0.00	0.00
125.789	355.22	355.20	-0.02	0.00
132.425	433.56	433.46	-0.10	-0.02
139.121	525.86	525.80	-0.06	-0.01
145.866	633.99	633.97	-0.02	0.00
152.659	760.00	759.96	-0.04	0.00
159.507	906.06	906.11	0.05	0.01
166.403	1074.60	1074.64	0.04	0.00
173.351	1268.00	1268.14	0.14	0.01
180.349	1489.10	1489.25	0.15	0.01
187.397	1740.80	1740.81	0.01	0.00
194.494	2026.00	2025.76	-0.24	-0.01

STANDARD DEVIATION = 0.08

2-METHYL-2-PENTANETHIOL C6H14S

OSBORN A.N.,DOUSLIN D.R.: J.CHEM.ENG.DATA 11,502(1966).

A = 6.85847 B = 1343.792 C = 212.802

B.P.(760) = 125.033

T	P EXPTL.	P CALCD.	DEV.	PERCENT
55.855	71.87	71.87	0.00	0.01
58.860	81.64	81.64	0.00	0.00
61.877	92.52	92.52	0.00	0.00
64.907	104.63	104.62	-0.01	-0.01
67.949	118.06	118.04	-0.02	-0.01
71.008	132.95	132.93	-0.02	-0.01
74.089	149.41	149.45	0.04	0.02
80.269	187.57	187.61	0.04	0.02
86.502	233.72	233.74	0.02	0.01
92.787	289.13	289.12	-0.01	0.00
99.127	355.22	355.19	-0.03	-0.01
105.521	433.56	433.50	-0.06	-0.01
111.972	525.86	525.83	-0.03	-0.01
118.475	633.99	633.96	-0.03	0.00
125.032	760.00	759.97	-0.03	0.00
131.646	906.06	906.10	0.04	0.00
138.314	1074.60	1074.64	0.04	0.00
145.037	1268.00	1268.14	0.14	0.01
151.815	1489.10	1489.28	0.18	0.01
158.645	1740.80	1740.77	-0.03	0.00
165.531	2026.00	2025.76	-0.24	-0.01

STANDARD DEVIATION = 0.08

4,5-DITHIAOCTANE C6H14S2

HUBBARD W.N.,ET AL.: J.AM.CHEM.SOC. 80,3547(1958).

A = 6.95981 B = 1593.177 C = 194.713

B.P.(100) = 126.504

T	P EXPTL.	P CALCD.	DEV.	PERCENT
117.472	71.87	71.86	-0.01	-0.01
120.869	81.64	81.55	-0.09	-0.11
124.340	92.52	92.54	0.02	0.03
127.798	104.63	104.69	0.06	0.05
131.251	118.06	118.09	0.03	0.03
134.732	132.95	133.00	0.05	0.04
138.207	149.41	149.40	-0.01	-0.01
145.223	187.57	187.55	-0.02	-0.01
152.297	233.72	233.69	-0.03	-0.01
159.426	289.13	289.12	-0.01	-0.01
166.604	355.22	355.18	-0.04	-0.01
173.848	433.56	433.62	0.06	0.01

STANDARD DEVIATION = 0.05

TRIETHYLALUMINIUM C6H15AL

FIC V.,DVORAK J.: CHEM.PRUM. 15,732(1965).

A = 11.64611 B = 4466.578 C = 322.871

B.P.(10) = 96.679

T	P EXPTL.	P CALCD.	DEV.	PERCENT
57.000	0.80	0.77	-0.03	-3.42
59.000	0.90	0.89	-0.01	-1.08
60.000	1.00	0.96	-0.04	-4.48
61.500	1.10	1.06	-0.04	-3.57
62.500	1.20	1.14	-0.06	-5.25
63.500	1.30	1.22	-0.08	-6.28
66.000	1.50	1.45	-0.05	-3.62
70.000	2.00	1.89	-0.11	-5.38
82.000	4.00	4.11	0.11	2.79
96.000	10.00	9.61	-0.39	-3.90
109.000	20.00	20.12	0.12	0.62
116.500	30.00	30.22	0.22	0.73
122.000	40.00	40.36	0.36	0.90
126.000	50.00	49.59	-0.41	-0.82

STANDARD DEVIATION = 0.22

TRIETHYLBORINE C6H15B

STOCK A.,ZEIDLER F.: BER.54,531(1921).

A = 5.65073 B = 694.266 C = 151.874

B.P.(100) = 38.298

T	P EXPTL.	P CALCD.	DEV.	PERCENT
0.000	12.50	12.01	-0.49	-3.95
49.500	158.00	159.62	1.62	1.02
59.000	229.00	228.25	-0.75	-0.33
64.900	282.00	280.55	-1.45	-0.51
69.900	332.00	331.30	-0.70	-0.21
74.700	386.00	385.95	-0.05	-0.01
77.800	423.00	424.52	1.52	0.36

STANDARD DEVIATION = 1.44

TRIETHYL BORATE C6H15BO3

CHRISTOPHER P.M.,SHILMAN A.:
J.CHEM.ENG.DATA 12,333(1967).

A = 7.51113 B = 1641.651 C = 236.283

B.P.(100) = 61.596

T	P EXPTL.	P CALCD.	DEV.	PERCENT
29.100	20.30	21.14	0.84	4.15
39.000	36.00	35.29	-0.71	-1.98
44.900	47.30	47.07	-0.23	-0.48
55.900	79.00	78.08	-0.92	-1.16
60.300	93.30	94.60	1.30	1.40
71.200	149.30	148.64	-0.66	-0.44
80.000	212.00	209.25	-2.75	-1.30
84.600	247.00	248.36	1.36	0.55
88.900	288.00	290.22	2.22	0.77
96.000	369.00	372.05	3.05	0.83
104.200	488.00	489.32	1.32	0.27
108.700	570.80	565.56	-5.24	-0.92

STANDARD DEVIATION = 2.52

TRIETHYLAMINE C6H15N

BITTRICH H.J.,KAUER E.:
Z.PHYS.CHEM. 219,226(1962).

A = 5.85879 B = 695.666 C = 144.832

B.P.(760) = 88.772

T	P EXPTL.	P CALCD.	DEV.	PERCENT
50.000	195.00	194.17	-0.83	-0.42
55.000	237.50	238.52	1.02	0.43
60.000	289.50	290.07	0.57	0.20
65.000	350.00	349.49	-0.51	-0.15
68.500	400.00	396.12	-3.88	-0.97
72.350	450.00	452.52	2.52	0.56
75.500	500.00	502.84	2.84	0.57
78.200	550.00	549.10	-0.90	-0.16
81.000	600.00	600.24	0.24	0.04
83.550	650.00	649.71	-0.29	-0.04
86.000	700.00	699.92	-0.08	-0.01
88.500	755.00	753.95	-1.05	-0.14
88.850	760.00	761.74	1.74	0.23
90.550	800.00	800.40	0.40	0.05
92.650	850.00	850.04	0.04	0.01
94.600	900.00	898.05	-1.95	-0.22

STANDARD DEVIATION = 1.77

TRIETHANOLAMINE C6H15NO3

MCDONALD R.A.,SHRADER S.A.,STULL D.R.:
J.CHEM.ENG.DATA 4,311(1959).

A = 10.06754 B = 4542.777 C = 297.763

B.P.(100) = 265.330

T	P EXPTL.	P CALCD.	DEV.	PERCENT
252.700	65.31	65.30	-0.01	-0.02
256.900	75.44	75.40	-0.04	-0.05
260.600	85.71	85.44	-0.27	-0.32
264.000	95.27	95.70	0.43	0.45
266.900	105.38	105.30	-0.08	-0.08
305.600	345.51	345.50	-0.01	0.00

STANDARD DEVIATION = 0.30

PENTAMETHYLDISILANYL CYANIDE C6H15NSI2

CRAIG A.D.,ET AL.:

J.CHEM.SOC. 1962,548.

A = 8.36127 B = 2472.700 C = 274.839

B.P.(100) = 113.872

T	P EXPTL.	P CALCD.	DEV.	PERCENT
61.900	10.40	10.43	0.03	0.27
73.100	18.00	17.97	-0.03	-0.16
82.800	28.00	28.01	0.01	0.04
87.400	34.30	34.29	-0.01	-0.04
90.400	39.00	39.01	0.01	0.03
96.100	49.60	49.57	-0.03	-0.06
101.400	61.50	61.54	0.04	0.06
105.000	71.00	71.03	0.03	0.04
108.000	79.90	79.88	-0.02	-0.03
115.400	105.90	105.90	0.00	0.00
129.100	173.70	173.70	0.00	0.00

STANDARD DEVIATION = 0.03

DIETHYL ETHYLPHOSPHONATE C6H15O3P

KOSOLAPOFF G.M.:

J.CHEM.SOC. 1955,2964.

A = 4.10159 B = 315.169 C = 15.503

B.P.(10) = 86.112

T	P EXPTL.	P CALCD.	DEV.	PERCENT
76.500	4.30	4.74	0.44	10.28
85.500	9.30	9.58	0.28	2.97
95.500	19.30	18.29	-1.01	-5.21
105.500	31.30	31.40	0.10	0.33
116.000	50.30	50.69	0.39	0.78
125.000	71.80	72.19	0.39	0.54
133.500	97.30	96.92	-0.38	-0.39

STANDARD DEVIATION = 0.66

TRIETHYLSILANOL C6H16OSI

GRUBB W.T.,OSTHOFF R.C.:
J.AM.CHEM.SOC. 75,2230(1953).

A = 7.79373 B = 1756.081 C = 202.368

B.P.(100) = 100.731

T	P EXPTL.	P CALCD.	DEV.	PERCENT
24.500	1.80	1.13	-0.67	-37.20
41.000	4.60	3.78	-0.82	-17.73
48.200	6.50	6.10	-0.40	-6.15
55.200	9.20	9.46	0.26	2.81
60.800	12.40	13.21	0.81	6.53
67.900	19.10	19.78	0.68	3.55
74.400	28.30	28.11	-0.20	-0.69
84.600	48.00	47.24	-0.76	-1.58
93.600	73.70	72.51	-1.19	-1.62
100.500	99.00	98.98	-0.02	-0.02
104.900	122.20	119.84	-2.36	-1.93
116.000	189.00	189.61	0.61	0.32
120.500	218.50	226.33	7.83	3.58
135.200	390.40	390.47	0.07	0.02
140.200	471.20	465.06	-6.14	-1.30

STANDARD DEVIATION = 3.02

N-PROPYLAMINE TRIMETHYLBORON C6H18BN

BROWN H.C.,TAYLOR M.D.,SUJISHI S.:
J.AM.CHEM.SOC. 73,2464(1951).

A = 9.08472 B = 2306.377 C = 222.106

B.P.(10) = 63.170

T	P EXPTL.	P CALCD.	DEV.	PERCENT
0.000	0.09	0.05	-0.04	-44.24
5.500	0.12	0.09	-0.03	-25.47
10.700	0.18	0.15	-0.03	-16.33
14.900	0.24	0.23	-0.01	-5.98
19.800	0.37	0.36	-0.01	-3.99
25.100	0.60	0.57	-0.03	-5.20
34.700	1.26	1.27	0.01	0.77
39.900	1.94	1.91	-0.03	-1.34
49.600	3.98	3.95	-0.03	-0.84
60.800	8.49	8.56	0.07	0.78
71.100	16.50	16.54	0.04	0.27
80.700	29.50	29.38	-0.12	-0.41
85.800	39.20	39.28	0.08	0.21
90.200	50.10	50.09	-0.01	-0.03

STANDARD DEVIATION = 0.06

TRIS-DIMETHYLAMINOFLUOROSILANE C6H18FN3SI

VONGROSSE-RUYKEN H.,KLEESAAT R.:
Z.ANORG.ALLG.CHEM. 308,122(1961).

A = 7.68606 B = 1962.741 C = 264.645

B.P.(100) = 80.540

T	P EXPTL.	P CALCD.	DEV.	PERCENT
24.000	8.00	7.70	-0.31	-3.81
39.600	17.00	17.17	0.17	1.02
60.700	45.00	45.00	0.00	0.01
79.000	94.00	94.30	0.30	0.32
90.900	147.00	146.45	-0.55	-0.38
99.200	196.00	195.71	-0.29	-0.15
109.000	270.00	271.08	1.08	0.40
120.200	386.00	385.46	-0.54	-0.14

STANDARD DEVIATION = 0.64

HEXAMETHYLDISILOXANE C6H18OSI2

SCOTT D.W.,ET AL.:
J.PHYS.CHEM. 65,1320(1961).

A = 6.77379 B = 1202.029 C = 208.251

B.P.(760) = 100.518

T	P EXPTL.	P CALCD.	DEV.	PERCENT
36.206	71.87	71.88	0.01	0.02
38.984	81.64	81.64	0.00	0.00
41.777	92.52	92.51	-0.01	-0.01
44.585	104.63	104.61	-0.02	-0.01
47.407	118.06	118.05	-0.01	-0.01
50.241	132.95	132.93	-0.02	-0.02
53.099	149.41	149.44	0.03	0.02
58.837	187.57	187.62	0.05	0.03
64.625	233.72	233.74	0.02	0.01
70.469	289.13	289.14	0.01	0.00
76.363	355.22	355.15	-0.07	-0.02
82.320	433.56	433.50	-0.06	-0.01
88.330	525.86	525.80	-0.06	-0.01
94.399	633.99	634.01	0.02	0.00
100.520	760.00	760.03	0.03	0.00
106.696	906.06	906.09	0.03	0.00
112.932	1074.60	1074.67	0.07	0.01
119.222	1268.00	1268.12	0.12	0.01
125.567	1489.10	1489.11	0.01	0.00
131.971	1740.80	1740.63	-0.17	-0.01
* 138.417	2026.00	2024.90	-1.10	-0.05

STANDARD DEVIATION = 0.06

HEXAMETHYLCYCLOTRISILOXANE C6H18O3SI3

OSTHOFF R.C.,GRUBB W.T.,BURKHARD C.A.:
J.AM.CHEM.SOC. 75,2227(1953).

A = 4.36135 B = 319.331 C = 58.973

B.P.(10) = 36.028

T	P EXPTL.	P CALCD.	DEV.	PERCENT
24.200	4.20	3.33	-0.87	-20.80
27.800	6.10	4.80	-1.30	-21.31
34.100	8.90	8.52	-0.38	-4.28
39.800	12.60	13.44	0.84	6.66
45.900	20.10	20.72	0.62	3.09
51.000	27.10	28.68	1.58	5.84
56.000	39.20	38.36	-0.84	-2.14
62.400	54.10	53.75	-0.35	-0.65

STANDARD DEVIATION = 1.18

DIMETHYLAMINOMETHYLBORANE CYCLIC DIMER C6H20B2N2

MILLER N.E.,MURPHY M.D.,REZNICEK D.L.:
INORG.CHEM. 5,1832(1966).

A = -6.37260 B = 1611.817 C = -299.505

B.P.(10) = 80.882

T	P EXPTL.	P CALCD.	DEV.	PERCENT
37.600	0.52	0.60	0.08	16.31
50.500	1.13	1.26	0.13	11.52
54.100	1.48	1.57	0.09	5.96
60.300	2.31	2.32	0.01	0.47
61.600	2.56	2.53	-0.03	-1.32
66.500	3.38	3.51	0.13	3.76
72.500	5.44	5.34	-0.10	-1.79
78.400	8.44	8.26	-0.18	-2.08
84.500	13.20	13.31	0.11	0.80

STANDARD DEVIATION = 0.13

PERFLUOROMETHYLCYCLOHEXANE C7F14

GOOD W.D.,ET AL.:

J.PHYS.CHEM. 63,1133(1959).

A = 6.82406 B = 1133.758 C = 211.223

B.P.(760) = 76.295

T	P EXPTL.	P CALCD.	DEV.	PERCENT
32.618	149.41	149.45	0.04	0.02
37.909	187.57	187.60	0.03	0.02
43.247	233.72	233.72	0.00	0.00
48.634	289.13	289.10	-0.03	-0.01
54.069	355.22	355.17	-0.05	-0.01
59.551	433.56	433.47	-0.09	-0.02
65.084	525.86	525.79	-0.07	-0.01
70.665	633.99	633.94	-0.06	-0.01
76.296	760.00	760.00	0.00	0.00
81.975	906.06	906.14	0.08	0.01
87.703	1074.60	1074.71	0.11	0.01
93.481	1268.00	1268.27	0.27	0.02
99.305	1489.10	1489.35	0.25	0.02
105.176	1740.80	1740.78	-0.02	0.00
111.094	2026.00	2025.53	-0.47	-0.02

STANDARD DEVIATION = 0.18

PERFLUOROHEPTANE C7F16

OLIVER G.D.,GRISARD J.W.: J.AM.CHEM.SOC. 73,1688(1951).

A = 6.93772 B = 1181.140 C = 208.657

B.P.(760) = 82.486

	T	P EXPTL.	P CALCD.	DEV.	PERCENT
	-1.890	16.88	16.80	-0.08	-0.48
	1.570	20.94	20.86	-0.08	-0.38
*	6.930	29.09	28.77	-0.32	-1.09
	13.540	41.72	41.88	0.16	0.38
	19.190	56.64	56.73	0.09	0.15
	23.950	72.52	72.42	-0.10	-0.13
	30.490	99.68	99.71	0.03	0.03
	35.910	128.39	128.29	-0.10	-0.08
	41.800	166.62	166.64	0.02	0.01
	46.860	206.57	206.62	0.05	0.02
	52.560	260.63	260.63	0.00	0.00
	58.990	334.70	334.70	0.00	0.00
	64.030	403.91	403.85	-0.06	-0.01
	68.960	482.15	482.11	-0.05	-0.01
	74.270	579.51	579.42	-0.09	-0.02
	79.710	694.49	694.62	0.13	0.02
	85.270	830.09	830.29	0.20	0.02
	90.080	963.72	963.68	-0.04	0.00
	95.030	1118.00	1117.83	-0.17	-0.02
	99.930	1288.70	1288.65	-0.05	0.00
	104.540	1467.00	1467.15	0.15	0.01
	106.030	1528.80	1528.73	-0.07	0.00

STANDARD DEVIATION = 0.10

3,4-DICHLOROPHENYLISOCYANATE C7H3CL2NO

KONSTANTINOV I.I.,ZHURAVLEV E.Z.,TENLOVA Z.G.:
ZH.PRIKL.KHIM. 40,1084(1967).

A = 8.67931 B = 3312.281 C = 333.987

B.P.(100) = 161.914

T	P EXPTL.	P CALCD.	DEV.	PERCENT
60.000	1.50	1.87	0.37	24.78
70.000	3.00	3.02	0.02	0.74
80.000	5.00	4.77	-0.23	-4.63
90.000	7.00	7.36	0.36	5.19
100.000	11.00	11.15	0.15	1.32
110.000	16.50	16.56	0.06	0.35
120.000	24.50	24.17	-0.33	-1.33
130.000	35.00	34.72	-0.28	-0.80
140.000	49.50	49.11	-0.39	-0.78
150.000	68.50	68.48	-0.02	-0.03
160.000	93.50	94.21	0.71	0.76
170.000	127.50	127.98	0.48	0.38
180.000	172.50	171.80	-0.70	-0.41
190.000	228.00	228.04	0.04	0.02

STANDARD DEVIATION = 0.42

PENTAFLUOROTOLUENE C7H3F5

AMBROSE D.:

J.CHEM.SOC.(A) 1968,1381.

A = 7.08478 B = 1392.203 C = 213.670

B.P.(760) = 117.494

T	P EXPTL.	P CALCD.	DEV.	PERCENT
39.164	37.92	37.88	-0.04	-0.11
50.578	65.53	65.50	-0.03	-0.05
58.357	92.66	92.66	0.00	0.00
62.734	111.65	111.67	0.02	0.02
66.375	129.82	129.84	0.02	0.02
72.174	163.75	163.78	0.03	0.02
76.551	193.92	193.97	0.05	0.02
81.824	236.21	236.23	0.02	0.01
85.651	271.34	271.37	0.03	0.01
92.625	346.32	346.31	-0.01	0.00
97.287	405.18	405.15	-0.03	-0.01
102.370	478.22	478.22	0.00	0.00
107.382	560.29	560.26	-0.03	0.00
112.608	657.47	657.43	-0.04	-0.01
116.158	730.81	730.76	-0.05	-0.01
117.884	768.88	768.68	-0.20	-0.03
118.979	793.66	793.54	-0.12	-0.01
123.337	898.81	898.86	0.05	0.01
127.924	1021.25	1021.33	0.08	0.01
133.176	1177.21	1177.27	0.06	0.01
138.225	1344.05	1344.22	0.17	0.01

STANDARD DEVIATION = 0.08

1-CHLOROPHENYLISOCYANATE C7H4CLNO

KONSTANTINOV I.I.,ZHURAVLEV E.Z.,TENLOVA Z.G.:
ZH.PRIKL.KHIM. 40,1084(1967).

A = 12.26590 B = 6532.551 C = 499.588

B.P.(100) = 136.747

T	P EXPTL.	P CALCD.	DEV.	PERCENT
50.000	2.50	2.40	-0.10	-4.13
60.000	4.00	3.91	-0.09	-2.28
70.000	6.50	6.27	-0.23	-3.60
80.000	10.50	9.88	-0.62	-5.88
90.000	15.50	15.35	-0.15	-0.98
100.000	23.00	23.49	0.49	2.12
110.000	35.50	35.45	-0.06	-0.15
120.000	53.00	52.79	-0.21	-0.40
130.000	78.50	77.62	-0.88	-1.12
140.000	112.00	112.77	0.77	0.69
150.000	160.00	161.97	1.97	1.23
160.000	232.00	230.09	-1.91	-0.82

STANDARD DEVIATION = 1.04

M-CHLOROPHENYLISOCYANATE C7H4CLNO

GOLDBERG N.A.,GORBUSHCHENKOV V.A.,TEPLOVA Z.G.:
ZH.PRIKL.KHIM. 37,745(1964).

A = 6.79729 B = 1512.425 C = 180.903

B.P.(100) = 134.364

T	P EXPTL.	P CALCD.	DEV.	PERCENT
71.000	6.00	6.21	0.21	3.55
84.800	13.00	12.74	-0.26	-2.01
97.700	23.50	23.37	-0.13	-0.55
110.300	40.00	40.14	0.14	0.35
121.000	61.50	61.33	-0.17	-0.28
129.000	82.00	82.60	0.60	0.73
136.200	107.00	106.60	-0.40	-0.37
144.800	142.50	142.47	-0.03	-0.02
158.300	218.00	218.03	0.03	0.02

STANDARD DEVIATION = 0.34

M-FLUOROBENZOTRIFLUORIDE C7H4F4

GOOD W.D.,ET AL.:
J.PHYS.CHEM. 63,1133(1959).

A = 7.00659 B = 1304.346 C = 215.666

B.P.(760) = 100.600

T	P EXPTL.	P CALCD.	DEV.	PERCENT
40.448	81.64	81.62	-0.02	-0.02
43.210	92.52	92.50	-0.02	-0.02
45.984	104.63	104.62	-0.01	-0.01
48.764	118.06	118.04	-0.02	-0.02
51.560	132.95	132.94	-0.01	-0.01
54.372	149.41	149.46	0.05	0.03
60.008	187.57	187.63	0.06	0.03
65.684	233.72	233.77	0.05	0.02
71.398	289.13	289.14	0.01	0.00
77.155	355.22	355.20	-0.02	-0.01
82.954	433.56	433.52	-0.04	-0.01
88.794	525.86	525.80	-0.06	-0.01
94.675	633.99	633.93	-0.06	-0.01
100.598	760.00	759.94	-0.06	-0.01
106.563	906.06	906.06	0.00	0.00
112.569	1074.60	1074.62	0.02	0.00
118.615	1268.00	1268.10	0.10	0.01
124.703	1489.10	1489.27	0.17	0.01
130.830	1740.80	1740.85	0.05	0.00
136.996	2026.00	2025.83	-0.17	-0.01

STANDARD DEVIATION = 0.07

ALPHA,ALPHA,ALPHA-TRIFLUOROTOLUENE C7H5F3

SCOTT D.W.,ET AL.:
J.AM.CHEM.SOC. 81,1015(1959).

A = 6.97045 B = 1306.352 C = 217.381

B.P.(760) = 102.048

T	P EXPTL.	P CALCD.	DEV.	PERCENT
55.001	149.41	149.42	0.01	0.01
60.729	187.57	187.59	0.02	0.01
66.501	233.72	233.73	0.01	0.00
72.316	289.13	289.12	-0.01	0.00
78.176	355.22	355.21	-0.01	0.00
84.075	433.56	433.49	-0.07	-0.02
90.023	525.86	525.81	-0.05	-0.01
96.014	633.99	633.96	-0.03	0.00
102.050	760.00	760.03	0.03	0.00
108.127	906.06	906.11	0.05	0.01
114.249	1074.60	1074.65	0.05	0.00
120.414	1268.00	1268.12	0.12	0.01
126.624	1489.10	1489.26	0.16	0.01
132.875	1740.80	1740.78	-0.02	0.00
139.169	2026.00	2025.74	-0.26	-0.01

STANDARD DEVIATION = 0.10

PHENYL ISOTHIOCYANATE C7H5NS

BAUER H.,BURSCHKIES K.:
BER. 68,1243(1935).

A = -0.70801 B = 106.418 C = -146.591

B.P.(10) = 84.286

T	P EXPTL.	P CALCD.	DEV.	PERCENT
10.000	1.20	1.18	-0.02	-1.85
15.000	1.30	1.26	-0.04	-3.01
20.000	1.40	1.36	-0.04	-3.06
25.000	1.50	1.47	-0.03	-2.03
30.000	1.60	1.60	0.00	0.14
35.000	1.70	1.76	0.06	3.56
40.000	1.80	1.95	0.15	8.42
50.000	2.30	2.48	0.18	7.65
55.000	2.70	2.84	0.14	5.32
60.000	3.30	3.32	0.02	0.57
65.000	4.10	3.95	-0.15	-3.73
70.000	5.00	4.80	-0.20	-3.96
75.000	6.10	6.00	-0.10	-1.57
80.000	7.60	7.76	0.16	2.16

STANDARD DEVIATION = 0.13

2,4,6-TRINITROTOLUENE $C_7H_5N_3O_6$

MAKSIMOV YU.YA.:
ZH.FIZ.KHIM. 42,2921(1968).

A = 7.67152 B = 2669.367 C = 205.630

B.P.(10) = 194.483

T	P EXPTL.	P CALCD.	DEV.	PERCENT
230.000	35.00	34.99	-0.01	-0.03
240.000	48.00	48.02	0.02	0.04
250.000	65.00	65.00	0.00	-0.01
250.000	65.00	65.00	0.00	-0.01

STANDARD DEVIATION = 0.02

2,4-DINITROTOLUENE $C_7H_6N_2O_4$

MAKSIMOV YU.YA.:
ZH.FIZ.KHIM. 42,2921(1968).

A = 5.79801 B = 1117.894 C = 61.826

B.P.(100) = 232.511

T	P EXPTL.	P CALCD.	DEV.	PERCENT
200.000	38.00	33.76	-4.24	-11.16
200.000	41.00	33.76	-7.24	-17.66
210.000	54.00	48.47	-5.53	-10.24
220.000	71.00	67.83	-3.17	-4.47
230.000	95.00	92.75	-2.25	-2.37
240.000	117.00	124.23	7.23	6.18
240.000	120.00	124.23	4.23	3.53
240.000	120.00	124.23	4.23	3.53
250.000	168.00	163.31	-4.69	-2.79
260.000	212.00	211.06	-0.94	-0.44
270.000	257.00	268.59	11.59	4.51
280.000	320.00	337.01	17.01	5.32
290.000	422.00	417.45	-4.55	-1.08
299.000	522.00	501.02	-20.98	-4.02

STANDARD DEVIATION =10.05

2,6-DINITROTOLUENE C7H6N2O4

MAKSIMOV YU.YA.:

ZH.FIZ.KHIM. 42,2921(1968).

A = 4.37195 B = 379.744 C = -43.562

B.P.(100) = 203.660

T	P EXPTL.	P CALCD.	DEV.	PERCENT
150.000	15.00	6.37	-8.63	-57.53
150.000	15.00	6.37	-8.63	-57.53
160.000	25.00	12.90	-12.10	-48.40
170.000	31.00	23.36	-7.64	-24.63
180.000	39.00	38.79	-0.21	-0.55
200.000	90.00	88.00	-2.00	-2.22
220.000	141.00	165.83	24.83	17.61
230.000	191.00	216.33	25.33	13.26
240.000	249.00	274.66	25.66	10.31
250.000	365.00	340.75	-24.25	-6.64
260.000	439.00	414.41	-24.59	-5.60

STANDARD DEVIATION =20.82

3,5-DINITROTOLUENE C7H6N2O4

MAKSIMOV YU.YA.:

ZH.FIZ.KHIM. 42,2921(1968).

A = 1.55566 B = 30.587 C = -301.717

B.P.(100) = 232.880

T	P EXPTL.	P CALCD.	DEV.	PERCENT
220.000	65.00	85.11	20.11	30.93
230.000	88.00	95.97	7.97	9.06
240.000	118.00	112.53	-5.47	-4.64
250.000	146.00	140.31	-5.69	-3.90
250.000	152.00	140.31	-11.69	-7.69
260.000	205.00	194.48	-10.52	-5.13
270.000	267.00	331.17	64.17	24.03

STANDARD DEVIATION =34.98

O-CHLOROANISOLE C7H7CLO

DREISBACH R.R.,SHRADER S.A.: IND.ENG.CHEM. 41,2879(1949).

A = 7.12136 B = 1655.802 C = 188.774

B.P.(100) = 134.539

T	P EXPTL.	P CALCD.	DEV.	PERCENT
115.130	47.16	47.09	-0.07	-0.15
119.810	57.04	56.96	-0.08	-0.14
123.630	66.39	66.25	-0.14	-0.22
127.270	75.86	76.24	0.38	0.51
140.470	123.76	123.67	-0.09	-0.08
169.440	315.52	315.48	-0.04	-0.01
186.190	507.50	507.53	0.03	0.01

STANDARC DEVIATION = 0.22

4-FLUOROTOLUENE C7H7F

SCOTT D.W.,ET AL.: J.CHEM.PHYS.37,867(1962).

A = 6.99426 B = 1374.055 C = 217.399

B.P.(760) = 116.641

T	P EXPTL.	P CALCD.	DEV.	PERCENT
67.679	149.41	149.39	-0.02	-0.01
73.649	187.57	187.59	0.02	0.01
79.661	233.72	233.75	0.03	0.01
85.716	289.13	289.16	0.03	0.01
91.810	355.22	355.21	-0.01	0.00
97.949	433.56	433.50	-0.06	-0.01
104.136	525.86	525.82	-0.04	-0.01
110.365	633.99	633.94	-0.05	-0.01
116.641	760.00	760.00	0.00	0.00
122.958	906.06	906.06	0.00	0.00
129.321	1074.60	1074.60	0.00	0.00
135.730	1268.00	1268.16	0.16	0.01
142.179	1489.10	1489.22	0.12	0.01
148.673	1740.80	1740.79	-0.01	0.00
155.210	2026.00	2025.81	-0.19	-0.01

STANDARD DEVIATION = 0.08

O-NITROTOLUENE C7H7NO2

DREISBACH R.R.,SHRADER S.A.: IND.ENG.CHEM.41,2879(1949).

A = 5.85126 B = 946.255 C = 96.072

B.P.(760) = 222.484

T	P EXPTL.	P CALCD.	DEV.	PERCENT
129.310	47.16	44.96	-2.20	-4.67
134.510	57.04	55.91	-1.13	-1.98
138.750	66.39	66.31	-0.08	-0.12
142.430	75.86	76.52	0.66	0.87
156.610	123.76	127.77	4.01	3.24
185.480	315.52	309.32	-6.20	-1.97
205.480	507.50	516.78	9.28	1.83
222.150	760.00	754.55	-5.45	-0.72

STANDARD DEVIATION = 5.95

P-NITROTOLUENE C7H7NO2

DREISBACH R.R.,SHRADER S.A.:
IND.ENG.CHEM.41,2879(1949).

A = 6.99477 B = 1720.385 C = 184.925

B.P.(760) = 233.257

T	P EXPTL.	P CALCD.	DEV.	PERCENT
147.710	66.39	66.49	0.10	0.16
151.430	75.86	75.86	-0.01	-0.01
165.980	123.76	123.61	-0.15	-0.12
197.750	315.52	315.57	0.05	0.02
216.170	507.50	507.64	0.14	0.03
233.250	760.00	759.88	-0.12	-0.02

STANDARD DEVIATION = 0.15

CYCLOHEPTATRIENE C_7H_8

FINKE H.L.,ET AL.:
J.AM.CHEM.SOC. 78,5469(1956).

A = 6.97433 B = 1376.843 C = 220.754

B.P.(100) = 56.036

T	P EXPTL.	P CALCD.	DEV.	PERCENT
0.000	5.50	5.46	-0.04	-0.70
15.000	13.61	13.62	0.01	0.07
20.000	18.00	18.01	0.01	0.04
25.000	23.52	23.54	0.02	0.08
30.000	30.44	30.44	0.00	0.01
35.000	38.99	38.98	-0.01	-0.02
40.000	49.44	49.44	0.00	0.00
45.000	62.15	62.15	0.00	0.00
50.000	77.46	77.46	0.00	0.01
55.000	95.81	95.79	-0.02	-0.02
60.000	117.55	117.55	0.00	0.00
65.000	143.22	143.23	0.01	0.01

STANDARD DEVIATION = 0.02

TOLUENE C_7H_8

WILLINGHAM C.J.,TAYLOR W.J.,PIGNOCCO J.M.,ROSSINI F.D.:
J.RESEARCH NATL.BUR.STANDARDS 35,219(1945).

A = 6.95805 B = 1346.773 C = 219.693

B.P.(760) = 110.622

T	P EXPTL.	P CALCD.	DEV.	PERCENT
35.366	47.68	47.62	-0.06	-0.12
39.343	57.41	57.40	-0.02	-0.03
42.810	67.22	67.23	0.01	0.01
45.948	77.28	77.29	0.01	0.02
48.867	87.75	87.75	0.00	0.00
52.802	103.64	103.67	0.03	0.03
57.293	124.67	124.68	0.01	0.01
61.851	149.43	149.46	0.03	0.02
66.079	175.89	175.91	0.02	0.01
71.738	217.16	217.17	0.01	0.01
76.942	261.75	261.75	0.00	0.00
83.202	324.93	324.87	-0.06	-0.02
89.667	402.43	402.38	-0.05	-0.01
96.559	500.68	500.62	-0.06	-0.01
104.037	627.93	627.88	-0.05	-0.01
109.312	732.08	732.11	0.03	0.00
109.879	744.05	744.08	0.03	0.00
110.403	755.24	755.28	0.04	0.01
110.991	767.95	768.00	0.05	0.01
111.509	779.34	779.35	0.01	0.00

STANDARD DEVIATION = 0.04

TOLUENE C_7H_8

PITZER K.S.,SCOTT D.W.:
J.AM.CHEM.SOC. 65,803(1943).

A = 6.93325 B = 1335.533 C = 218.752

B.P.(10) = 6.341

T	P EXPTL.	P CALCD.	DEV.	PERCENT
0.000	6.75	6.73	-0.02	-0.30
12.500	14.40	14.39	-0.01	-0.08
25.000	28.40	28.46	0.06	0.20
37.500	52.70	52.66	-0.04	-0.08
50.000	92.00	92.01	0.01	0.01

STANDARD DEVIATION = 0.05

BENZYL ALCOHOL C7H8O

DREISBACH R.R.,SHRADER S.A.: IND.ENG.CHEM. 41,2879(1949).

A = 7.19817 B = 1632.593 C = 172.790

B.P.(760) = 205.357

T	P EXPTL.	P CALCD.	DEV.	PERCENT
122.520	47.16	46.75	-0.41	-0.87
127.120	57.04	56.83	-0.21	-0.37
130.900	66.39	66.42	0.03	0.05
134.280	75.86	76.12	0.26	0.34
147.090	123.76	124.28	0.52	0.42
174.640	315.52	315.58	0.06	0.02
190.500	507.50	506.12	-1.38	-0.27
205.410	760.00	761.06	1.06	0.14

STANDARD DEVIATION = 0.85

BENZYL ALCOHOL C7H8O

GARDNER G.S.,BREWER J.E.: IND.ENG.CHEM.29,179(1937).

A = 7.24083 B = 1638.405 C = 170.371

B.P.(100) = 142.252

T	P EXPTL.	P CALCD.	DEV.	PERCENT
38.800	0.20	0.26	0.06	27.93
42.500	0.60	0.35	-0.25	-41.66
46.800	0.80	0.50	-0.30	-37.85
50.500	0.90	0.67	-0.23	-26.10
85.200	6.80	6.76	-0.04	-0.56
85.700	6.90	6.96	0.06	0.86
97.200	13.10	13.11	0.01	0.07
113.500	29.10	29.46	0.36	1.22
128.800	58.40	58.12	-0.28	-0.48
151.600	141.90	141.96	0.06	0.04

STANDARD DEVIATION = 0.25

O-CRESOL C7H8O

DREISBACH R.R.,SHRADER S.A.:
IND.ENG.CHEM. 41,2879(1949).

A = 6.91172 B = 1435.503 C = 165.158

B.P.(760) = 190.966

T	P EXPTL.	P CALCD.	DEV.	PERCENT
120.220	75.86	76.13	0.27	0.35
132.630	123.76	123.36	-0.40	-0.33
160.170	315.52	315.67	0.15	0.05
176.140	507.50	507.81	0.31	0.06
190.950	760.00	759.67	-0.33	-0.04

STANDARD DEVIATION = 0.48

M-CRESOL C7H8O

GOLDBLUM K.B.,MARTIN R.W.,YOUNG R.B.:
IND.ENG.CHEM. 39,1474(1947).

A = 7.50798 B = 1856.356 C = 199.065

B.P.(760) = 202.121

T	P EXPTL.	P CALCD.	DEV.	PERCENT
149.500	152.00	152.15	0.15	0.10
149.600	153.00	152.68	-0.32	-0.21
154.100	178.00	178.50	0.50	0.28
164.300	251.00	250.72	-0.28	-0.11
172.000	321.00	320.03	-0.97	-0.30
176.400	366.00	366.29	0.29	0.08
179.600	404.00	403.28	-0.72	-0.18
183.400	451.00	451.14	0.14	0.03
187.400	505.00	506.46	1.46	0.29
192.100	577.00	578.45	1.45	0.25
196.300	649.00	649.65	0.65	0.10
201.100	742.00	739.60	-2.40	-0.32

STANDARD DEVIATION = 1.17

P-CRESOL C7H8O

DREISBACH R.R.,SHRADER S.A.: IND.ENG.CHEM. 41,2879(1949).

A = 7.03508 B = 1511.080 C = 161.854

B.P.(760) = 201.888

T	P EXPTL.	P CALCD.	DEV.	PERCENT
128.050	66.39	66.49	0.10	0.14
131.260	75.86	75.82	-0.04	-0.05
143.860	123.76	123.68	-0.08	-0.07
171.270	315.52	315.49	-0.03	-0.01
187.170	507.50	507.73	0.23	0.04
201.880	760.00	759.83	-0.17	-0.02

STANDARD DEVIATION = 0.18

METHYL PHENYL ETHER C7H8O

COLLERSON R.R., ET AL.: J.CHEM.SOC. 3697(1965).

A = 7.05269 B = 1489.986 C = 203.570

B.P.(760) = 153.580

T	P EXPTL.	P CALCD.	DEV.	PERCENT
109.876	199.12	199.12	0.00	0.00
116.255	247.70	247.71	0.01	0.00
122.045	299.77	299.76	-0.01	0.00
126.854	349.43	349.44	0.01	0.00
131.266	400.72	400.67	-0.05	-0.01
135.078	449.66	449.66	0.00	0.00
138.640	499.65	499.67	0.02	0.00
141.919	549.54	549.54	0.00	0.00
145.003	599.99	600.01	0.02	0.00
147.890	650.54	650.54	0.00	0.00
150.429	697.72	697.72	0.00	0.00
153.143	751.08	751.11	0.03	0.00
155.554	801.21	801.21	0.00	0.00
157.810	850.44	850.44	0.00	0.00
160.009	900.72	900.70	-0.02	0.00
162.087	950.32	950.32	0.00	0.00
164.114	1000.80	1000.78	-0.02	0.00

STANDARD DEVIATION = 0.02

QUAIACOL

C7H8O2

VONTERRES E.,ET AL.:

BRENNSTOFF CHEM. 36,272(1955).

A = 6.16122 B = 1051.203 C = 115.844

B.P.(760) = 204.605

T	P EXPTL.	P CALCD.	DEV.	PERCENT
82.000	10.00	7.05	-2.95	-29.54
102.000	25.00	21.66	-3.34	-13.35
119.000	50.00	48.42	-1.58	-3.16
129.500	75.00	75.27	0.27	0.36
136.000	100.00	97.09	-2.91	-2.91
149.000	150.00	155.62	5.62	3.75
155.500	200.00	193.71	-6.29	-3.14
164.800	250.00	260.34	10.34	4.13
169.000	300.00	295.64	-4.36	-1.45
173.000	325.00	332.56	7.56	2.33
175.900	350.00	361.45	11.45	3.27
179.000	400.00	394.39	-5.61	-1.40
184.700	450.00	460.84	10.84	2.41
187.000	500.00	489.90	-10.10	-2.02
191.000	550.00	543.70	-6.30	-1.15
194.000	600.00	586.86	-13.15	-2.19
198.000	650.00	648.29	-1.71	-0.26
201.000	700.00	697.41	-2.59	-0.37
205.000	760.00	767.09	7.09	0.93

STANDARD DEVIATION = 7.66

1-PHENYL-1-THIAETHANE C7H8S

OSBORN A.N.,DOUSLIN D.R.: J.CHEM.ENG.DATA 11,502(1966).

A = 7.00614 B = 1600.077 C = 193.563

B.P.(760) = 194.304

T	P EXPTL.	P CALCD.	DEV.	PERCENT
117.146	71.87	71.84	-0.03	-0.04
120.530	81.64	81.63	-0.01	-0.01
123.912	92.52	92.50	-0.02	-0.02
127.321	104.63	104.63	0.00	0.00
130.728	118.06	118.05	-0.01	-0.01
134.156	132.95	132.95	0.00	0.00
137.606	149.41	149.46	0.05	0.04
144.514	187.57	187.61	0.04	0.02
151.477	233.72	233.76	0.04	0.02
158.487	289.13	289.14	0.01	0.00
165.548	355.22	355.21	-0.01	0.00
172.658	433.56	433.49	-0.07	-0.02
179.821	525.86	525.78	-0.08	-0.02
187.036	633.99	633.93	-0.06	-0.01
194.306	760.00	760.04	0.04	0.01
201.622	906.06	906.17	0.11	0.01

STANDARD DEVIATION = 0.05

1-BUTYL PENTAFLUOROPROPIONATE C7H9F5O2

SHEEHAN R.J.,LANGER S.H.: J.CHEM.ENG.DATA 14,248(1969).

A = 6.65100 B = 1108.022 C = 177.044

B.P.(760) = 116.847

T	P EXPTL.	P CALCD.	DEV.	PERCENT
81.700	234.20	233.71	-0.49	-0.21
96.800	400.50	402.55	2.05	0.51
106.000	547.10	544.92	-2.18	-0.40
116.000	740.60	741.17	0.57	0.08

STANDARD DEVIATION = 3.08

2,3-LUTIDINE

C7H9N

COULSON E.A.,COX J.D.,HERINGTON E.F.G.,MARTIN J.F.: J.CHEM.SOC. 1959,1934.

A = 7.44781 B = 1832.608 C = 240.114

B.P.(760) = 161.157

T	P EXPTL.	P CALCD.	DEV.	PERCENT
155.326	650.82	650.82	0.00	0.00
157.940	698.06	698.07	0.01	0.00
158.603	710.48	710.48	0.00	0.00
159.141	720.70	720.68	-0.02	0.00
159.586	729.20	729.22	0.02	0.00
160.125	739.67	739.65	-0.02	0.00
160.668	750.29	750.30	0.01	0.00
161.199	760.85	760.83	-0.02	0.00
161.672	770.32	770.30	-0.02	0.00
162.077	778.48	778.49	0.01	0.00
162.412	785.29	785.32	0.03	0.00

STANDARD DEVIATION = 0.02

2,4-LUTIDINE

C7H9N

COULSON E.A.,COX J.D.,HERINGTON E.F.G.,MARTIN J.F.: J.CHEM.SOC. 1959,1934.

A = 7.33898 B = 1733.387 C = 230.407

B.P.(760) = 158.405

T	P EXPTL.	P CALCD.	DEV.	PERCENT
150.386	612.21	612.25	0.04	0.01
152.789	653.85	653.85	0.00	0.00
155.232	698.46	698.45	-0.01	0.00
155.739	708.01	708.00	-0.01	0.00
155.900	711.07	711.06	-0.01	0.00
156.107	715.08	715.01	-0.07	-0.01
156.690	726.23	726.21	-0.02	0.00
156.693	726.30	726.27	-0.03	0.00
157.202	736.16	736.17	0.01	0.00
157.712	746.21	746.20	-0.01	0.00
157.801	748.00	747.96	-0.04	-0.01
158.333	758.50	758.56	0.06	0.01
159.092	773.86	773.89	0.03	0.00
159.825	788.89	788.93	0.04	0.01

STANDARD DEVIATION = 0.04

2,5-LUTIDINE C7H9N

HERINGTON E.F.G.,MARTIN J.F.: TRANS.FARADAY SOC.49,154(1953).

A = 7.08100 B = 1539.636 C = 209.561

B.P.(760) = 157.002

T	P EXPTL.	P CALCD.	DEV.	PERCENT
85.098	71.76	71.76	0.00	-0.01
86.101	74.79	74.75	-0.04	-0.06
90.342	88.60	88.56	-0.04	-0.05
97.185	115.25	115.28	0.03	0.02
101.938	137.45	137.51	0.06	0.04
109.268	178.59	178.64	0.05	0.03
111.895	195.55	195.63	0.08	0.04
114.862	216.41	216.39	-0.02	-0.01
119.232	250.27	250.22	-0.05	-0.02
123.279	285.27	285.26	0.00	0.00
126.322	314.18	314.17	-0.01	0.00
130.857	361.59	361.60	0.01	0.00
131.669	370.70	370.67	-0.03	-0.01
134.062	398.50	398.48	-0.02	0.00
136.358	426.72	426.72	0.00	0.00
138.935	460.49	460.32	-0.17	-0.04
141.497	495.78	495.79	0.01	0.00
142.806	514.77	514.75	-0.02	0.00
145.250	551.57	551.69	0.12	0.02
147.417	586.11	586.18	0.07	0.01
148.991	612.38	612.30	-0.08	-0.01
150.972	646.67	646.49	-0.18	-0.03
152.371	671.33	671.53	0.20	0.03
* 155.155	732.48	723.67	-8.81	-1.20
155.273	725.98	725.95	-0.03	0.00
155.815	736.48	736.48	0.00	0.00
157.299	765.90	765.96	0.06	0.01

STANDARD DEVIATION = 0.08

2,6-LUTIDINE C7H9N

HERINGTON E.F.G., MARTIN J.F.:
TRANS.FARADAY SOC.49,154(1953).

A = 7.05674 B = 1470.172 C = 208.015

B.P.(760) = 144.044

T	P EXPTL.	P CALCD.	DEV.	PERCENT
79.290	87.06	87.02	-0.04	-0.05
83.938	105.01	104.97	-0.04	-0.03
95.180	161.30	161.36	0.06	0.04
100.326	194.32	194.41	0.09	0.05
105.723	234.75	234.82	0.07	0.03
108.211	255.69	255.63	-0.06	-0.02
112.351	293.55	293.56	0.01	0.00
113.949	309.25	309.36	0.11	0.04
116.979	341.36	341.23	-0.13	-0.04
120.305	379.30	379.20	-0.10	-0.03
122.877	410.92	410.84	-0.08	-0.02
125.637	447.22	447.12	-0.10	-0.02
127.495	473.01	472.97	-0.04	-0.01
129.190	497.55	497.57	0.02	0.00
131.655	535.01	535.17	0.17	0.03
133.468	564.22	564.25	0.03	0.01
135.162	592.55	592.55	0.00	0.00
136.638	618.10	618.12	0.02	0.00
138.750	656.24	656.23	-0.01	0.00
139.909	678.16	677.92	-0.24	-0.04
141.532	709.02	709.24	0.22	0.03
142.850	735.65	735.53	-0.13	-0.02
143.917	757.30	757.36	0.06	0.01
144.300	765.22	765.32	0.10	0.01

STANDARD DEVIATION = 0.11

3,4-LUTIDINE C7H9N

COULSON E.A.,COX J.D.,HERINGTON E.F.G.,MARTIN J.F.: J.CHEM.SOC. 1959,1934.

A = 7.36204 B = 1840.122 C = 231.496

B.P.(760) = 179.133

T	P EXPTL.	P CALCD.	DEV.	PERCENT
173.104	651.65	651.67	0.02	0.00
175.761	697.77	697.75	-0.02	0.00
176.327	707.89	707.90	0.01	0.00
177.035	720.78	720.76	-0.02	0.00
177.384	727.20	727.17	-0.03	0.00
177.931	737.34	737.31	-0.03	0.00
178.487	747.75	747.73	-0.02	0.00
178.933	756.18	756.18	0.00	0.00
179.274	762.64	762.68	0.04	0.01
179.401	765.09	765.12	0.03	0.00
179.860	773.99	773.97	-0.02	0.00
180.349	783.49	783.50	0.01	0.00

STANDARD DEVIATION = 0.03

3,5-LUTIDINE C7H9N

COULSON E.A.,COX J.D.,HERINGTON E.F.G.,MARTIN J.F.: J.CHEM.SOC. 1959,1934.

A = 7.33306 B = 1783.625 C = 228.704

B.P.(760) = 171.909

T	P EXPTL.	P CALCD.	DEV.	PERCENT
162.850	599.51	599.52	0.01	0.00
165.941	650.87	650.85	-0.02	0.00
168.620	698.15	698.16	0.01	0.00
169.344	711.39	711.41	0.02	0.00
169.790	719.67	719.68	0.01	0.00
170.306	729.32	729.33	0.01	0.00
170.862	739.87	739.85	-0.02	0.00
171.265	747.62	747.55	-0.07	-0.01
171.961	761.03	761.01	-0.02	0.00
172.307	767.76	767.77	0.01	0.00
172.729	776.02	776.08	0.06	0.01

STANDARD DEVIATION = 0.04

N-METHYLANILINE C7H9N

NELSON O.A.,WALES H.: J.AM.CHEM.SOC. 47,867(1927).

A = 7.08188 B = 1631.333 C = 192.440

B.P.(760) = 195.874

	T	P EXPTL.	P CALCD.	DEV.	PERCENT
*	50.000	3.40	2.25	-1.15	-33.69
	60.000	5.20	4.17	-1.04	-19.90
*	70.000	7.30	7.34	0.04	0.58
	80.000	13.00	12.42	-0.58	-4.49
	90.100	22.10	20.33	-1.77	-8.03
	99.660	31.70	31.41	-0.29	-0.92
	103.530	37.60	37.16	-0.44	-1.18
	111.960	53.50	52.81	-0.69	-1.29
	122.080	77.60	78.55	0.95	1.22
	134.180	121.00	122.26	1.26	1.04
	135.250	124.70	126.94	2.24	1.79
	141.680	158.00	158.27	0.27	0.17
	146.840	185.10	187.78	2.68	1.45
	149.920	206.50	207.45	0.95	0.46
	154.770	248.40	241.81	-6.59	-2.65
	164.210	325.10	321.98	-3.12	-0.96
	171.320	397.90	395.58	-2.32	-0.58
	175.790	444.80	448.41	3.61	0.81
	181.280	518.50	520.90	2.40	0.46
	183.830	555.00	557.62	2.62	0.47
	188.090	623.60	623.55	-0.05	-0.01
	190.740	665.50	667.61	2.11	0.32
	190.960	671.50	671.38	-0.13	-0.02
	194.490	733.60	734.14	0.54	0.07
	195.720	761.30	757.08	-4.22	-0.55
	199.610	834.70	833.38	-1.32	-0.16

STANDARD DEVIATION = 2.48

O-TOLUIDINE C7H9N

DREISBACH R.R.,SHRADER S.A.: IND.ENG.CHEM. 41,2879(1949).

A = 7.08203 B = 1627.718 C = 187.127

B.P.(760) = 200.313

T	P EXPTL.	P CALCD.	DEV.	PERCENT
118.460	57.04	56.95	-0.09	-0.16
122.220	66.39	66.11	-0.28	-0.43
125.990	75.80	76.49	0.69	0.91
139.000	123.76	123.30	-0.46	-0.37
168.060	315.52	315.73	0.21	0.07
184.800	507.50	507.67	0.17	0.03
200.300	760.00	759.75	-0.25	-0.03

STANDARD DEVIATION = 0.48

M-TOLUIDINE C7H9N

DREISBACH R.R.,SHRADER S.A.: IND.ENG.CHEM. 41,2879(1949).

A = 7.09367 B = 1631.428 C = 183.906

B.P.(760) = 203.344

T	P EXPTL.	P CALCD.	DEV.	PERCENT
121.770	57.04	57.09	0.05	0.08
125.570	66.39	66.39	0.00	-0.01
129.030	75.86	75.92	0.06	0.08
142.270	123.76	123.59	-0.17	-0.14
171.180	315.52	315.65	0.13	0.04
187.870	507.50	507.53	0.03	0.01
203.340	760.00	759.92	-0.08	-0.01

STANDARD DEVIATION = 0.12

CHLOROHEXYLISOCYANATE C7H12CLNO

ZHURAVLEV E.Z.,KONSTANTINOV I.I.:
ZH.PRIKL.KHIM. 41,1170(1968).

A = 7.74095 B = 2340.501 C = 241.902

B.P.(100) = 165.783

T	P EXPTL.	P CALCD.	DEV.	PERCENT
90.000	5.00	4.89	-0.11	-2.23
100.000	8.00	7.86	-0.14	-1.75
110.000	12.50	12.30	-0.20	-1.59
120.000	18.50	18.78	0.28	1.52
130.000	28.00	28.03	0.03	0.11
140.000	41.00	40.97	-0.04	-0.09
150.000	59.00	58.72	-0.28	-0.47
160.000	82.00	82.68	0.68	0.83
170.000	115.00	114.49	-0.51	-0.44
180.000	156.00	156.12	0.12	0.08

STANDARD DEVIATION = 0.37

CYCLOHEPTANE C_7H_{14}

FINKE H.L.,ET AL.: J.AM.CHEM.SOC. 78,5469(1956).

A = 6.85395 B = 1331.567 C = 216.349

B.P.(760) = 118.794

T	P EXPTL.	P CALCD.	DEV.	PERCENT
68.204	149.41	149.43	0.02	0.01
74.338	187.57	187.58	0.01	0.01
80.529	233.72	233.73	0.01	0.00
86.771	289.13	289.12	-0.01	0.00
93.068	355.22	355.20	-0.02	0.00
99.416	433.56	433.50	-0.06	-0.01
105.820	525.86	525.79	-0.07	-0.01
112.281	633.99	633.98	-0.01	0.00
118.793	760.00	759.98	-0.02	0.00
125.364	906.06	906.16	0.10	0.01
131.985	1074.60	1074.67	0.07	0.01
138.665	1268.00	1268.24	0.24	0.02
145.387	1489.10	1489.01	-0.09	-0.01
152.178	1740.80	1740.73	-0.07	0.00
159.022	2026.00	2025.86	-0.14	-0.01

STANDARD DEVIATION = 0.10

1,1-DIMETHYLCYCLOPENTANE C7H14

FORZIATI A.F.,NORRIS W.R.,ROSSINI F.D.:
J.RESEARCH NATL.BUR.STANDARDS 43,555(1949).

A = 6.82323 B = 1222.858 C = 222.336

B.P.(760) = 87.844

T	P EXPTL.	P CALCD.	DEV.	PERCENT
15.498	48.12	48.04	-0.08	-0.17
19.262	57.72	57.77	0.05	0.08
22.527	67.43	67.48	0.05	0.07
25.476	77.47	77.37	-0.10	-0.12
28.300	87.92	87.94	0.02	0.03
32.069	103.81	103.87	0.06	0.06
36.361	124.81	124.80	-0.01	-0.01
40.744	149.60	149.62	0.02	0.01
44.815	176.08	176.12	0.04	0.02
50.257	217.36	217.37	0.01	0.00
61.312	325.17	325.12	-0.05	-0.02
67.558	402.72	402.65	-0.07	-0.02
74.228	500.99	500.96	-0.03	-0.01
81.470	628.20	628.19	-0.01	0.00
86.585	732.40	732.38	-0.02	0.00
87.137	744.40	744.39	-0.01	0.00
87.647	755.59	755.62	0.03	0.00
88.227	768.49	768.55	0.06	0.01
88.736	779.99	780.03	0.04	0.00

STANDARD DEVIATION = 0.05

CIS-1,2-DIMETHYLCYCLOPENTANE C7H14

FORZIATI A.F.,NORRIS W.R.,ROSSINI F.D.:
J.RESEARCH NATL.BUR.STANDARDS 43,555(1949).

A = 6.85112 B = 1269.738 C = 220.277

B.P.(760) = 99.532

T	P EXPTL.	P CALCD.	DEV.	PERCENT
25.347	48.11	48.05	-0.06	-0.13
29.195	57.72	57.73	0.01	0.02
32.555	67.42	67.46	0.04	0.06
35.616	77.46	77.47	0.01	0.01
38.482	87.93	87.92	-0.01	-0.01
42.346	103.80	103.82	0.02	0.02
46.770	124.81	124.85	0.04	0.03
51.253	149.60	149.58	-0.02	-0.01
55.426	176.08	176.06	-0.02	-0.01
61.012	217.36	217.33	-0.03	-0.02
66.155	261.88	261.92	0.04	0.01
72.358	325.17	325.18	0.01	0.00
78.755	402.71	402.67	-0.04	-0.01
85.589	500.99	500.99	0.00	0.00
93.005	628.21	628.19	-0.02	0.00
98.244	732.40	732.41	0.01	0.00
98.806	744.40	744.35	-0.05	-0.01
99.329	755.60	755.60	0.00	0.00
99.922	768.49	768.51	0.02	0.00
100.446	780.00	780.06	0.06	0.01

STANDARD DEVIATION = 0.03

TRANS-1,2-DIMETHYLCYCLOPENTANE C_7H_{14}

FORZIATI A.F.,NORRIS W.R.,ROSSINI F.D.:
J.RESEARCH NATL.BUR.STANDARDS 43,555(1949).

A = 6.83744 B = 1238.935 C = 221.256

B.P.(760) = 91.873

	T	P EXPTL.	P CALCD.	DEV.	PERCENT
	26.113	67.42	67.45	0.03	0.05
	29.113	77.46	77.45	-0.01	-0.02
	31.934	87.93	87.93	0.00	0.00
	35.719	103.80	103.80	0.00	0.00
	40.055	124.81	124.80	-0.01	-0.01
	44.460	149.60	149.56	-0.04	-0.03
	48.564	176.08	176.09	0.01	0.00
	54.047	217.36	217.36	0.00	0.00
	59.094	261.88	261.94	0.06	0.02
	65.178	325.17	325.13	-0.04	-0.01
	71.466	402.71	402.69	-0.02	0.00
	78.181	500.99	501.06	0.07	0.01
	85.463	628.22	628.23	0.01	0.00
	90.608	732.40	732.42	0.02	0.00
	91.160	744.40	744.35	-0.05	-0.01
	91.673	755.61	755.58	-0.03	0.00
*	92.255	768.90	768.47	-0.43	-0.06
	92.769	780.01	780.01	0.00	0.00

STANDARD DEVIATION = 0.03

CIS-1,3-DIMETHYLCYCLOPENTANE C_7H_{14}

FORZIATI A.F.,NORRIS W.R.,ROSSINI F.D.:
J.RESEARCH NATL.BUR.STANDARDS 43,555(1949).

A = 6.84069 B = 1241.480 C = 221.790

B.P.(760) = 91.724

T	P EXPTL.	P CALCD.	DEV.	PERCENT
25.977	67.60	67.61	0.01	0.02
28.970	77.64	77.59	-0.05	-0.06
31.784	88.06	88.06	0.00	0.00
35.582	104.00	104.00	0.00	0.00
39.912	124.95	124.98	0.03	0.02
44.323	149.76	149.78	0.02	0.01
48.422	176.26	176.29	0.03	0.02
53.902	217.54	217.55	0.01	0.00
58.949	262.13	262.12	-0.01	0.00
65.039	325.41	325.39	-0.02	-0.01
71.323	402.93	402.89	-0.04	-0.01
78.030	501.15	501.12	-0.03	-0.01
85.315	628.32	628.30	-0.03	0.00
90.465	732.53	732.55	0.02	0.00
91.018	744.49	744.50	0.01	0.00
91.533	755.74	755.77	0.03	0.00
92.115	768.63	768.66	0.03	0.00
92.628	780.20	780.17	-0.03	0.00

STANDARD DEVIATION = 0.03

TRANS-1,3-DIMETHYLCYCLOPENTANE C7H14

FORZIATI A.F.,NORRIS W.R.,ROSSINI F.D.:
J.RESEARCH NATL.BUR.STANDARDS 43,555(1949).

A = 6.83973 B = 1238.914 C = 222.170

B.P.(760) = 90.773

T	P EXPTL.	P CALCD.	DEV.	PERCENT
18.005	48.11	48.01	-0.10	-0.20
21.782	57.72	57.71	-0.01	-0.03
25.081	67.42	67.45	0.03	0.04
28.099	77.46	77.52	0.06	0.07
30.906	87.93	87.96	0.03	0.04
34.685	103.80	103.83	0.03	0.03
39.010	124.81	124.80	-0.01	-0.01
43.417	149.60	149.59	-0.01	0.00
47.510	176.08	176.08	0.00	0.00
52.985	217.36	217.33	-0.03	-0.01
58.029	261.88	261.93	0.05	0.02
64.105	325.18	325.10	-0.08	-0.02
70.388	402.71	402.68	-0.03	-0.01
77.092	500.99	500.99	0.00	0.00
84.368	628.22	628.19	-0.03	0.00
90.062	744.40	744.39	-0.01	0.00
90.575	755.61	755.63	0.02	0.00
91.156	768.50	768.52	0.02	0.00
91.670	780.01	780.06	0.05	0.01

STANDARD DEVIATION = 0.04

ETHYLCYCLOPENTANE C7H14

FORZIATI A.F.,NORRIS W.R.,ROSSINI F.D.:
J.RESEARCH NATL.BUR.STANDARDS 43,555(1949).

A = 6.88302 B = 1296.119 C = 220.385

B.P.(760) = 103.466

	T	P EXPTL.	P CALCD.	DEV.	PERCENT
*	28.778	48.11	47.99	-0.12	-0.26
	32.673	57.72	57.70	-0.02	-0.03
	36.058	67.42	67.42	0.00	0.00
	39.165	77.46	77.50	0.04	0.05
	42.040	87.93	87.90	-0.03	-0.03
	45.935	103.80	103.81	0.01	0.01
	50.386	124.81	124.81	0.00	0.00
	54.908	149.60	149.58	-0.02	-0.01
	59.114	176.08	176.08	0.00	0.00
	64.739	217.36	217.37	0.01	0.01
	69.909	261.88	261.92	0.04	0.01
	76.143	325.17	325.11	-0.06	-0.02
	82.588	402.71	402.73	0.02	0.00
	89.453	500.98	500.96	-0.02	0.00
	96.910	628.21	628.22	0.01	0.00
	102.173	732.40	732.43	0.03	0.00
	102.739	744.39	744.40	0.01	0.00
	103.264	755.60	755.63	0.03	0.00
	103.855	768.50	768.44	-0.06	-0.01
	104.382	780.00	780.00	0.00	0.00

STANDARD DEVIATION = 0.03

METHYLCYCLOHEXANE C7H14

WILLINGHAM C.J.,TAYLOR W.J.,PIGNOCCO J.M.,ROSSINI F.D.: J.RESEARCH NATL.BUR.STANDARDS 35,219(1945).

A = 6.82827 B = 1273.673 C = 221.723

B.P.(760) = 100.934

T	P EXPTL.	P CALCD.	DEV.	PERCENT
25.586	47.66	47.66	0.00	-0.01
29.533	57.42	57.42	0.00	-0.01
32.976	67.22	67.23	0.01	0.01
36.089	77.28	77.26	-0.02	-0.03
38.998	87.73	87.71	-0.02	-0.02
42.929	103.66	103.66	0.00	0.00
47.407	124.65	124.65	0.00	0.00
51.964	149.40	149.45	0.05	0.03
56.194	175.91	175.93	0.02	0.01
61.857	217.20	217.20	0.00	0.00
67.067	261.74	261.75	0.01	0.00
73.349	324.95	324.92	-0.03	-0.01
79.840	402.46	402.43	-0.03	-0.01
86.771	500.74	500.70	-0.04	-0.01
94.299	627.96	627.96	0.00	0.00
99.614	732.12	732.14	0.02	0.00
100.185	744.10	744.09	-0.01	0.00
100.715	755.29	755.32	0.03	0.00
101.312	768.05	768.12	0.07	0.01
101.832	779.46	779.41	-0.05	-0.01

STANDARD DEVIATION = 0.03

1-HEPTENE C7H14

FORZIATI A.F.,CAMIN D.L.,ROSSINI F.D.:
J.RESEARCH NATL.BUR.STANDARDS 45,406(1950).

A = 6.91381 B = 1265.120 C = 220.051

B.P.(760) = 93.642

T	P EXPTL.	P CALCD.	DEV.	PERCENT
21.609	47.89	47.72	-0.17	-0.36
25.492	57.69	57.74	0.05	0.08
28.768	67.44	67.50	0.06	0.09
34.525	87.91	87.96	0.05	0.06
38.281	103.85	103.88	0.03	0.03
42.564	124.84	124.85	0.01	0.01
46.923	149.60	149.64	0.04	0.03
50.970	176.13	176.12	-0.01	0.00
56.384	217.43	217.39	-0.04	-0.02
67.366	325.27	325.17	-0.10	-0.03
73.563	402.82	402.73	-0.09	-0.02
80.179	501.05	501.14	0.09	0.02
92.391	732.44	732.26	-0.18	-0.02
92.941	744.31	744.35	0.04	0.01
93.444	755.56	755.56	0.00	0.00
94.022	768.53	768.58	0.05	0.01
94.531	780.08	780.21	0.13	0.02

STANDARD DEVIATION = 0.09

1-HEPTENE C7H14

BENT H.E.,CUTHBERTSON G.R.,DORFMAN M.,LEARY R.E.:
J.AM.CHEM.SOC. 58,165(1936).

A = 6.86778 B = 1272.447 C = 225.412

B.P.(100) = 35.990

T	P EXPTL.	P CALCD.	DEV.	PERCENT
0.040	17.00	16.74	-0.26	-1.52
0.140	17.30	16.84	-0.46	-2.67
0.140	17.30	16.84	-0.46	-2.67
0.140	17.40	16.84	-0.56	-3.23
21.140	49.70	50.91	1.21	2.44
21.140	50.50	50.91	0.41	0.81
21.140	50.70	50.91	0.21	0.42
21.140	51.00	50.91	-0.09	-0.17
21.240	51.40	51.16	-0.24	-0.47
41.740	127.40	127.28	-0.12	-0.09
41.740	127.50	127.28	-0.22	-0.17
42.040	129.00	128.86	-0.14	-0.11
42.240	129.30	129.92	0.62	0.48
62.040	276.60	276.14	-0.46	-0.16
62.340	279.30	279.09	-0.21	-0.07
62.640	282.40	282.07	-0.33	-0.12
62.840	284.30	284.07	-0.23	-0.08
88.040	643.50	643.15	-0.35	-0.05
88.740	655.60	656.69	1.09	0.17

STANDARD DEVIATION = 0.54

2,4-DIMETHYL-3-PENTANONE $C_7H_{14}O$

DREISBACH R.R.,SHRADER S.A.: IND.ENG.CHEM. 41,2879(1949).

A = 6.96853 B = 1382.841 C = 213.061

B.P.(760) = 125.231

T	P EXPTL.	P CALCD.	DEV.	PERCENT
48.040	47.16	47.02	-0.14	-0.29
52.190	57.04	56.91	-0.13	-0.23
55.730	66.39	66.66	0.27	0.40
58.660	75.86	75.74	-0.12	-0.16
70.580	123.76	123.94	0.18	0.14
96.340	315.52	315.58	0.06	0.02
111.280	507.50	506.97	-0.53	-0.10
125.250	760.00	760.39	0.39	0.05

STANDARD DEVIATION = 0.34

BUTYL PROPIONATE $C_7H_{14}O_2$

USANOVICH M.,DEMBICKIJ A.: ZH.OBSHCH.KHIM.29,1771(1959).

A = 9.48489 B = 2852.580 C = 296.980

B.P.(100) = 84.132

T	P EXPTL.	P CALCD.	DEV.	PERCENT
32.400	6.50	6.67	0.17	2.69
36.300	8.50	8.43	-0.07	-0.84
43.900	13.00	13.08	0.08	0.61
50.500	19.00	18.86	-0.14	-0.73
57.500	28.00	27.39	-0.61	-2.16
64.300	39.00	38.83	-0.17	-0.45
70.100	52.00	51.75	-0.25	-0.49
76.800	70.00	71.31	1.31	1.88
81.300	88.00	87.89	-0.11	-0.12
86.700	112.00	112.23	0.23	0.20
92.900	148.00	147.34	-0.66	-0.44

STANDARD DEVIATION = 0.59

HEPTANOIC ACID $C_7H_{14}O_2$

KAHLBAUM G.W.A.:

Z.PHYS.CHEM. 13,14(1894).

A = 5.28736 B = 665.543 C = 42.068

B.P.(10) = 113.166

	T	P EXPTL.	P CALCD.	DEV.	PERCENT
	112.600	10.00	9.65	-0.35	-3.55
	113.500	10.40	10.21	-0.19	-1.78
	114.300	10.80	10.74	-0.06	-0.53
	114.500	10.80	10.88	0.08	0.72
	114.800	11.20	11.08	-0.12	-1.04
	116.300	12.10	12.16	0.06	0.48
	118.400	13.30	13.80	0.50	3.75
	118.500	14.30	13.88	-0.42	-2.93
	118.600	14.50	13.96	-0.54	-3.70
	119.400	14.50	14.64	0.14	0.96
	119.200	14.90	14.47	-0.43	-2.90
	119.900	15.40	15.08	-0.33	-2.11
	120.400	15.40	15.52	0.12	0.78
	120.500	15.80	15.61	-0.19	-1.20
	121.300	16.50	16.35	-0.15	-0.92
	122.200	17.40	17.21	-0.19	-1.09
	122.800	18.00	17.80	-0.20	-1.08
	123.200	18.50	18.21	-0.29	-1.57
	124.200	18.80	19.25	0.45	2.42
	124.300	19.50	19.36	-0.14	-0.71
	125.600	20.00	20.79	0.79	3.97
	125.600	20.90	20.79	-0.11	-0.50
*	127.000	21.40	22.43	1.03	4.81
	126.800	22.20	22.19	-0.01	-0.05
	127.800	23.10	23.41	0.31	1.33
	127.900	23.30	23.53	0.23	1.00
	128.500	24.20	24.29	0.09	0.37
	129.700	25.60	25.86	0.26	1.03
	129.700	25.90	25.86	-0.04	-0.14
	130.700	27.20	27.24	0.04	0.13
	131.600	28.30	28.52	0.22	0.76
	132.100	29.10	29.25	0.15	0.51
	132.900	30.60	30.45	-0.15	-0.50
	134.000	32.00	32.16	0.16	0.50
	134.500	33.00	32.96	-0.04	-0.11
	135.200	34.10	34.11	0.01	0.04
	136.800	36.70	36.86	0.16	0.42
	137.900	38.80	38.84	0.04	0.09
	138.800	40.20	40.52	0.32	0.79
	139.700	42.40	42.25	-0.15	-0.35
	140.000	42.70	42.84	0.14	0.34
	140.600	44.20	44.05	-0.15	-0.35
	142.300	47.60	47.59	-0.01	-0.02
	142.800	49.10	48.67	-0.43	-0.88
	143.600	50.00	50.44	0.44	0.88
	144.200	51.50	51.80	0.30	0.58
	145.200	54.10	54.13	0.03	0.05
	145.800	55.40	55.56	0.16	0.29
	147.900	61.40	60.80	-0.60	-0.98
	149.500	65.30	65.04	-0.26	-0.40

STANDARD DEVIATION = 0.29

ISOAMYL ACETATE C7H14O2

USANOVICH M.,DEMBICKIJ A.: ZH.OBSHCH.KHIM.29,1771(1959).

A = 7.43566 B = 1606.581 C = 215.658

B.P.(100) = 79.905

T	P EXPTL.	P CALCD.	DEV.	PERCENT
40.600	15.00	14.66	-0.34	-2.23
47.500	22.00	21.41	-0.59	-2.67
58.000	37.00	36.72	-0.28	-0.76
64.000	49.00	49.07	0.07	0.15
71.800	70.00	70.26	0.26	0.38
77.500	89.50	90.24	0.74	0.83
83.400	114.00	115.75	1.75	1.54
88.000	142.00	139.61	-2.39	-1.69
94.500	180.00	180.21	0.21	0.12

STANDARD DEVIATION = 1.29

METHYL CAPROATE C7H14O2

ROSE A.,SUPINA W.R.: J.CHEM.ENG.DATA6,173(1961).

A = 7.40932 B = 1672.742 C = 218.980

B.P.(100) = 90.253

T	P EXPTL.	P CALCD.	DEV.	PERCENT
44.500	11.80	11.50	-0.30	-2.55
55.700	20.80	20.87	0.07	0.34
60.300	26.20	26.29	0.09	0.36
64.200	31.70	31.79	0.09	0.29
69.000	39.80	39.88	0.08	0.21
73.800	49.70	49.66	-0.04	-0.08
75.900	54.40	54.54	0.14	0.25
77.600	58.90	58.78	-0.12	-0.21
81.500	69.70	69.57	-0.13	-0.19
84.900	80.30	80.30	0.00	0.00
88.500	93.10	93.14	0.04	0.05
93.900	115.60	115.62	0.02	0.02
99.600	144.00	144.11	0.11	0.08
101.300	154.00	153.66	-0.34	-0.22
104.800	174.80	175.00	0.20	0.11

STANDARD DEVIATION = 0.17

METHYL CAPROATE C7H14O2

ALTHOUSE P.M.,TRIEBOLD H.O.:
IND.ENG.CHEM.ANAL.ED.16,605(1944).

A = 6.57998 B = 1300.053 C = 191.174

B.P.(10) = 41.812

T	P EXPTL.	P CALCD.	DEV.	PERCENT
15.000	2.00	1.88	-0.12	-5.96
26.000	4.00	3.92	-0.08	-1.90
33.000	6.00	6.03	0.03	0.58
38.000	8.00	8.08	0.08	0.95
42.000	10.00	10.10	0.10	1.04
55.000	20.00	19.90	-0.10	-0.48
70.000	40.00	40.02	0.02	0.04

STANDARD DEVIATION = 0.11

1-CHLOROHEPTANE C7H15CL

KEMME R.H.,KREPS S.I.:
J.CHEM.ENG.DATA 14,98(1969).

A = 6.91670 B = 1453.960 C = 199.825

B.P.(760) = 160.433

T	P EXPTL.	P CALCD.	DEV.	PERCENT
34.400	5.10	5.12	0.02	0.37
46.600	10.50	10.39	-0.11	-1.08
53.500	15.20	15.04	-0.16	-1.07
59.100	20.10	20.01	-0.09	-0.43
67.500	30.10	30.04	-0.06	-0.18
73.600	39.80	39.73	-0.07	-0.18
81.100	55.00	55.09	0.09	0.16
90.900	81.90	82.33	0.43	0.52
101.100	121.60	121.63	0.03	0.03
115.600	203.00	202.85	-0.15	-0.08
127.900	302.60	302.11	-0.49	-0.16
145.100	502.50	502.81	0.31	0.06
160.100	753.40	753.49	0.09	0.01

STANDARD DEVIATION = 0.25

2,2-DIMETHYLPENTANE C7H16

WILLINGHAM C.J.,TAYLOR W.J.,PIGNOCCO J.M.,ROSSINI F.D.: J.RESEARCH NATL.BUR.STANDARDS 35,219(1945).

A = 6.81262 B = 1188.892 C = 223.176

B.P.(760) = 79.202

T	P EXPTL.	P CALCD.	DEV.	PERCENT
15.325	67.27	67.26	-0.01	-0.01
18.253	77.31	77.31	0.00	0.00
20.979	87.74	87.74	0.00	0.00
24.670	103.68	103.69	0.01	0.01
28.879	124.68	124.69	0.01	0.01
33.160	149.46	149.48	0.02	0.02
37.129	175.94	175.92	-0.02	-0.01
42.454	217.21	217.20	-0.01	0.00
47.359	261.79	261.83	0.04	0.02
53.261	324.99	324.97	-0.02	-0.01
59.362	402.46	402.45	-0.01	0.00
65.879	500.72	500.71	-0.01	0.00
72.958	627.94	627.92	-0.02	0.00
77.959	732.11	732.12	0.01	0.00
78.496	744.07	744.06	-0.01	0.00
78.995	755.28	755.29	0.01	0.00
79.550	768.04	767.94	-0.10	-0.01
80.050	779.35	779.48	0.13	0.02

STANDARD DEVIATION = 0.04

2,3-DIMETHYLPENTANE C7H16

FORZIATI A.F.,NORRIS W.R.,ROSSINI F.D.:
J.RESEARCH NATL.BUR.STANDARDS 43,555(1949).

A = 6.85695 B = 1239.787 C = 222.024

B.P.(760) = 89.783

T	P EXPTL.	P CALCD.	DEV.	PERCENT
17.523	48.07	48.02	-0.05	-0.11
21.293	57.76	57.75	-0.01	-0.01
24.575	67.49	67.51	0.03	0.04
27.563	77.54	77.55	0.01	0.01
30.342	87.96	87.96	0.00	0.00
34.106	103.88	103.87	-0.01	-0.01
38.407	124.83	124.86	0.03	0.02
42.786	149.65	149.67	0.02	0.02
46.849	176.14	176.15	0.01	0.01
52.290	217.42	217.45	0.03	0.01
57.295	262.03	262.02	-0.01	0.00
63.333	325.30	325.28	-0.02	-0.01
69.562	402.83	402.78	-0.05	-0.01
76.209	501.07	500.99	-0.08	-0.02
83.429	628.23	628.19	-0.04	-0.01
88.531	732.46	732.45	-0.01	0.00
89.080	744.39	744.43	0.04	0.00
89.588	755.60	755.65	0.05	0.01
90.167	768.58	768.59	0.01	0.00
90.678	780.12	780.16	0.04	0.01

STANDARD DEVIATION = 0.04

2,4-DIMETHYLPENTANE C7H16

FORZIATI A.F.,NORRIS W.R.,ROSSINI F.D.:
J.RESEARCH NATL.BUR.STANDARDS 43,555(1949).

A = 6.83043 B = 1194.370 C = 221.902

B.P.(760) = 80.500

T	P EXPTL.	P CALCD.	DEV.	PERCENT
13.714	57.76	57.72	-0.05	-0.08
16.911	67.49	67.48	-0.01	-0.02
19.823	77.54	77.52	-0.02	-0.03
22.540	87.96	87.97	0.01	0.01
26.213	103.88	103.91	0.03	0.03
30.399	124.83	124.89	0.06	0.05
34.661	149.65	149.68	0.03	0.02
38.623	176.14	176.18	0.04	0.02
43.919	217.42	217.42	0.00	0.00
48.801	262.03	262.01	-0.02	-0.01
54.688	325.30	325.26	-0.04	-0.01
60.764	402.83	402.76	-0.07	-0.02
67.251	501.07	501.01	-0.06	-0.01
74.297	628.23	628.21	-0.02	0.00
79.277	732.46	732.45	-0.01	0.00
79.813	744.39	744.43	0.04	0.00
80.308	755.60	755.63	0.03	0.00
80.874	768.58	768.59	0.01	0.00
81.374	780.12	780.18	0.06	0.01

STANDARD DEVIATION = 0.04

3,3-DIMETHYLPENTANE C7H16

FORZIATI A.F.,NORRIS W.R.,ROSSINI F.D.:
J.RESEARCH NATL.BUR.STANDARDS 43,555(1949).

A = 6.82779 B = 1229.297 C = 225.389

B.P.(760) = 86.063

T	P EXPTL.	P CALCD.	DEV.	PERCENT
13.484	48.06	48.04	-0.02	-0.05
17.252	57.75	57.74	-0.01	-0.02
20.547	67.49	67.51	0.02	0.03
23.545	77.54	77.55	0.01	0.01
30.109	103.88	103.85	-0.03	-0.03
34.432	124.83	124.87	0.04	0.04
38.824	149.65	149.67	0.02	0.01
42.904	176.14	176.15	0.01	0.00
48.365	217.42	217.41	-0.01	0.00
53.396	262.02	262.01	-0.01	0.00
59.466	325.30	325.29	-0.01	0.00
65.727	402.83	402.80	-0.03	-0.01
72.411	501.06	501.03	-0.03	-0.01
79.672	628.23	628.23	0.00	0.00
84.803	732.45	732.45	0.00	0.00
85.355	744.38	744.42	0.04	0.00
85.866	755.60	755.63	0.03	0.00
86.447	768.57	768.54	-0.03	0.00
86.962	780.12	780.13	0.01	0.00

STANDARD DEVIATION = 0.02

3-ETHYLPENTANE C7H16

FORZIATI A.F.,NORRIS W.R.,ROSSINI F.D.:
J.RESEARCH NATL.BUR.STANDARDS 43,555(1949).

A = 6.87833 B = 1253.316 C = 220.050

B.P.(760) = 93.474

	T	P EXPTL.	P CALCD.	DEV.	PERCENT
	21.126	48.07	48.04	-0.03	-0.05
	24.900	57.76	57.77	0.01	0.02
	28.182	67.50	67.51	0.01	0.01
	31.179	77.55	77.55	0.00	0.00
	33.971	87.97	87.99	0.02	0.02
	37.734	103.89	103.86	-0.03	-0.03
	42.045	124.84	124.86	0.02	0.01
	46.423	149.66	149.61	-0.05	-0.03
	50.511	176.15	176.21	0.06	0.04
	55.954	217.43	217.47	0.04	0.02
	60.962	262.03	262.01	-0.02	-0.01
	73.242	402.84	402.77	-0.07	-0.02
	87.119	628.23	628.22	-0.01	0.00
*	92.238	732.46	732.81	0.35	0.05
	92.771	744.39	744.44	0.05	0.01
	93.277	755.61	755.61	0.00	0.00
	93.856	768.58	768.56	-0.02	0.00
	94.367	780.12	780.13	0.01	0.00

STANDARD DEVIATION = 0.04

HEPTANE C7H16

WILLINGHAM C.J.,TAYLOR W.J.,PIGNOCCO J.M.,ROSSINI F.D.:
J. RESEARCH NATL. BUR. STANDARDS 35, 219(1945).

A = 6.90253 B = 1267.828 C = 216.823

B.P.(760) = 98.423

T	P EXPTL.	P CALCD.	DEV.	PERCENT
25.925	47.78	47.83	0.05	0.11
29.699	57.49	57.50	0.01	0.02
33.024	67.33	67.31	-0.02	-0.02
36.017	77.34	77.30	-0.04	-0.05
38.822	87.76	87.74	-0.02	-0.02
42.599	103.67	103.61	-0.06	-0.06
46.929	124.54	124.63	0.09	0.07
51.320	149.41	149.40	-0.01	-0.01
55.394	175.84	175.83	-0.01	0.00
60.862	217.14	217.18	0.04	0.02
65.882	261.75	261.75	0.00	0.00
71.930	324.97	324.95	-0.02	-0.01
78.169	402.39	402.42	0.03	0.01
84.823	500.66	500.60	-0.06	-0.01
92.053	627.85	627.87	0.02	0.00
97.154	731.99	732.08	0.09	0.01
97.702	743.96	744.03	0.07	0.01
98.207	755.23	755.19	-0.04	-0.01
98.773	768.01	767.84	-0.17	-0.02
99.285	779.37	779.43	0.06	0.01

STANDARD DEVIATION = 0.07

2-METHYLHEXANE C7H16

FORZIATI A.F.,NORRIS W.R.,ROSSINI F.D.:
J.RESEARCH NATL.BUR.STANDARDS 43,555(1949).

A = 6.87689 B = 1238.122 C = 219.783

B.P.(760) = 90.052

T	P EXPTL.	P CALCD.	DEV.	PERCENT
18.528	48.08	48.03	-0.05	-0.11
22.260	57.77	57.76	-0.01	-0.03
25.518	67.51	67.54	0.03	0.04
28.469	77.55	77.54	-0.01	-0.01
31.235	87.97	88.00	0.03	0.03
34.956	103.90	103.88	-0.02	-0.02
39.219	124.84	124.89	0.05	0.04
43.549	149.67	149.67	0.00	0.00
47.579	176.15	176.20	0.05	0.03
52.960	217.44	217.45	0.01	0.01
57.913	262.04	262.02	-0.02	-0.01
63.889	325.31	325.28	-0.03	-0.01
70.051	402.84	402.77	-0.07	-0.02
76.628	501.07	501.00	-0.07	-0.01
83.769	628.24	628.21	-0.03	-0.01
88.814	732.46	732.46	0.00	0.00
89.357	744.40	744.45	0.05	0.01
89.860	755.62	755.68	0.06	0.01
90.430	768.58	768.57	-0.01	0.00
90.936	780.13	780.16	0.03	0.00

STANDARD DEVIATION = 0.04

3-METHYLHEXANE C7H16

FORZIATI A.F.,NORRIS W.R.,ROSSINI F.D.: J.RESEARCH NATL.BUR.STANDARDS 43,555(1949).

A = 6.86919 B = 1241.069 C = 219.323

B.P.(760) = 91.849

T	P EXPTL.	P CALCD.	DEV.	PERCENT
19.915	48.08	48.04	-0.04	-0.08
23.662	57.77	57.76	-0.01	-0.03
26.932	67.50	67.52	0.02	0.03
29.911	77.55	77.56	0.01	0.02
32.684	87.97	87.99	0.02	0.03
36.428	103.89	103.88	-0.01	-0.01
40.713	124.87	124.88	0.01	0.01
45.068	149.66	149.66	0.00	0.00
49.119	176.15	176.18	0.03	0.02
54.535	217.43	217.46	0.03	0.02
59.509	262.04	261.96	-0.08	-0.03
65.533	325.41	325.35	-0.06	-0.02
71.732	402.84	402.86	0.02	0.00
78.347	501.07	501.08	0.01	0.00
85.529	628.24	628.25	0.01	0.00
90.602	732.46	732.44	-0.02	0.00
91.148	744.40	744.41	0.01	0.00
91.655	755.61	755.67	0.06	0.01
92.229	768.58	768.57	-0.01	0.00
92.737	780.13	780.13	0.00	0.00

STANDARD DEVIATION = 0.03

2,2,3-TRIMETHYLBUTANE C7H16

FORZIATI A.F.,NORRIS W.R.,ROSSINI F.D.:
J.RESEARCH NATL.BUR.STANDARDS 43,555(1949).

A = 6.79481 B = 1201.965 C = 226.213

B.P.(760) = 80.881

T	P EXPTL.	P CALCD.	DEV.	PERCENT
12.555	57.68	57.65	-0.03	-0.06
15.833	67.43	67.45	0.02	0.02
18.804	77.47	77.48	0.01	0.01
21.571	87.90	87.89	-0.01	-0.01
25.313	103.81	103.78	-0.03	-0.03
29.599	124.75	124.79	0.04	0.03
33.960	149.58	149.60	0.02	0.01
38.012	176.06	176.11	0.05	0.03
43.424	217.34	217.31	-0.03	-0.01
48.422	261.95	261.95	0.00	0.00
54.445	325.22	325.19	-0.03	-0.01
60.668	402.77	402.75	-0.02	0.00
67.308	501.01	500.97	-0.04	-0.01
74.523	628.17	628.14	-0.03	0.00
79.627	732.41	732.42	0.01	0.00
80.174	744.32	744.34	0.02	0.00
80.682	755.50	755.56	0.06	0.01
81.262	768.54	768.52	-0.02	0.00
81.772	780.07	780.07	0.00	0.00

STANDARD DEVIATION = 0.03

1-HEPTANOL C7H16O

KEMME R.H.,KREPS S.I.:
J.CHEM.ENG.DATA 14,98(1969).

A = 6.97899 B = 1321.126 C = 145.985

B.P.(760) = 176.384

T	P EXPTL.	P CALCD.	DEV.	PERCENT
63.600	4.70	4.74	0.04	0.78
76.100	10.90	10.72	-0.18	-1.64
81.500	15.00	14.84	-0.16	-1.06
87.600	21.10	21.04	-0.06	-0.27
94.300	30.20	30.26	0.06	0.19
99.900	40.20	40.37	0.17	0.42
106.400	55.50	55.52	0.02	0.03
114.500	80.50	80.76	0.26	0.32
124.100	122.40	122.31	-0.09	-0.07
136.500	201.00	200.54	-0.46	-0.23
147.800	303.50	303.45	-0.06	-0.02
162.800	501.00	501.80	0.80	0.16
176.400	760.80	760.35	-0.45	-0.06

STANDARD DEVIATION = 0.35

1-HEPTANETHIOL $C_7H_{16}S$

OSBORN A.N.,DOUSLIN D.R.:
J.CHEM.ENG.DATA 11,502(1966).

A = 6.95238 B = 1525.232 C = 197.687

B.P.(760) = 176.919

T	P EXPTL.	P CALCD.	DEV.	PERCENT
101.627	71.87	71.88	0.01	0.01
104.908	81.64	81.63	-0.01	-0.01
108.205	92.52	92.51	-0.01	-0.01
111.517	104.63	104.62	-0.01	-0.01
114.840	118.06	118.05	-0.01	-0.01
118.182	132.95	132.95	0.00	0.00
121.546	149.41	149.48	0.07	0.04
128.269	187.57	187.55	-0.02	-0.01
135.066	233.72	233.72	0.00	0.00
141.911	289.13	289.13	0.00	0.00
148.807	355.22	355.20	-0.02	-0.01
155.759	433.56	433.57	0.01	0.00
162.758	525.86	525.83	-0.03	0.00
169.812	633.99	633.97	-0.02	0.00
176.919	760.00	759.99	-0.01	0.00
184.082	906.06	906.16	0.10	0.01
191.292	1074.60	1074.63	0.03	0.00
198.551	1268.00	1267.92	-0.08	-0.01

STANDARD DEVIATION = 0.04

DIETHYL PROPYLPHOSPHONATE $C_7H_{17}O_3P$

KOSOLAPOFF G.M.:
J.CHEM.SOC. 1955,2964.

A = 4.55811 B = 446.503 C = 26.165

B.P.(10) = 99.324

T	P EXPTL.	P CALCD.	DEV.	PERCENT
87.300	3.90	4.20	0.30	7.61
95.500	7.90	7.73	-0.17	-2.16
105.000	14.90	14.26	-0.65	-4.33
115.000	24.90	24.84	-0.06	-0.25
124.000	37.40	38.43	1.03	2.76
134.000	59.40	58.93	-0.47	-0.79

STANDARD DEVIATION = 0.78

DIISOPROPYL METHYLPHOSPHONATE $C_7H_{17}O_3P$

KOSOLAPOFF G.M.:

J.CHEM.SOC. 1955,2964.

A = 4.17979 B = 331.976 C = 22.964

B.P.(100) = 129.333

T	P EXPTL.	P CALCD.	DEV.	PERCENT
70.000	2.00	4.06	2.06	103.11
77.000	7.00	7.22	0.22	3.21
89.000	18.50	16.40	-2.10	-11.37
98.000	28.50	27.25	-1.25	-4.39
107.000	41.00	42.21	1.21	2.95
116.000	60.50	61.78	1.28	2.12
125.000	84.50	86.33	1.83	2.17
133.000	114.50	112.52	-1.98	-1.73

STANDARD DEVIATION = 2.03

1-BUTYL TRIMETHYLSILYL ETHER $C_7H_{18}SiO$

SHEEHAN R.J.,LANGER S.H.:

J.CHEM.ENG.DATA 14,248(1969).

A = 7.76300 B = 1884.675 C = 261.313

B.P.(760) = 124.718

T	P EXPTL.	P CALCD.	DEV.	PERCENT
71.200	124.40	124.46	0.06	0.05
95.200	300.00	299.63	-0.37	-0.12
110.600	495.30	495.99	0.70	0.14
124.100	746.80	746.42	-0.38	-0.05

STANDARD DEVIATION = 0.88

N-BUTYLAMINE TRIMETHYLBORON C7H20BN

BROWN H.C.,TAYLOR M.D.,SUJISHI S.: J.AM.CHEM.SOC. 73,2464(1951).

A = 8.46521 B = 1980.984 C = 193.597

B.P.(10) = 71.765

T	P EXPTL.	P CALCD.	DEV.	PERCENT
0.000	0.03	0.02	-0.01	-43.04
10.100	0.07	0.06	-0.02	-21.48
19.800	0.22	0.15	-0.07	-30.87
30.400	0.43	0.42	-0.01	-2.74
41.000	1.10	1.05	-0.05	-4.58
49.700	2.17	2.10	-0.07	-3.06
59.300	4.24	4.29	0.05	1.08
70.200	8.99	9.03	0.04	0.45
79.800	16.70	16.57	-0.13	-0.76
86.300	24.30	24.42	0.12	0.47
91.200	32.30	32.32	0.02	0.05
94.900	39.60	39.69	0.09	0.22
97.600	46.20	45.95	-0.25	-0.54
99.000	49.40	49.53	0.13	0.26

STANDARD DEVIATION = 0.11

PERFLUORODIMETHYLCYCLOHEXANE C8F16

FOWLER R.D.,ET AL.: IND.ENG.CHEM.39,375(1947).

A = 6.79965 B = 1224.595 C = 209.883

B.P.(760) = 102.607

T	P EXPTL.	P CALCD.	DEV.	PERCENT
2.890	12.30	11.07	-1.23	-9.98
11.820	19.70	18.88	-0.82	-4.15
20.810	31.40	31.00	-0.40	-1.29
30.040	49.30	49.60	0.30	0.62
39.920	78.40	78.96	0.56	0.71
49.930	121.10	121.97	0.87	0.72
59.800	180.70	181.46	0.76	0.42
69.710	262.50	262.86	0.36	0.14
80.100	377.00	377.27	0.27	0.07
90.100	524.00	521.71	-2.29	-0.44
95.200	612.00	610.48	-1.52	-0.25
100.100	712.00	706.51	-5.49	-0.77
106.800	854.00	856.44	2.44	0.29
114.500	1053.00	1058.01	5.01	0.48

STANDARD DEVIATION = 2.58

PERFLUOROETHYLCYCLOHEXANE C8F16

GOOD W.D.,ET AL.: J.PHYS.CHEM. 63,1133(1959).

A = 6.84993 B = 1216.651 C = 205.557

B.P.(760) = 100.973

T	P EXPTL.	P CALCD.	DEV.	PERCENT
38.098	71.87	71.88	0.01	0.01
40.834	81.64	81.66	0.02	0.03
43.570	92.52	92.53	0.01	0.01
46.326	104.63	104.64	0.01	0.01
49.086	118.06	118.05	-0.01	-0.01
51.864	132.95	132.93	-0.02	-0.02
54.667	149.41	149.46	0.05	0.03
60.275	187.57	187.56	-0.01	0.00
65.941	233.72	233.70	-0.02	-0.01
71.653	289.13	289.07	-0.06	-0.02
77.418	355.22	355.16	-0.07	-0.02
83.233	433.56	433.50	-0.06	-0.01
89.099	525.86	525.85	-0.01	0.00
95.011	633.99	633.99	0.00	0.00
100.975	760.00	760.04	0.04	0.01
106.990	906.06	906.20	0.14	0.02
113.055	1074.60	1074.79	0.19	0.02
119.166	1268.00	1268.19	0.19	0.02
125.326	1489.10	1489.15	0.05	0.00
131.532	1740.80	1740.34	-0.46	-0.03
* 137.786	2026.00	2024.78	-1.22	-0.06

STANDARD DEVIATION = 0.14

PERFLUOROOCTANE C8F18

KREGLEWSKI A.: BULL.ACAD.POL.SCI. 10,629(1962).

A = 5.90248 B = 1225.928 C = 198.993

B.P.(10) = 51.069

T	P EXPTL.	P CALCD.	DEV.	PERCENT
37.320	5.17	5.19	0.02	0.29
37.880	5.33	5.33	0.00	0.07
46.670	8.15	8.17	0.02	0.24
62.420	16.31	16.33	0.02	0.10
74.700	26.62	26.50	-0.12	-0.44
82.820	35.66	35.67	0.01	0.04
92.710	50.00	50.10	0.10	0.20
105.760	75.86	75.82	-0.04	-0.05

STANDARD DEVIATION = 0.07

ALPHA-BROMBENZYL CYANIDE C8H6BRN

GOULD C.,HOLZMAN G.,NIEMANN C.:
ANAL.CHEM. 19,204(1947).

A = 5.04459 B = 734.821 C = 59.273

B.P.(10) = 122.407

T	P EXPTL.	P CALCD.	DEV.	PERCENT
85.000	1.00	0.89	-0.11	-10.60
105.800	4.00	3.92	-0.08	-2.04
121.600	10.00	9.59	-0.41	-4.07
143.000	25.00	25.81	0.81	3.24
151.800	37.00	36.58	-0.42	-1.14

STANDARD DEVIATION = 0.71

O-BROMOVINYLBENZENE C8H7BR

DREISBACH R.R.,SHRADER S.A.:
IND.ENG.CHEM. 41,2879(1949).

A = 0.56497 B = 82.913 C = -191.706

B.P.(10) = 1.114

T	P EXPTL.	P CALCD.	DEV.	PERCENT
110.190	37.58	38.20	0.62	1.66
116.530	47.16	46.55	-0.61	-1.30
121.450	57.04	55.61	-1.43	-2.51
125.600	66.39	65.95	-0.44	-0.66
129.270	75.86	78.15	2.29	3.02

STANDARD DEVIATION = 2.03

P-BROMOVINYLBENZENE C8H7BR

DREISBACH R.R.,SHRADER S.A.: IND.ENG.CHEM. 41,2879(1949).

A = 12.50417 B = 7348.996 C = 559.042

B.P.(100) = 140.584

T	P EXPTL.	P CALCD.	DEV.	PERCENT
119.510	47.16	47.18	0.02	0.04
124.710	57.04	57.03	-0.01	-0.01
128.950	66.39	66.43	0.04	0.06
132.660	75.86	75.80	-0.06	-0.08
146.810	123.76	123.78	0.02	0.01

STANDARD DEVIATION = 0.06

O-CHLOROVINYLBENZENE C8H7CL

DREISBACH R.R.,SHRADER S.A.: IND.ENG.CHEM. 41,2879(1949).

A = 6.95663 B = 1602.248 C = 204.503

B.P.(100) = 118.751

T	P EXPTL.	P CALCD.	DEV.	PERCENT
98.510	47.16	46.66	-0.50	-1.07
103.570	57.04	56.98	-0.06	-0.10
111.410	75.86	76.70	0.84	1.11
124.820	123.76	123.41	-0.35	-0.28
154.940	315.52	315.53	0.01	0.00

STANDARD DEVIATION = 0.74

P-CHLOROVINYLBENZENE C8H7CL

DREISBACH R.R.,SHRADER S.A.:
IND.ENG.CHEM. 41,2879(1949).

A = 9.96907 B = 4093.480 C = 392.413

B.P.(100) = 121.258

T	P EXPTL.	P CALCD.	DEV.	PERCENT
100.910	47.16	46.91	-0.25	-0.52
106.090	57.04	57.22	0.18	0.31
109.970	66.39	66.21	-0.18	-0.27
113.760	75.86	76.20	0.34	0.45
127.270	123.76	123.65	-0.11	-0.09

STANDARD DEVIATION = 0.35

P-CHLOROACETOPHENONE C8H7CLO

DREISBACH R.R.,SHRADER S.A.:
IND.ENG.CHEM. 41,2879(1949).

A = 7.08457 B = 1693.631 C = 190.952

B.P.(760) = 211.933

T	P EXPTL.	P CALCD.	DEV.	PERCENT
122.010	47.16	47.09	-0.07	-0.14
126.800	57.04	56.82	-0.22	-0.38
130.900	66.39	66.44	0.05	0.08
134.540	75.86	76.08	0.22	0.29
148.330	123.76	123.81	0.05	0.04
178.380	315.52	315.44	-0.08	-0.03
195.790	507.50	507.39	-0.11	-0.02
211.940	760.00	760.13	0.13	0.02

STANDARD DEVIATION = 0.17

CYCLOOCTATETRAEN C_8H_8

SCOTT D.W., GROSS M.E., OLIVER G.D., HUFFMAN H.M.: J.AM.CHEM.SOC. 71, 1634 (1944).

A = 7.00669 B = 1472.112 C = 215.843

B.P.(10) = 29.235

T	P EXPTL.	P CALCD.	DEV.	PERCENT
0.000	1.51	1.54	0.03	1.73
15.000	4.29	4.26	-0.03	-0.66
20.000	5.84	5.82	-0.02	-0.37
25.000	7.84	7.84	0.00	0.01
30.000	10.43	10.44	0.01	0.09
35.000	13.72	13.74	0.02	0.16
40.000	17.91	17.90	-0.01	-0.08
45.000	23.06	23.07	0.01	0.04
50.000	29.45	29.46	0.01	0.02
55.000	37.2[illegible]	37.27	0.01	0.04
60.000	46.79	46.77	-0.03	-0.05
65.000	58.20	58.20	0.00	0.00
70.000	71.89	71.88	-0.01	-0.01
75.000	88.13	88.14	0.01	0.01

STANDARD DEVIATION = 0.02

STYRENE C8H8

DREYER R.,MARTIN W.,VON WEBER U.:
J.PRAKT.CHEM. 273,324(1954/55).

A = 7.06623 B = 1507.434 C = 214.985

B.P.(760) = 145.178

T	P EXPTL.	P CALCD.	DEV.	PERCENT
29.920	8.20	8.15	-0.05	-0.63
39.210	13.70	13.68	-0.02	-0.16
60.040	38.30	38.47	0.17	0.45
63.230	44.30	44.46	0.16	0.37
74.420	72.30	72.03	-0.27	-0.38
80.170	90.50	90.99	0.49	0.54
85.530	112.50	112.22	-0.28	-0.25
99.510	188.30	187.52	-0.78	-0.42
110.060	269.10	268.30	-0.80	-0.30
113.100	295.40	296.20	0.80	0.27
120.810	377.50	377.64	0.14	0.04
125.380	432.50	433.87	1.37	0.32
125.410	434.40	434.26	-0.14	-0.03
134.830	571.60	571.48	-0.12	-0.02
137.230	610.00	611.46	1.46	0.24
144.770	753.90	751.73	-2.17	-0.29

STANDARD DEVIATION = 0.92

STYRENE C8H8

CHAIYAVECH P.,VAN WINKLE M.:
J.CHEM.ENG.DATA 4,53(1959).

A = 7.14016 B = 1574.511 C = 224.087

B.P.(100) = 82.229

T	P EXPTL.	P CALCD.	DEV.	PERCENT
32.400	10.00	10.03	0.03	0.32
45.600	20.00	20.04	0.04	0.19
53.860	30.00	29.88	-0.12	-0.40
60.050	40.00	39.70	-0.30	-0.75
65.450	50.00	50.36	0.36	0.73
76.600	80.00	80.13	0.13	0.16
82.190	100.00	99.85	-0.15	-0.15

STANDARD DEVIATION = 0.26

2-(2,4-DICHLOROPHENOXY)-ETHANOL C8H8CL2O2

MCDONALD R.A.,SHRADER S.A.,STULL D.R.:
J.CHEM.ENG.DATA 4,311(1959).

A = 7.24009 B = 2004.308 C = 157.251

B.P.(100) = 225.244

T	P EXPTL.	P CALCD.	DEV.	PERCENT
211.480	63.86	63.74	-0.12	-0.19
215.870	73.84	73.85	0.01	0.01
219.680	83.61	83.69	0.08	0.09
223.100	93.34	93.42	0.08	0.09
236.460	141.07	141.02	-0.05	-0.04
267.660	333.52	333.49	-0.03	-0.01
286.300	526.37	526.40	0.03	0.00

STANDARD DEVIATION = 0.09

M-ETHYLACETOPHENONE C8H8O

KHOREVSKAYA A.S.,BYK S.SH.:
ZH.PRIKL.KHIM. 41,2566(1968).

A = 3.76715 B = 708.052 C = 182.638

B.P.(10) = 73.240

T	P EXPTL.	P CALCD.	DEV.	PERCENT
19.100	2.15	1.81	-0.34	-15.87
29.200	3.11	2.66	-0.45	-14.50
38.500	3.81	3.68	-0.13	-3.54
46.400	5.24	4.74	-0.50	-9.55
58.200	6.61	6.72	0.11	1.62
68.100	8.18	8.78	0.60	7.28
78.310	11.50	11.32	-0.18	-1.58
87.800	13.40	14.09	0.69	5.17
95.750	17.63	16.74	-0.89	-5.05
112.000	22.39	23.12	0.73	3.27
127.000	29.85	30.23	0.38	1.27
135.000	34.28	34.52	0.24	0.69
143.000	39.81	39.16	-0.65	-1.64

STANDARD DEVIATION = 0.59

P-ETHYLACETOPHENONE C8H80

KHOREVSKAYA A.S.,BYK S.SH.: ZH.PRIKL.KHIM. 41,2566(1968).

A = 4.27459 B = 629.339 C = 120.896

B.P.(10) = 71.293

T	P EXPTL.	P CALCD.	DEV.	PERCENT
21.400	0.86	0.71	-0.15	-17.33
36.400	2.00	1.88	-0.12	-6.12
45.200	3.00	3.06	0.06	1.96
58.420	5.74	5.82	0.08	1.39
67.850	8.84	8.72	-0.13	-1.42
77.100	12.55	12.47	-0.08	-0.60
87.250	17.36	17.83	0.47	2.68
94.050	22.49	22.22	-0.27	-1.21

STANDARD DEVIATION = 0.27

METHYL BENZOATE C8H802

DREISBACH R.R.,SHRADER S.A.: IND.ENG.CHEM. 41,2879(1949).

A = 7.27349 B = 1846.562 C = 220.906

B.P.(760) = 199.467

T	P EXPTL.	P CALCD.	DEV.	PERCENT
110.880	57.04	51.05	-5.99	-10.51
115.550	66.39	60.98	-5.41	-8.14
123.230	75.86	80.85	4.99	6.58
136.800	123.76	129.20	5.44	4.39
166.280	315.52	319.37	3.85	1.22
183.320	507.50	507.39	-0.11	-0.02
199.100	760.00	753.31	-6.69	-0.88

STANDARD DEVIATION = 6.70

METHYL SALICYLATE C8H8O3

MATTHEWS J.B.,SUMNER J.F.,MOELWYN-HUGHES E.A.:
TRANS.FARADAY SOC. 46,797(1950).

A = 7.08329 B = 1712.797 C = 187.066

B.P.(760) = 220.503

T	P EXPTL.	P CALCD.	DEV.	PERCENT
79.000	4.28	4.42	0.14	3.36
84.200	5.68	5.88	0.20	3.48
87.900	7.27	7.15	-0.12	-1.68
89.200	7.37	7.65	0.28	3.75
92.900	9.06	9.23	0.17	1.92
93.100	9.27	9.33	0.06	0.62
96.000	10.55	10.77	0.22	2.13
100.900	12.94	13.66	0.72	5.54
103.500	15.04	15.44	0.40	2.64
104.500	16.32	16.17	-0.15	-0.90
107.000	17.93	18.14	0.21	1.20
110.700	21.01	21.44	0.43	2.02
114.400	24.60	25.22	0.62	2.51
114.600	26.61	25.44	-1.17	-4.40
117.600	28.77	28.93	0.16	0.57
123.400	37.34	36.85	-0.49	-1.32
130.100	48.28	48.19	-0.09	-0.19
132.400	53.57	52.70	-0.87	-1.62
135.700	62.82	59.79	-3.03	-4.82
141.000	73.43	72.84	-0.59	-0.80
144.100	81.91	81.52	-0.39	-0.48
148.400	94.84	94.96	0.12	0.13
152.200	107.70	108.33	0.63	0.58
156.100	123.10	123.63	0.53	0.43
159.700	136.80	139.30	2.50	1.82
166.100	173.00	171.18	-1.82	-1.05
170.400	195.30	195.79	0.49	0.25
178.000	248.80	246.34	-2.46	-0.99
184.200	293.40	295.04	1.64	0.56
191.000	352.80	357.16	4.36	1.24
193.000	374.60	377.31	2.71	0.72
195.800	402.00	407.06	5.06	1.26
200.100	456.80	456.40	-0.40	-0.09
200.100	460.10	456.40	-3.70	-0.80
202.000	489.30	479.68	-9.62	-1.97
203.900	499.40	503.90	4.50	0.90
206.000	529.50	531.80	2.30	0.43
207.800	556.50	556.69	0.19	0.03
* 208.200	579.80	562.35	-17.45	-3.01
210.300	593.90	592.80	-1.11	-0.19
214.000	642.20	649.64	7.44	1.16
220.000	762.00	750.96	-11.04	-1.45
220.500	760.50	759.94	-0.56	-0.07

STANDARD DEVIATION = 3.18

2-BROMOETHYLBENZENE C8H9BR

MCDONALD R.A.,SHRADER S.A.,STULL D.R.:
J.CHEM.ENG.DATA 4,311(1959).

A = 7.79977 B = 2235.368 C = 238.729

B.P.(760) = 215.711

T	P EXPTL.	P CALCD.	DEV.	PERCENT
127.450	49.90	49.57	-0.33	-0.67
143.770	90.38	90.29	-0.09	-0.10
166.190	189.27	190.21	0.94	0.50
190.310	389.53	388.68	-0.85	-0.22
213.880	729.11	725.97	-3.14	-0.43
216.080	765.50	767.02	1.52	0.20
217.180	786.41	788.25	1.84	0.23

STANDARD DEVIATION = 2.08

P-ETHYLCHLOROBENZENE C8H9CL

DREISBACH R.R.,SHRADER S.A.:
IND.ENG.CHEM. 41,2879(1949).

A = 6.95112 B = 1557.129 C = 198.124

B.P.(760) = 184.434

T	P EXPTL.	P CALCD.	DEV.	PERCENT
108.980	75.86	75.99	0.13	0.17
122.330	123.76	123.59	-0.17	-0.14
151.620	315.52	315.45	-0.07	-0.02
168.660	507.50	507.88	0.38	0.08
184.420	760.00	759.73	-0.27	-0.04

STANDARD DEVIATION = 0.37

P-CHLOROPHENETOLE C8H9CLO

DREISBACH R.R.,SHRADER S.A.:
IND.ENG.CHEM. 41,2879(1949).

A = 7.08457 B = 1693.631 C = 190.952

B.P.(760) = 211.933

T	P EXPTL.	P CALCD.	DEV.	PERCENT
122.010	47.16	47.09	-0.07	-0.14
126.800	57.04	56.82	-0.22	-0.38
130.900	66.39	66.44	0.05	0.08
134.540	75.86	76.08	0.22	0.29
148.330	123.76	123.81	0.05	0.04
178.380	315.52	315.44	-0.08	-0.03
195.790	507.50	507.39	-0.11	-0.02
211.940	760.00	760.13	0.13	0.02

STANDARD DEVIATION = 0.17

BETA-CHLORO-BETA-PHENYL ETHYL ALCOHOL C8H9CLO

DREISBACH R.R.,SHRADER S.A.:
IND.ENG.CHEM. 41,2879(1949).

A = 6.91733 B = 1635.630 C = 145.870

B.P.(760) = 259.337

T	P EXPTL.	P CALCD.	DEV.	PERCENT
165.980	47.16	47.03	-0.13	-0.27
171.000	57.04	56.95	-0.09	-0.16
175.220	66.39	66.58	0.19	0.29
178.780	75.86	75.72	-0.14	-0.19
193.200	123.76	124.01	0.25	0.20
224.310	315.52	315.40	-0.12	-0.04
242.450	507.50	507.30	-0.20	-0.04
259.350	760.00	760.21	0.21	0.03

STANDARD DEVIATION = 0.22

2-METHYL-5-VINYLPYRIDINE C8H9N

FROLOV A.F.,LOGINOVA M.A.,KISELEVA M.M.:
ZH.FIZ.KHIM. 35,1784(1961).

A = 6.15609 B = 1022.917 C = 129.176

B.P.(760) = 183.139

T	P EXPTL.	P CALCD.	DEV.	PERCENT
69.620	11.00	10.25	-0.75	-6.86
85.750	25.00	24.93	-0.07	-0.29
94.200	39.50	37.73	-1.77	-4.47
120.100	112.00	112.86	0.86	0.77
143.920	251.00	257.31	6.31	2.51
154.520	357.00	355.14	-1.86	-0.52
163.150	462.00	453.79	-8.21	-1.78
183.190	757.00	760.93	3.93	0.52

STANDARD DEVIATION = 5.11

ETHYLBENZENE C8H10

WILLINGHAM C.J.,TAYLOR W.J.,PIGNOCCO J.M.,ROSSINI F.D.:
J.RESEARCH NATL.BUR.STANDARDS 35,219(1945).

A = 6.95650 B = 1423.543 C = 213.091

B.P.(760) = 136.185

T	P EXPTL.	P CALCD.	DEV.	PERCENT
56.589	47.68	47.63	-0.05	-0.11
60.796	57.41	57.40	-0.01	-0.01
64.463	67.22	67.24	0.02	0.03
67.775	77.28	77.29	0.01	0.01
70.862	87.75	87.74	-0.01	-0.01
75.027	103.65	103.67	0.02	0.02
79.777	124.67	124.68	0.01	0.01
84.599	149.44	149.46	0.02	0.02
89.071	175.89	175.92	0.03	0.02
95.056	217.17	217.18	0.01	0.00
100.561	261.75	261.75	0.00	0.00
107.183	324.93	324.89	-0.04	-0.01
114.020	402.44	402.38	-0.06	-0.02
121.312	500.70	500.65	-0.05	-0.01
129.221	627.93	627.90	-0.03	-0.01
134.800	732.08	732.11	0.04	0.00
135.399	744.05	744.07	0.02	0.00
135.954	755.24	755.28	0.04	0.01
136.574	767.95	767.96	0.01	0.00
137.124	779.34	779.35	0.01	0.00

STANDARD DEVIATION = 0.03

O-XYLENE C8H10

WILLINGHAM C.J.,TAYLOR W.J.,PIGNOCCO J.M.,ROSSINI F.D.:
J.RESEARCH NATL.BUR.STANDARDS 35,219(1945).

A = 7.00154 B = 1476.393 C = 213.872

B.P.(760) = 144.413

T	P EXPTL.	P CALCD.	DEV.	PERCENT
63.460	47.66	47.64	-0.02	-0.04
67.746	57.41	57.41	0.00	0.00
71.481	67.23	67.24	0.01	0.01
74.857	77.28	77.29	0.01	0.01
77.993	87.75	87.71	-0.04	-0.04
82.242	103.65	103.67	0.02	0.02
87.081	124.67	124.68	0.01	0.01
91.987	149.44	149.45	0.01	0.01
96.541	175.89	175.92	0.03	0.02
102.632	217.17	217.19	0.02	0.01
108.227	261.75	261.74	-0.01	0.00
114.965	324.94	324.93	-0.01	0.00
121.909	402.44	402.38	-0.06	-0.01
129.318	500.71	500.68	-0.03	-0.01
137.346	627.93	627.91	-0.02	0.00
143.007	732.09	732.11	0.02	0.00
143.614	744.06	744.05	-0.01	0.00
144.176	755.25	755.24	-0.01	0.00
144.809	767.95	768.00	0.05	0.01
145.367	779.36	779.39	0.03	0.00

STANDARD DEVIATION = 0.03

O-XYLENE C8H10

PITZER K.S.,SCOTT D.W.:
J.AM.CHEM.SOC. 65,803(1943).

A = 6.81499 B = 1386.228 C = 206.313

B.P.(10) = 32.076

T	P EXPTL.	P CALCD.	DEV.	PERCENT
0.000	1.30	1.25	-0.05	-4.06
12.500	3.00	3.02	0.02	0.61
25.000	6.60	6.64	0.04	0.60
37.500	13.50	13.47	-0.03	-0.22
50.000	25.50	25.51	0.01	0.03

STANDARD DEVIATION = 0.05

M-XYLENE

C8H10

WILLINGHAM C.J.,TAYLOR W.J.,PIGNOCCO J.M.,ROSSINI F.D.:
J.RESEARCH NATL.BUR.STANDARDS 35,219(1945).

A = 7.00646 B = 1460.183 C = 214.827

B.P.(760) = 139.101

T	P EXPTL.	P CALCD.	DEV.	PERCENT
59.203	47.67	47.63	-0.04	-0.08
63.436	57.41	57.41	0.00	-0.01
67.123	67.23	67.23	0.00	0.00
70.458	77.28	77.29	0.01	0.01
73.558	87.74	87.73	-0.01	-0.01
77.747	103.65	103.67	0.02	0.02
82.522	124.67	124.68	0.01	0.01
87.367	149.44	149.46	0.02	0.01
91.860	175.89	175.92	0.03	0.02
97.870	217.17	217.18	0.01	0.00
103.396	261.75	261.76	0.01	0.00
110.041	324.94	324.91	-0.03	-0.01
116.896	402.44	402.39	-0.05	-0.01
124.205	500.70	500.66	-0.04	-0.01
132.128	627.93	627.90	-0.03	0.00
137.713	732.09	732.09	0.00	0.00
138.314	744.06	744.07	0.01	0.00
138.869	755.25	755.27	0.02	0.00
139.493	767.95	768.02	0.07	0.01
140.041	779.36	779.35	-0.01	0.00

STANDARD DEVIATION = 0.03

M-XYLENE

C8H10

PITZER K.S.,SCOTT D.W.:
J.AM.CHEM.SOC. 65,803(1943).

A = 7.33324 B = 1639.046 C = 230.686

B.P.(10) = 28.114

T	P EXPTL.	P CALCD.	DEV.	PERCENT
0.000	1.75	1.69	-0.06	-3.37
12.500	3.90	3.92	0.02	0.53
25.000	8.35	8.37	0.02	0.27
37.500	16.65	16.66	0.01	0.05
50.000	31.20	31.18	-0.02	-0.08
60.000	49.50	49.51	0.01	0.02

STANDARD DEVIATION = 0.04

P-XYLENE C8H10

WILLINGHAM C.J.,TAYLOR W.J.,PIGNOCCO J.M.,ROSSINI F.D.: J.RESEARCH NATL.BUR.STANDARDS 35,219(1945).

A = 6.98820 B = 1451.792 C = 215.111

B.P.(760) = 138.347

T	P EXPTL.	P CALCD.	DEV.	PERCENT
58.288	47.66	47.65	-0.01	-0.03
62.523	57.43	57.42	-0.01	-0.02
66.216	67.25	67.25	0.00	0.00
69.549	77.30	77.29	-0.01	-0.02
72.657	87.73	87.74	0.01	0.01
76.852	103.67	103.68	0.01	0.01
81.636	124.68	124.69	0.01	0.01
86.488	149.45	149.47	0.02	0.01
90.990	175.92	175.93	0.01	0.01
97.013	217.19	217.21	0.02	0.01
102.546	261.78	261.75	-0.03	-0.01
109.211	324.97	324.95	-0.02	-0.01
116.083	402.45	402.43	-0.02	0.00
123.409	500.71	500.68	-0.03	-0.01
131.355	627.93	627.94	0.01	0.00
136.956	732.10	732.11	0.01	0.00
137.558	744.07	744.07	0.00	0.00
138.114	755.27	755.26	-0.01	0.00
138.742	768.00	768.05	0.05	0.01
139.289	779.36	779.34	-0.02	0.00

STANDARD DEVIATION = 0.02

P-XYLENE C8H10

PITZER K.S.,SCOTT D.W.: J.AM.CHEM.SOC. 65,803(1943).

A = 7.06992 B = 1505.942 C = 220.995

B.P.(10) = 27.104

T	P EXPTL.	P CALCD.	DEV.	PERCENT
25.000	8.90	8.87	-0.03	-0.30
37.500	17.50	17.54	0.04	0.25
50.000	32.60	32.57	-0.03	-0.09
60.000	51.35	51.36	0.01	0.02

STANDARD DEVIATION = 0.06

O-ETHYLPHENOL $C_8H_{10}O$

VONTERRES E.,ET AL.:
BRENNSTOFF CHEM. 36,272(1955).

A = 7.80035 B = 2140.360 C = 227.401

B.P.(760) = 207.672

T	P EXPTL.	P CALCD.	DEV.	PERCENT
86.000	10.00	9.35	-0.65	-6.48
106.500	25.00	24.56	-0.44	-1.77
123.000	50.00	49.21	-0.79	-1.58
134.000	75.00	75.50	0.50	0.67
141.600	100.00	99.99	-0.01	-0.01
153.400	150.00	151.24	1.24	0.83
161.300	200.00	196.75	-3.25	-1.63
169.000	250.00	251.70	1.70	0.68
175.000	300.00	302.96	2.96	0.99
177.200	325.00	323.82	-1.18	-0.36
179.900	350.00	351.05	1.05	0.30
184.200	400.00	398.36	-1.64	-0.41
188.700	450.00	453.43	3.43	0.76
191.600	500.00	492.17	-7.83	-1.57
196.000	550.00	556.15	6.15	1.12
199.000	600.00	603.62	3.62	0.60
201.600	650.00	647.42	-2.58	-0.40
204.500	700.00	699.33	-0.67	-0.10
207.500	760.00	756.60	-3.40	-0.45

STANDARD DEVIATION = 3.29

M-ETHYLPHENOL C8H10O

VONTERRES E.,ET AL.: BRENNSTOFF CHEM. 36,272(1955).

A = 7.46793 B = 1855.639 C = 187.073

B.P.(760) = 217.460

T	P EXPTL.	P CALCD.	DEV.	PERCENT
97.200	10.00	8.71	-1.29	-12.85
* 106.000	25.00	13.69	-11.31	-45.26
134.000	50.00	48.80	-1.20	-2.40
144.500	75.00	74.38	-0.62	-0.83
152.000	100.00	98.91	-1.09	-1.09
163.500	150.00	149.54	-0.46	-0.30
172.500	200.00	202.89	2.89	1.44
179.000	250.00	250.54	0.54	0.22
184.900	300.00	301.50	1.50	0.50
187.600	325.00	327.52	2.52	0.77
189.900	350.00	351.12	1.12	0.32
194.100	400.00	397.82	-2.18	-0.54
199.000	450.00	458.65	8.65	1.92
202.000	500.00	499.51	-0.49	-0.10
205.000	550.00	543.29	-6.71	-1.22
208.500	600.00	598.29	-1.71	-0.29
211.000	650.00	640.28	-9.72	-1.50
213.800	700.00	690.13	-9.87	-1.41
218.200	760.00	774.80	14.80	1.95

STANDARD DEVIATION = 6.12

P-ETHYLPHENOL C8H10O

VONTERRES E.,ET AL.:
BRENNSTOFF CHEM. 36,272(1955).

A = 8.29091 B = 2423.197 C = 229.000

B.P.(760) = 218.902

	T	P EXPTL.	P CALCD.	DEV.	PERCENT
	101.000	10.00	8.87	-1.13	-11.31
*	109.800	25.00	13.76	-11.24	-44.96
	137.500	50.00	47.77	-2.23	-4.45
	148.000	75.00	73.00	-2.00	-2.66
	155.500	100.00	97.43	-2.57	-2.57
	167.000	150.00	148.50	-1.50	-1.00
	177.200	200.00	211.54	11.54	5.77
	182.000	250.00	248.35	-1.65	-0.66
	187.700	300.00	299.02	-0.98	-0.33
	190.500	325.00	326.98	1.98	0.61
	192.600	350.00	349.38	-0.62	-0.18
	197.000	400.00	400.55	0.55	0.14
	201.700	450.00	462.09	12.09	2.69
	204.000	500.00	495.01	-4.99	-1.00
	207.400	550.00	547.29	-2.71	-0.49
	210.800	600.00	604.15	4.15	0.69
	213.100	650.00	645.37	-4.63	-0.71
	216.100	700.00	702.67	2.67	0.38
	218.200	760.00	745.27	-14.73	-1.94

STANDARD DEVIATION = 6.33

ETHYL PHENYL ETHER C8H10O

COLLERSON R.R., ET AL.: J.CHEM.SOC. 3697(1965).

A = 7.02138 B = 1508.391 C = 194.490

B.P.(760) = 169.806

T	P EXPTL.	P CALCD.	DEV.	PERCENT
117.431	153.32	153.31	-0.01	-0.01
124.908	198.96	198.96	0.00	0.00
131.478	247.70	247.72	0.02	0.01
137.430	299.86	299.87	0.01	0.00
142.338	349.27	349.26	-0.01	0.00
146.908	400.94	400.95	0.01	0.00
150.795	449.59	449.61	0.02	0.00
154.432	499.32	499.31	-0.01	0.00
157.829	549.63	549.61	-0.02	0.00
160.982	599.84	599.83	-0.01	0.00
163.972	650.77	650.76	-0.01	0.00
166.622	698.70	698.72	0.02	0.00
169.315	750.31	750.28	-0.03	0.00
171.821	800.98	800.92	-0.06	-0.01
174.190	851.22	851.23	0.01	0.00
176.407	900.51	900.54	0.03	0.00
178.511	949.37	949.38	0.01	0.00
180.608	1000.09	1000.12	0.03	0.00

STANDARD DEVIATION = 0.02

ALPHA-PHENYL ETHYL ALCOHOL C8H10O

DREISBACH R.R.,SHRADER S.A.: IND.ENG.CHEM. 41,2879(1949).

A = 1.50821 B = 91.426 C = -263.295

B.P.(100) = 77.390

T	P EXPTL.	P CALCD.	DEV.	PERCENT
81.990	75.95	102.91	26.96	35.50
84.780	85.65	104.80	19.15	22.36
147.430	233.55	198.28	-35.27	-15.10
174.080	324.90	341.19	16.29	5.01
189.760	518.85	564.30	45.45	8.76

STANDARD DEVIATION =48.31

BETA-PHENYL ETHYL ALCOHOL C8H10O

DREISBACH R.R.,SHRADER S.A.:
IND.ENG.CHEM. 41,2879(1949).

A = 6.60421 B = 1306.481 C = 132.043

B.P.(760) = 218.841

T	P EXPTL.	P CALCD.	DEV.	PERCENT
133.050	47.16	47.40	0.24	0.52
137.150	57.04	56.35	-0.69	-1.21
141.100	66.39	66.23	-0.16	-0.24
144.610	75.86	76.17	0.31	0.40
157.600	123.76	124.04	0.28	0.22
186.300	315.52	316.38	0.86	0.27
202.910	507.50	505.50	-2.00	-0.40
218.900	760.00	761.09	1.09	0.14

STANDARD DEVIATION = 1.15

5-OXIHYDRINDENE C8H10O

VONTERRES E.,ET AL.:
BRENNSTOFF CHEM. 36,272(1955).

A = 9.21371 B = 3665.771 C = 326.350

B.P.(760) = 252.495

T	P EXPTL.	P CALCD.	DEV.	PERCENT
120.500	10.00	10.24	0.24	2.36
143.000	25.00	25.32	0.32	1.27
161.500	50.00	50.07	0.07	0.14
172.900	75.00	74.33	-0.67	-0.89
181.000	100.00	97.36	-2.64	-2.64
194.900	150.00	151.73	1.73	1.15
204.000	200.00	200.32	0.32	0.16
211.800	250.00	252.30	2.30	0.92
218.000	300.00	301.65	1.65	0.55
220.500	325.00	323.81	-1.19	-0.37
223.200	350.00	349.32	-0.68	-0.19
228.000	400.00	399.01	-0.99	-0.25
232.200	450.00	447.41	-2.59	-0.58
236.500	500.00	502.16	2.16	0.43
240.100	550.00	552.38	2.38	0.43
243.500	600.00	603.74	3.74	0.62
246.200	650.00	647.42	-2.58	-0.40
249.000	700.00	695.57	-4.43	-0.63
* 251.000	760.00	731.84	-28.16	-3.71

STANDARD DEVIATION = 2.29

2,3-XYLENOL C8H10O

ANDON R.J.L.,ET AL.: J.CHEM.SOC. 5246(1960).

A = 7.05397 B = 1617.566 C = 170.742

B.P.(760) = 216.871

	T	P EXPTL.	P CALCD.	DEV.	PERCENT
*	149.346	99.79	100.11	0.32	0.32
	160.940	150.35	150.35	0.00	0.00
	169.634	200.30	200.30	0.00	0.00
	176.314	247.24	247.25	0.01	0.01
	182.500	298.37	298.37	0.00	0.00
	187.374	344.43	344.42	-0.01	0.00
	192.094	394.33	394.32	-0.01	0.00
	196.778	449.43	449.43	0.00	0.00
	200.752	500.88	500.89	0.01	0.00
	204.000	546.36	546.37	0.01	0.00
	207.544	599.68	599.68	0.00	0.00
	210.612	649.12	649.12	0.00	0.00
	213.454	697.75	697.75	0.00	0.00
	214.137	709.89	709.86	-0.03	0.00
	214.768	721.19	721.19	0.00	0.00
	215.091	727.03	727.05	0.02	0.00
	215.646	737.18	737.20	0.02	0.00
	216.144	746.41	746.40	-0.01	0.00
	216.987	762.16	762.19	0.03	0.00
	217.323	768.59	768.56	-0.03	0.00
	217.928	780.14	780.13	-0.01	0.00
	218.467	790.54	790.55	0.01	0.00

STANDARD DEVIATION = 0.02

2,3-XYLENOL C8H100

VONTERRES E.,ET AL.:
BRENNSTOFF CHEM. 36,272(1955).

A = 6.76034 B = 1547.289 C = 181.013

B.P.(760) = 217.822

T	P EXPTL.	P CALCD.	DEV.	PERCENT
84.000	10.00	8.35	-1.65	-16.48
105.600	25.00	23.00	-2.00	-7.99
124.000	50.00	48.69	-1.31	-2.61
135.200	75.00	73.64	-1.36	-1.81
144.000	100.00	99.91	-0.09	-0.09
157.000	150.00	152.31	2.31	1.54
166.100	200.00	200.79	0.79	0.39
174.000	250.00	252.31	2.31	0.92
180.200	300.00	299.73	-0.27	-0.09
183.600	325.00	328.61	3.61	1.11
186.100	350.00	351.22	1.22	0.35
191.000	400.00	399.11	-0.89	-0.22
196.000	450.00	453.16	3.16	0.70
200.000	500.00	500.43	0.43	0.09
203.900	550.00	550.16	0.16	0.03
207.000	600.00	592.39	-7.61	-1.27
210.100	650.00	637.11	-12.89	-1.98
213.500	700.00	689.14	-10.86	-1.55
219.000	760.00	780.25	20.25	2.66

STANDARD DEVIATION = 7.07

2,4-XYLENOL C8H10O

ANDON R.J.L.,ET AL.: J.CHEM.SOC. 5246(1960).

A = 7.05539 B = 1587.459 C = 169.339

B.P.(760) = 210.929

T	P EXPTL.	P CALCD.	DEV.	PERCENT
144.382	98.97	98.92	-0.05	-0.05
156.053	150.19	150.24	0.05	0.03
164.129	197.18	197.21	0.03	0.02
171.241	247.90	247.94	0.04	0.01
176.700	293.69	293.68	-0.01	0.00
182.116	345.61	345.59	-0.02	-0.01
186.619	394.21	394.19	-0.02	-0.01
191.348	451.02	451.00	-0.02	-0.01
194.795	496.45	496.41	-0.04	-0.01
198.444	548.41	548.39	-0.02	0.00
201.747	599.13	599.11	-0.02	0.00
204.809	649.43	649.41	-0.02	0.00
207.513	696.56	696.57	0.01	0.00
208.299	710.79	710.77	-0.02	0.00
208.898	721.74	721.75	0.01	0.00
209.314	729.44	729.46	0.02	0.00
209.847	739.41	739.42	0.01	0.00
210.450	750.83	750.83	0.00	0.00
210.951	760.42	760.41	-0.01	0.00
211.453	770.06	770.10	0.04	0.01
211.888	778.59	778.59	0.00	0.00
212.320	787.06	787.08	0.02	0.00

STANDARD DEVIATION = 0.03

2,5-XYLENOL

$C_8H_{10}O$

ANDON R.J.L.,ET AL.: J.CHEM.SOC. 5246(1960).

A = 7.05156 B = 1592.698 C = 170.742

B.P.(760) = 211.131

	T	P EXPTL.	P CALCD.	DEV.	PERCENT
*	143.924	97.43	97.72	0.29	0.30
	154.805	144.27	144.27	0.00	0.00
	164.433	199.40	199.40	0.00	0.00
	171.549	250.33	250.33	0.00	0.00
	176.902	295.23	295.22	-0.01	0.00
	182.455	348.48	348.48	0.00	0.00
	187.328	401.35	401.37	0.03	0.01
	191.460	451.10	451.12	0.02	0.00
	194.984	497.38	497.35	-0.03	-0.01
	198.830	552.06	552.06	0.00	0.00
	201.834	598.06	598.04	-0.02	0.00
	204.603	643.08	643.09	0.01	0.00
	207.748	697.47	697.47	0.00	0.00
	208.661	713.92	713.92	0.00	0.00
	209.138	722.68	722.64	-0.04	-0.01
	209.556	730.36	730.35	-0.01	0.00
	210.117	740.79	740.80	0.01	0.00
	210.711	751.99	751.99	0.00	0.00
	211.232	761.91	761.92	0.01	0.00
	211.736	771.59	771.62	0.03	0.00

STANDARD DEVIATION = 0.02

2,6-XYLENOL $C_8H_{10}O$

ANDON R.J.L.,ET AL.: J.CHEM.SOC. 5246(1960).

A = 7.07070 B = 1628.323 C = 187.603

B.P.(760) = 201.029

T	P EXPTL.	P CALCD.	DEV.	PERCENT
144.798	148.61	148.60	-0.01	-0.01
153.385	197.43	197.42	-0.01	-0.01
160.624	248.11	248.12	0.01	0.00
166.759	298.92	298.96	0.04	0.01
171.831	347.08	347.10	0.02	0.01
176.276	394.26	394.27	0.01	0.00
180.369	442.14	442.14	0.00	0.00
184.792	499.01	499.03	0.02	0.00
188.694	553.93	553.94	0.01	0.00
191.754	600.37	600.30	-0.07	-0.01
194.810	649.67	649.64	-0.03	0.00
196.676	681.38	681.31	-0.07	-0.01
198.414	711.94	711.91	-0.03	0.00
198.937	721.38	721.33	-0.05	-0.01
199.432	730.36	730.33	-0.03	0.00
199.983	740.47	740.46	-0.01	0.00
200.450	749.17	749.13	-0.04	-0.01
201.059	760.56	760.56	0.00	0.00
201.616	771.12	771.13	0.01	0.00
202.519	788.44	788.52	0.08	0.01
203.525	808.10	808.25	0.15	0.02

STANDARD DEVIATION = 0.05

3,4-XYLENOL $C_8H_{10}O$

ANDON R.J.L.,ET AL.: J.CHEM.SOC. 5246(1960).

A = 7.07919 B = 1621.451 C = 159.261

B.P.(760) = 226.947

T	P EXPTL.	P CALCD.	DEV.	PERCENT
171.933	152.57	152.55	-0.02	-0.01
180.131	200.29	200.30	0.01	0.00
186.548	245.64	245.66	0.02	0.01
193.380	302.79	302.81	0.02	0.01
197.879	346.04	346.02	-0.02	-0.01
202.370	393.98	393.98	0.00	0.00
207.376	453.62	453.63	0.01	0.00
211.126	502.91	502.89	-0.02	0.00
214.126	545.32	545.32	0.00	0.00
217.587	597.79	597.76	-0.03	0.00
220.898	651.67	651.63	-0.04	-0.01
223.649	699.26	699.27	0.01	0.00
224.292	710.79	710.80	0.01	0.00
224.799	720.03	719.99	-0.04	-0.01
225.276	728.71	728.72	0.01	0.00
225.928	740.84	740.80	-0.04	-0.01
226.392	749.46	749.49	0.03	0.00
226.921	759.38	759.49	0.11	0.01
227.397	768.62	768.59	-0.03	0.00
228.487	789.73	789.73	0.00	0.00
228.899	797.85	797.85	0.00	0.00

STANDARD DEVIATION = 0.04

3,5-XYLENOL C8H10O

ANDON R.J.L.,ET AL.: J.CHEM.SOC. 5246(1960).

A = 7.13076 B = 1639.856 C = 164.162

B.P.(760) = 221.691

T	P EXPTL.	P CALCD.	DEV.	PERCENT
154.720	97.39	97.33	-0.06	-0.06
166.472	148.21	148.26	0.05	0.03
175.126	198.35	198.39	0.04	0.02
182.389	250.46	250.51	0.05	0.02
187.937	297.45	297.43	-0.02	-0.01
192.743	343.64	343.64	0.00	0.00
197.298	392.67	392.64	-0.03	-0.01
201.925	448.10	448.06	-0.04	-0.01
205.601	496.49	496.45	-0.04	-0.01
209.151	547.11	547.07	-0.04	-0.01
212.159	593.10	593.14	0.04	0.01
215.799	653.08	652.99	-0.09	-0.01
218.298	696.84	696.79	-0.05	-0.01
219.146	712.20	712.18	-0.02	0.00
219.584	720.26	720.23	-0.03	0.00
220.097	729.78	729.75	-0.03	0.00
220.692	740.84	740.92	0.08	0.01
221.309	752.57	752.65	0.08	0.01
221.709	760.28	760.33	0.05	0.01
222.279	771.40	771.39	-0.01	0.00
222.724	780.12	780.10	-0.02	0.00
223.321	791.86	791.92	0.06	0.01

STANDARD DEVIATION = 0.05

CYCLOHEXYL TRIFLUOROACETATE C8H11F3O2

SHEEHAN R.J.,LANGER S.H.: J.CHEM.ENG.DATA 14,248(1969).

A = 7.80235 B = 1954.662 C = 249.327

B.P.(760) = 147.837

T	P EXPTL.	P CALCD.	DEV.	PERCENT
72.000	52.30	52.39	0.09	0.18
93.500	126.30	126.11	-0.19	-0.15
117.800	300.50	300.71	0.21	0.07
146.900	740.00	739.89	-0.11	-0.01

STANDARD DEVIATION = 0.31

DIMETHYLANILINE C8H11N

NELSON O.A.,WALES H.: J.AM.CHEM.SOC.47,867(1925).

A = 7.36768 B = 1857.076 C = 220.355

B.P.(760) = 193.536

T	P EXPTL.	P CALCD.	DEV.	PERCENT
* 71.020	10.90	9.87	-1.03	-9.48
80.750	16.30	15.85	-0.45	-2.73
91.670	26.00	26.06	0.06	0.23
95.890	31.50	31.29	-0.21	-0.67
101.800	39.90	40.10	0.20	0.50
114.890	67.20	67.33	0.13	0.20
118.500	77.70	77.13	-0.57	-0.73
128.610	110.60	111.18	0.58	0.52
134.950	138.00	138.35	0.35	0.25
140.510	166.80	166.53	-0.27	-0.16
143.910	185.70	186.01	0.31	0.17
152.090	240.70	240.72	0.02	0.01
156.220	273.20	273.02	-0.18	-0.07
162.860	333.40	332.38	-1.02	-0.31
166.160	364.60	365.61	1.01	0.28
168.370	388.60	389.34	0.74	0.19
171.870	430.10	429.50	-0.60	-0.14
174.100	456.30	456.80	0.50	0.11
179.420	528.60	527.69	-0.91	-0.17
182.010	566.50	565.30	-1.20	-0.21
190.300	700.90	700.57	-0.33	-0.05
190.910	711.50	711.47	-0.03	0.00
* 193.980	764.90	768.45	3.55	0.46
195.400	796.30	796.02	-0.28	-0.04
196.780	821.70	823.57	1.87	0.23

STANDARD DEVIATION = 0.72

P-ETHYLAMINOBENZENE C8H11N

DREISBACH R.R.,SHRADER S.A.: IND.ENG.CHEM. 41,2879(1949).

A = 7.07504 B = 1688.340 C = 185.253

B.P.(760) = 217.286

T	P EXPTL.	P CALCD.	DEV.	PERCENT
140.410	76.64	77.75	1.11	1.45
153.960	127.30	125.26	-2.04	-1.60
183.880	315.30	317.13	1.83	0.58
201.170	507.80	508.03	0.23	0.05
217.220	760.00	758.79	-1.21	-0.16

STANDARD DEVIATION = 2.27

N-ETHYLANILINE

C8H11N

NELSON O.A.,WALES H.: J.AM.CHEM.SOC. 47,867 (1927).

A = 7.42283 B = 1903.351 C = 214.318

B.P.(760) = 204.736

T	P EXPTL.	P CALCD.	DEV.	PERCENT
50.000	2.40	1.67	-0.73	-30.56
60.000	4.00	3.05	-0.95	-23.74
70.000	6.10	5.35	-0.75	-12.29
81.000	10.30	9.50	-0.80	-7.77
91.550	16.90	15.85	-1.05	-6.21
101.520	24.20	24.92	0.72	2.96
107.420	32.00	32.14	0.14	0.42
110.660	38.50	36.81	-1.69	-4.39
116.580	46.50	46.85	0.35	0.76
122.990	60.20	60.26	0.06	0.11
128.880	74.60	75.32	0.72	0.96
134.420	91.90	92.26	0.36	0.39
135.080	92.50	94.47	1.97	2.13
143.470	126.70	126.78	0.08	0.06
146.580	141.60	140.90	-0.71	-0.50
152.110	167.50	169.23	1.73	1.04
156.340	194.00	193.98	-0.02	-0.01
159.260	215.20	212.77	-2.43	-1.13
162.660	235.70	236.51	0.81	0.34
169.970	295.80	295.05	-0.75	-0.25
172.670	319.90	319.49	-0.41	-0.13
182.410	424.30	421.89	-2.41	-0.57
184.080	439.80	441.89	2.09	0.48
189.060	509.00	506.17	-2.83	-0.56
190.300	521.90	523.31	1.41	0.27
194.600	586.10	586.44	0.34	0.06
197.000	624.00	624.28	0.28	0.05
200.690	685.50	686.32	0.82	0.12
200.900	688.80	690.00	1.20	0.17
203.850	745.70	743.34	-2.36	-0.32
205.200	769.30	768.83	-0.47	-0.06
206.700	795.80	797.99	2.19	0.27

STANDARD DEVIATION = 1.39

2-METHYL-5-ETHYLPYRIDINE C8H11N

FROLOV A.F.,LOGINOVA M.A.,KISELEVA M.M.: ZH.FIZ.KHIM. 35,1784(1961).

A = 5.04999 B = 516.848 C = 58.997

B.P.(760) = 179.272

T	P EXPTL.	P CALCD.	DEV.	PERCENT
51.900	6.00	2.45	-3.55	-59.15
75.940	24.00	16.58	-7.42	-30.90
86.340	37.00	31.17	-5.83	-15.75
94.180	52.00	47.40	-4.60	-8.84
100.560	92.00	64.67	-27.33	-29.70
114.570	116.00	118.08	2.08	1.79
121.590	144.00	154.15	10.15	7.05
130.700	186.00	211.54	25.54	13.73
137.100	242.00	259.60	17.60	7.27
145.060	303.00	328.95	25.95	8.56
158.100	461.00	466.94	5.94	1.29
162.570	521.00	521.55	0.55	0.10
169.600	633.00	615.22	-17.78	-2.81
176.660	758.00	719.06	-38.94	-5.14

STANDARD DEVIATION =20.18

2-ETHYLHEXEN-2-AL C8H14O

DYKYJ J.,SEPRAKOVA M.,PAULECH J.: CHEM.ZVESTI 15,465(1962).

A = 6.86133 B = 1457.379 C = 190.565

B.P.(760) = 175.563

T	P EXPTL.	P CALCD.	DEV.	PERCENT
53.750	7.87	7.87	0.00	0.04
53.500	7.97	7.76	-0.21	-2.59
58.180	10.10	10.06	-0.04	-0.44
65.500	14.93	14.79	-0.14	-0.96
72.200	21.00	20.65	-0.35	-1.65
83.950	36.09	35.68	-0.41	-1.14
96.530	61.00	60.96	-0.04	-0.07
101.090	73.17	73.18	0.01	0.02
105.240	86.00	86.00	0.00	0.00
110.520	104.60	104.93	0.33	0.32
120.720	150.35	151.19	0.84	0.56
130.980	212.13	213.26	1.13	0.53
133.420	230.43	230.70	0.27	0.12
144.700	328.40	326.88	-1.52	-0.46
152.400	411.44	409.24	-2.20	-0.53
164.630	573.70	573.18	-0.52	-0.09
174.950	746.10	748.41	2.31	0.31

STANDARD DEVIATION = 1.04

2-ETHYL-4-METHYLPENTEN-2-AL $C_8H_{14}O$

DYKYJ J.,SEPRAKOVA M.,PAULECH J.:
CHEM.ZVESTI 15,465(1962).

A = 7.00997 B = 1511.469 C = 202.955

B.P.(760) = 163.092

T	P EXPTL.	P CALCD.	DEV.	PERCENT
38.860	5.68	5.75	0.07	1.18
46.980	9.06	9.17	0.11	1.25
58.580	17.03	17.01	-0.02	-0.11
67.000	25.80	25.76	-0.04	-0.14
74.410	36.35	36.36	0.01	0.02
82.730	52.60	52.40	-0.21	-0.39
90.800	72.90	73.22	0.32	0.44
98.700	100.50	99.86	-0.64	-0.64
111.520	159.20	159.83	0.63	0.39
119.360	209.20	209.20	0.00	0.00
140.550	407.40	407.22	-0.18	-0.04
152.360	570.10	570.28	0.18	0.03
161.850	738.80	735.78	-3.02	-0.41
162.310	741.90	744.67	2.77	0.37

STANDARD DEVIATION = 1.27

CYCLOHEXYL ACETATE $C_8H_{14}O_2$

SHEEHAN R.J.,LANGER S.H.:
J.CHEM.ENG.DATA 14,248(1969).

A = 7.97586 B = 2167.994 C = 252.299

B.P.(760) = 173.211

T	P EXPTL.	P CALCD.	DEV.	PERCENT
95.500	55.10	55.26	0.16	0.29
109.700	97.30	97.03	-0.27	-0.28
129.500	198.20	198.38	0.18	0.09
172.200	739.10	739.05	-0.05	-0.01

STANDARD DEVIATION = 0.36

CYCLOOCTANE C_8H_{16}

FINKE H.L.,ET AL.:
J.AM.CHEM.SOC. 78,5469(1956).

A = 6.86187 B = 1437.785 C = 210.016

B.P.(760) = 151.140

T	P EXPTL.	P CALCD.	DEV.	PERCENT
96.711	149.41	149.41	0.00	0.00
103.318	187.57	187.59	0.02	0.01
109.977	233.72	233.72	0.00	0.00
116.694	289.13	289.12	-0.01	0.00
123.472	355.22	355.24	0.02	0.00
130.301	433.56	433.54	-0.02	0.00
137.190	525.86	525.84	-0.02	0.00
144.133	633.99	633.93	-0.06	-0.01
151.146	760.00	760.10	0.10	0.01
158.203	906.06	906.09	0.03	0.00
165.321	1074.60	1074.53	-0.07	-0.01
172.502	1268.00	1268.04	0.04	0.00
179.738	1489.10	1489.07	-0.03	0.00
187.040	1740.80	1740.83	0.03	0.00
194.397	2026.00	2025.96	-0.04	0.00

STANDARD DEVIATION = 0.05

1,1-DIMETHYLCYCLOHEXANE

C8H16

FORZIATI A.F.,NORRIS W.R.,ROSSINI F.D.:
J.RESEARCH NATL.BUR.STANDARDS 43,555(1949).

A = 6.81334 B = 1330.625 C = 218.822

B.P.(760) = 119.542

	T	P EXPTL.	P CALCD.	DEV.	PERCENT
	40.497	48.10	48.10	0.00	-0.01
	44.531	57.71	57.64	-0.07	-0.13
	48.153	67.40	67.49	0.09	0.14
	51.399	77.46	77.47	0.01	0.01
	54.450	87.94	87.92	-0.02	-0.02
	58.564	103.80	103.83	0.03	0.03
	63.260	124.81	124.79	-0.02	-0.02
	68.047	149.61	149.59	-0.02	-0.02
	72.496	176.09	176.09	0.00	0.00
	78.447	217.37	217.36	-0.02	-0.01
	83.925	261.87	261.91	0.04	-0.01
*	90.497	325.18	324.75	-0.43	0.02
	97.361	402.70	402.65	-0.05	-0.13
	104.658	501.00	501.03	0.03	-0.01
	112.575	628.24	628.25	0.01	0.01
	118.168	732.42	732.44	0.02	0.00
	118.768	744.42	744.37	-0.05	0.00
	119.327	755.65	755.63	-0.02	-0.01
	119.959	768.52	768.51	-0.01	0.00
	120.520	780.05	780.09	0.04	0.00
					0.00

STANDARD DEVIATION = 0.04

CIS-1,2-DIMETHYLCYCLOHEXANE C8H16

WILLINGHAM C.B.,TAYLOR W.J.,PIGNOCCO J.M.,ROSSINI F.D.:
J.RESEARCH NATL.BUR.STANDARDS 35,219(1945).

A = 6.84360 B = 1370.738 C = 216.177

B.P.(760) = 129.726

	T	P EXPTL.	P CALCD.	DEV.	PERCENT
	49.185	47.65	47.65	0.00	0.00
	53.413	57.42	57.42	0.00	0.00
	57.094	67.25	67.23	-0.02	-0.03
	60.429	77.29	77.27	-0.02	-0.02
	63.543	87.73	87.74	0.01	0.01
	67.742	103.66	103.67	0.01	0.01
	72.533	124.67	124.68	0.01	0.01
	77.402	149.44	149.46	0.02	0.01
	81.921	175.90	175.92	0.02	0.01
	87.974	217.18	217.18	0.00	0.00
	93.548	261.76	261.77	0.01	0.01
	100.258	324.96	324.92	-0.04	-0.01
	107.192	402.45	402.40	-0.05	-0.01
	114.600	500.73	500.72	-0.01	0.00
	122.639	627.94	627.94	0.00	0.00
	128.315	732.11	732.11	0.00	0.00
	128.926	744.08	744.09	0.01	0.00
	129.491	755.27	755.29	0.02	0.00
*	130.315	767.98	771.87	3.89	0.51
	130.684	779.38	779.39	0.01	0.00

STANDARD DEVIATION = 0.02

1-METHYL-1-ETHYLCYCLOPENTANE C8H16

FORZIATI A.F.,NORRIS W.R.,ROSSINI F.D.:
J.RESEARCH NATL.BUR.STANDARDS 43,555(1949).

A = 6.85920 B = 1347.602 C = 217.212

B.P.(760) = 121.519

	T	P EXPTL.	P CALCD.	DEV.	PERCENT
	43.056	48.10	48.02	-0.08	-0.16
	47.137	57.71	57.73	0.02	0.03
	50.691	67.41	67.45	0.04	0.06
	53.937	77.46	77.49	0.03	0.03
	56.967	87.94	87.93	-0.01	-0.01
	61.049	103.80	103.81	0.01	0.01
	65.724	124.81	124.82	0.01	0.01
	70.476	149.61	149.61	0.00	0.00
	74.888	176.09	176.08	-0.01	-0.01
	80.796	217.37	217.35	-0.02	-0.01
	86.232	261.87	261.92	0.05	0.02
	92.783	325.18	325.10	-0.08	-0.02
	99.559	402.71	402.73	0.02	0.00
*	106.835	501.00	501.80	0.80	0.16
	114.622	628.24	628.24	0.00	0.00
	120.159	732.41	732.44	0.03	0.00
	120.752	744.41	744.36	-0.05	-0.01
	121.307	755.64	755.65	0.01	0.00
	121.933	768.52	768.54	0.02	0.00
	122.484	780.04	780.03	-0.01	0.00

STANDARD DEVIATION = 0.04

1-METHYL-CIS-2-ETHYLCYCLOPENTANE C8H16

FORZIATI A.F.,NORRIS W.R.,ROSSINI F.D.:
J.RESEARCH NATL.BUR.STANDARDS 43,555(1949).

A = 6.90588 B = 1388.412 C = 216.892

B.P.(760) = 128.049

	T	P EXPTL.	P CALCD.	DEV.	PERCENT
	48.846	48.02	47.99	-0.03	-0.06
	52.975	57.71	57.69	-0.02	-0.04
*	56.539	67.45	67.32	-0.13	-0.19
	59.862	77.50	77.47	-0.03	-0.04
	62.935	87.93	87.94	0.01	0.01
	67.070	103.84	103.86	0.02	0.02
	71.792	124.78	124.86	0.08	0.06
	76.582	149.61	149.60	-0.01	-0.01
	81.050	176.09	176.15	0.06	0.03
	87.011	217.38	217.41	0.03	0.01
	92.490	261.98	261.93	-0.05	-0.02
	99.105	325.25	325.18	-0.07	-0.02
	105.929	402.79	402.72	-0.07	-0.02
	113.205	501.03	500.96	-0.07	-0.01
	121.105	628.19	628.22	0.03	0.01
	126.684	732.43	732.52	0.09	0.01
	127.281	744.34	744.43	0.09	0.01
	127.832	755.54	755.57	0.03	0.00
	128.463	768.55	768.48	-0.07	-0.01
	129.021	780.09	780.04	-0.05	-0.01

STANDARD DEVIATION = 0.06

1-OCTENE C8H16

FORZIATI A.F.,CAMIN D.L.,ROSSINI F.D.:
J.RESEARCH NATL.BUR.STANDARDS 45,406(1950).

A = 6.93637 B = 1355.779 C = 213.022

B.P.(760) = 121.279

T	P EXPTL.	P CALCD.	DEV.	PERCENT
44.893	47.87	47.83	-0.04	-0.09
48.975	57.68	57.75	0.07	0.13
52.410	67.46	67.38	-0.08	-0.12
55.581	77.48	77.42	-0.06	-0.08
58.557	87.91	87.93	0.02	0.03
62.557	103.84	103.90	0.06	0.06
67.096	124.84	124.84	-0.01	0.00
71.736	149.60	149.69	0.09	0.06
76.022	176.13	176.12	-0.01	-0.01
81.779	217.44	217.47	0.03	0.01
87.053	262.03	261.96	-0.07	-0.03
93.428	325.27	325.25	-0.02	0.00
106.997	501.09	500.97	-0.12	-0.02
119.967	732.50	732.53	0.03	0.00
120.539	744.38	744.40	0.02	0.00
121.075	755.64	755.66	0.02	0.00
121.685	768.62	768.64	0.02	0.00
122.223	780.21	780.23	0.02	0.00

STANDARD DEVIATION = 0.06

PROPYLCYCLOPENTANE C8H16

FORZIATI A.F.,NORRIS W.R.,ROSSINI F.D.:
J.RESEARCH NATL.BUR.STANDARDS 43,555(1949).

A = 6.91061 B = 1388.511 C = 213.615

B.P.(760) = 130.946

	T	P EXPTL.	P CALCD.	DEV.	PERCENT
	51.875	48.01	47.93	-0.08	-0.16
	56.030	57.70	57.70	0.00	0.01
	59.628	67.45	67.46	0.01	0.01
	62.911	77.49	77.51	0.02	0.02
	65.966	87.92	87.95	0.03	0.03
	70.088	103.83	103.84	0.01	0.01
	74.794	124.78	124.81	0.03	0.02
	79.591	149.61	149.62	0.01	0.01
	84.039	176.09	176.10	0.01	0.01
	89.990	217.37	217.37	0.00	0.00
	102.068	325.25	325.22	-0.03	-0.01
	108.877	402.79	402.76	-0.03	-0.01
	116.132	501.03	500.93	-0.10	-0.02
	124.015	628.19	628.19	0.00	0.00
	129.579	732.42	732.43	0.01	0.00
	130.176	744.34	744.37	0.03	0.00
	130.731	755.53	755.61	0.08	0.01
*	131.379	768.55	768.91	0.36	0.05
	131.917	780.08	780.08	0.00	0.00

STANDARD DEVIATION = 0.04

ISOPROPYLCYCLOPENTANE C8H16

FORZIATI A.F.,NORRIS W.R.,ROSSINI F.D.:
J.RESEARCH NATL.BUR.STANDARDS 43,555(1949).

A = 6.88736 B = 1380.115 C = 218.047

B.P.(760) = 126.418

T	P EXPTL.	P CALCD.	DEV.	PERCENT
47.033	48.01	47.97	-0.04	-0.09
51.183	57.70	57.70	0.00	0.01
54.798	67.45	67.47	0.02	0.03
58.088	77.49	77.52	0.03	0.03
61.122	87.92	87.84	-0.08	-0.09
65.296	103.83	103.88	0.05	0.05
70.012	124.77	124.82	0.05	0.04
74.827	149.60	149.64	0.04	0.02
79.275	176.08	176.01	-0.07	-0.04
85.265	217.37	217.38	0.01	0.00
90.769	261.98	262.00	0.02	0.01
97.396	325.25	325.23	-0.02	-0.01
104.235	402.79	402.75	-0.04	-0.01
111.532	501.03	501.01	-0.02	0.00
119.450	628.19	628.19	0.00	0.00
125.044	732.42	732.43	0.01	0.00
125.643	744.34	744.35	0.01	0.00
126.201	755.53	755.59	0.06	0.01
126.836	768.55	768.54	-0.01	0.00
127.394	780.08	780.07	-0.01	0.00

STANDARD DEVIATION = 0.04

1,1,2-TRIMETHYLCYCLOPENTANE C_8H_{16}

FORZIATI A.F., NORRIS W.R., ROSSINI F.D.:
J.RESEARCH NATL.BUR.STANDARDS 43,555(1949).

A = 6.82238 B = 1309.813 C = 218.580

B.P.(760) = 113.728

T	P EXPTL.	P CALCD.	DEV.	PERCENT
36.207	48.10	48.03	-0.07	-0.14
40.224	57.72	57.72	0.00	0.00
43.728	67.41	67.44	0.03	0.05
46.937	77.46	77.50	0.04	0.05
49.920	87.94	87.92	-0.02	-0.02
53.962	103.81	103.86	0.05	0.05
58.562	124.82	124.80	-0.02	-0.01
63.251	149.61	149.57	-0.04	-0.02
67.615	176.09	176.08	-0.01	0.00
73.455	217.37	217.39	0.02	0.01
85.302	325.18	325.16	-0.02	-0.01
91.992	402.71	402.67	-0.04	-0.01
99.144	501.00	501.05	0.05	0.01
106.900	628.24	628.23	-0.01	0.00
112.381	732.42	732.43	0.01	0.00
112.971	744.42	744.40	-0.02	0.00
113.517	755.64	755.62	-0.02	0.00
114.138	768.52	768.54	0.02	0.00
114.686	780.04	780.07	0.03	0.00

STANDARD DEVIATION = 0.03

1,1,3-TRIMETHYLCYCLOPENTANE C8H16

FORZIATI A.F.,NORRIS W.R.,ROSSINI F.D.:
J.RESEARCH NATL.BUR.STANDARDS 43,555(1949).

A = 6.80931 B = 1275.915 C = 219.892

B.P.(760) = 104.893

T	P EXPTL.	P CALCD.	DEV.	PERCENT
28.944	48.11	48.06	-0.05	-0.11
32.881	57.72	57.76	0.04	0.07
36.299	67.42	67.45	0.03	0.04
39.432	77.46	77.47	0.01	0.02
42.361	87.94	87.92	-0.02	-0.02
46.308	103.81	103.81	0.00	0.00
50.828	124.82	124.81	-0.01	-0.01
55.423	149.61	149.60	-0.01	-0.01
65.411	217.37	217.35	-0.02	-0.01
70.674	261.88	261.92	0.04	0.02
77.023	325.18	325.14	-0.04	-0.01
83.585	402.71	402.70	-0.01	0.00
90.594	501.00	501.07	0.07	0.01
98.197	628.23	628.23	0.00	0.00
103.572	732.42	732.44	0.02	0.00
104.148	744.42	744.36	-0.06	-0.01
104.688	755.63	755.67	0.05	0.01
105.292	768.52	768.49	-0.03	0.00
105.830	780.03	780.04	0.01	0.00

STANDARD DEVIATION = 0.04

CIS,CIS,TRANS-1,2,4-TRIMETHYLCYCLOPENTANE C_8H_{16}

FORZIATI A.F.,NORRIS W.R.,ROSSINI F.D.:
J.RESEARCH NATL.BUR.STANDARDS 43,555(1949).

A = 6.85738 B = 1335.686 C = 219.159

B.P.(760) = 116.730

	T	P EXPTL.	P CALCD.	DEV.	PERCENT
	38.907	48.12	48.04	-0.08	-0.16
	42.972	57.72	57.79	0.07	0.13
	46.426	67.43	67.32	-0.11	-0.16
	49.701	77.47	77.52	0.05	0.07
	52.704	87.93	87.96	0.03	0.04
	56.756	103.81	103.86	0.05	0.05
	61.380	124.82	124.81	-0.01	-0.01
	66.100	149.60	149.63	0.03	0.02
	70.477	176.08	176.11	0.03	0.01
	76.327	217.36	217.31	-0.05	-0.02
	81.725	261.89	261.92	0.03	0.01
	88.224	325.17	325.11	-0.06	-0.02
	94.942	402.72	402.69	-0.03	-0.01
	102.109	500.99	501.00	0.01	0.00
*	109.887	625.21	628.21	3.00	0.48
	115.378	732.41	732.38	-0.03	0.00
	115.969	744.40	744.35	-0.05	-0.01
	116.518	755.60	755.61	0.01	0.00
	117.140	768.49	768.52	0.03	0.00
	117.690	780.00	780.09	0.09	0.01

STANDARD DEVIATION = 0.06

CIS,TRANS,CIS-1,2,4-TRIMETHYLCYCLOPENTANE C8H16

FORZIATI A.F.,NORRIS W.R.,ROSSINI F.D.:
J.RESEARCH NATL.BUR.STANDARDS 43,555(1949).

A = 6.85128 B = 1307.098 C = 219.916

B.P.(760) = 109.289

T	P EXPTL.	P CALCD.	DEV.	PERCENT
32.948	48.11	48.10	-0.01	-0.03
36.878	57.72	57.71	-0.01	-0.02
40.338	67.43	67.44	0.01	0.01
43.487	77.47	77.44	-0.04	-0.05
46.454	87.93	87.94	0.01	0.02
50.433	103.81	103.86	0.05	0.04
54.976	124.82	124.83	0.01	0.01
59.599	149.60	149.62	0.02	0.01
63.892	176.08	176.08	0.00	0.00
69.641	217.36	217.35	-0.01	-0.01
74.929	261.89	261.89	0.00	0.00
81.310	325.18	325.11	-0.07	-0.02
87.906	402.72	402.72	0.00	0.00
94.940	500.99	501.04	0.05	0.01
102.571	628.22	628.21	-0.01	0.00
107.962	732.41	732.40	-0.01	0.00
108.541	744.41	744.34	-0.07	-0.01
109.082	755.60	755.65	0.05	0.01
109.690	768.50	768.50	0.00	0.00
110.229	780.01	780.05	0.04	0.00

STANDARD DEVIATION = 0.03

CAPRYLIC ACID $C_8H_{16}O_2$

ROSE A.,ACCIARRI J.A.,JOHNSON R.C.,SANDERS N.N.: IND.ENG.CHEM.49,104(1957).

A = 7.77064 B = 1933.048 C = 159.363

B.P.(100) = 175.617

T	P EXPTL.	P CALCD.	DEV.	PERCENT
130.200	12.00	12.44	0.44	3.68
131.900	13.10	13.61	0.51	3.89
140.100	20.90	20.68	-0.22	-1.05
142.800	24.20	23.62	-0.58	-2.40
147.800	29.80	30.02	0.22	0.74
150.200	33.50	33.59	0.09	0.26
154.000	40.60	39.99	-0.61	-1.51
159.600	51.20	51.31	0.11	0.22
161.500	55.60	55.73	0.13	0.24
169.300	77.90	77.46	-0.44	-0.56
169.400	78.00	77.78	-0.22	-0.28
176.500	103.40	103.55	0.15	0.15
176.600	103.50	103.96	0.46	0.45
186.800	153.00	153.61	0.61	0.40
193.800	198.40	198.20	-0.20	-0.10
200.300	248.70	248.90	0.20	0.08
206.300	305.50	304.94	-0.56	-0.18

STANDARD DEVIATION = 0.43

BUTYL ALPHA-HYDROXYISOBUTYRATE C8H16O3

FRIED V.,PICK J.,HALA E.,VILIM O.:
CHEM.LISTY 48,774(1954).

A = 8.42169 B = 2617.318 C = 287.093

B.P.(760) = 185.272

T	P EXPTL.	P CALCD.	DEV.	PERCENT
111.900	71.80	72.76	0.96	1.33
117.300	89.70	89.02	-0.68	-0.76
125.000	117.70	117.60	-0.10	-0.09
136.800	177.80	176.69	-1.11	-0.63
142.800	218.20	215.47	-2.73	-1.25
152.600	294.00	294.50	0.50	0.17
155.600	321.50	323.16	1.66	0.52
158.600	352.80	354.17	1.37	0.39
161.400	385.50	385.37	-0.13	-0.03
163.600	410.80	411.49	0.69	0.17
165.500	434.00	435.26	1.26	0.29
167.800	463.00	465.57	2.57	0.55
171.000	508.50	510.71	2.21	0.44
173.000	541.40	540.77	-0.63	-0.12
174.400	566.50	562.69	-3.81	-0.67
177.800	612.80	619.08	6.28	1.02
181.100	676.50	678.31	1.81	0.27
182.600	710.30	706.78	-3.52	-0.50
184.300	742.00	740.25	-1.75	-0.24
185.000	760.00	754.42	-5.58	-0.73

STANDARD DEVIATION = 2.80

1-CHLOROOCTANE C8H17CL

KEMME R.H.,KREPS S.I.:
J.CHEM.ENG.DATA 14,98(1969).

A = 7.05152 B = 1600.239 C = 200.283

B.P.(760) = 183.403

T	P EXPTL.	P CALCD.	DEV.	PERCENT
54.100	5.60	5.77	0.17	2.96
63.900	9.80	9.87	0.07	0.69
72.400	15.30	15.24	-0.06	-0.39
78.200	20.30	20.19	-0.11	-0.52
87.100	30.30	30.42	0.12	0.40
93.400	40.10	40.05	-0.05	-0.11
101.100	55.20	55.19	-0.01	-0.02
110.700	80.50	80.50	-0.01	-0.01
121.800	121.20	121.09	-0.11	-0.09
136.700	201.00	200.81	-0.19	-0.09
149.800	301.60	302.33	0.73	0.24
167.400	501.00	500.37	-0.63	-0.13
184.100	773.20	773.35	0.15	0.02

STANDARD DEVIATION = 0.33

DEUTERO-1-OCTANOL C8H17DO

GEISELER G.,FRUWERT J.,HUETTING R.:
CHEM.BER. 99,1594(1966).

A = 6.93838 B = 1417.843 C = 149.694

B.P.(10) = 89.065

T	P EXPTL.	P CALCD.	DEV.	PERCENT
20.000	0.04	0.04	0.00	-4.27
30.000	0.12	0.11	-0.01	-6.91
40.000	0.29	0.29	0.00	0.37
50.000	0.68	0.69	0.01	1.35
60.000	1.51	1.50	-0.01	-0.48
70.000	3.05	3.05	0.00	0.09
80.000	5.83	5.83	0.00	-0.01

STANDARD DEVIATION = 0.01

DEUTERO-2-OCTANOL C8H17DO

GEISELER G.,FRUWERT J.,HUETTING R.:
CHEM.BER. 99,1594(1966).

A = 7.43959 B = 1574.635 C = 169.144

B.P.(10) = 75.380

T	P EXPTL.	P CALCD.	DEV.	PERCENT
10.000	0.06	0.04	-0.02	-25.58
20.000	0.13	0.13	0.00	0.13
30.000	0.34	0.34	0.00	0.25
40.000	0.81	0.81	0.00	0.50
50.000	1.78	1.80	0.02	0.87
60.000	3.73	3.70	-0.03	-0.90
70.000	7.14	7.16	0.02	0.33
80.000	13.17	13.16	-0.01	-0.04

STANDARD DEVIATION = 0.02

DEUTERO-3-OCTANOL C8H17DO

GEISELER G.,FRUWERT J.,HUETTING R.:
CHEM.BER. 99,1594(1966).

A = 6.47520 B = 1130.797 C = 134.506

B.P.(10) = 72.024

T	P EXPTL.	P CALCD.	DEV.	PERCENT
10.000	0.05	0.04	-0.01	-10.67
20.000	0.15	0.14	-0.01	-4.43
30.000	0.41	0.40	-0.01	-2.61
40.000	0.96	0.99	0.03	3.03
50.000	2.25	2.22	-0.03	-1.32
60.000	4.57	4.59	0.02	0.37
70.000	8.83	8.83	0.00	-0.04
80.000	15.98	15.98	0.00	0.00

STANDARD DEVIATION = 0.02

DEUTERO-4-OCTANOL C8H17DO

GEISELER G.,FRUWERT J.,HUETTING R.:
CHEM.BER. 99,1594(1966).

A = 6.41661 B = 1102.659 C = 132.532

B.P.(10) = 71.038

T	P EXPTL.	P CALCD.	DEV.	PERCENT
10.000	0.06	0.05	-0.01	-20.15
20.000	0.18	0.15	-0.03	-14.43
30.000	0.45	0.43	-0.02	-4.69
40.000	1.01	1.06	0.05	5.02
50.000	2.42	2.38	-0.04	-1.85
60.000	4.86	4.89	0.03	0.65
70.000	9.39	9.38	-0.01	-0.10
80.000	16.92	16.92	0.00	0.00

STANDARD DEVIATION = 0.04

2-METHYLHEPTANE C8H18

WILLINGHAM C.J.,TAYLOR W.J.,PIGNOCCO J.M.,ROSSINI F.D.:
J.RESEARCH NATL.BUR.STANDARDS 35,219(1945).

A = 6.88814 B = 1319.529 C = 211.625

B.P.(760) = 117.654

	T	P EXPTL.	P CALCD.	DEV.	PERCENT
	41.707	47.84	47.80	-0.04	-0.08
	45.687	57.55	57.55	0.00	-0.01
*	49.165	67.17	67.36	0.19	0.28
	52.301	77.36	77.36	0.00	0.00
	55.229	87.78	87.78	0.00	0.00
	59.192	103.69	103.69	0.00	0.00
	63.711	124.59	124.66	0.07	0.05
	68.308	149.40	149.42	0.02	0.01
	72.580	175.81	175.90	0.09	0.05
	78.278	217.16	217.03	-0.13	-0.06
	83.549	261.73	261.70	-0.03	-0.01
	89.892	324.96	324.97	0.01	0.00
	96.422	402.38	402.36	-0.03	-0.01
	103.397	500.58	500.56	-0.02	0.00
	110.971	627.73	627.76	0.03	0.01
	116.318	731.92	731.96	0.04	0.01
	116.888	743.74	743.82	0.08	0.01
	117.426	755.00	755.15	0.15	0.02
	118.022	767.94	767.87	-0.07	-0.01
	118.544	779.31	779.14	-0.17	-0.02

STANDARD DEVIATION = 0.08

2-METHYLHEPTANE C8H18

NICOLINI E.,LAFFITTE P.:

COMPTES RENDUS 229,757(1949).

A = 6.85999 B = 1313.125 C = 230.020

B.P.(100) = 40.171

T	P EXPTL.	P CALCD.	DEV.	PERCENT
23.400	47.90	47.68	-0.22	-0.45
29.700	63.60	63.69	0.09	0.14
31.300	68.10	68.39	0.29	0.43
33.400	74.80	75.00	0.20	0.27
35.100	80.70	80.73	0.03	0.04
37.250	88.50	88.49	-0.01	-0.01
38.850	94.80	94.65	-0.15	-0.16
40.600	102.20	101.79	-0.41	-0.40
43.400	114.30	114.13	-0.17	-0.15
46.100	127.20	127.16	-0.04	-0.03
48.350	138.70	138.93	0.23	0.16
52.100	160.50	160.51	0.01	0.00
54.750	177.40	177.34	-0.06	-0.03
56.200	187.10	187.14	0.04	0.02
58.250	201.30	201.74	0.44	0.22
62.300	233.30	233.29	-0.01	0.00
65.250	258.70	258.69	-0.01	0.00
67.000	274.90	274.78	-0.12	-0.04
69.500	299.20	299.15	-0.05	-0.02
72.250	328.20	327.92	-0.28	-0.08
75.000	358.70	358.87	0.17	0.05

STANDARD DEVIATION = 0.21

3-METHYLHEPTANE C8H18

WILLINGHAM C.J.,TAYLOR W.J.,PIGNOCCO J.M.,ROSSINI F.D.: J.RESEARCH NATL.BUR.STANDARDS 35,219(1945).

A = 6.89127 B = 1326.154 C = 211.756

B.P.(760) = 118.917

T	P EXPTL.	P CALCD.	DEV.	PERCENT
42.672	47.78	47.75	-0.03	-0.06
46.672	57.48	57.50	0.02	0.03
50.171	67.33	67.33	0.00	0.00
53.317	77.33	77.32	-0.01	-0.01
56.265	87.76	87.77	0.01	0.01
60.243	103.67	103.68	0.01	0.01
64.775	124.53	124.62	0.09	0.07
69.372	149.41	149.28	-0.13	-0.08
73.676	175.84	175.85	0.01	0.01
79.418	217.14	217.16	0.02	0.01
84.698	261.75	261.75	0.00	0.00
91.057	324.98	324.96	-0.02	-0.01
97.615	402.39	402.41	0.02	0.00
104.616	500.66	500.64	-0.02	0.00
112.217	627.85	627.87	0.02	0.00
117.584	731.97	732.11	0.14	0.02
118.160	743.98	744.05	0.07	0.01
118.693	755.08	755.24	0.16	0.02
119.283	768.01	767.79	-0.22	-0.03
119.814	779.37	779.21	-0.16	-0.02

STANDARD DEVIATION = 0.09

OCTANE C8H18

WILLINGHAM C.J., TAYLOR W.J., PIGNOCCO J.M., ROSSINI F.D.: J.RESEARCH NATL.BUR.STANDARDS 35,219(1945).

A = 6.91874 B = 1351.756 C = 209.100

B.P.(760) = 125.665

T	P EXPTL.	P CALCD.	DEV.	PERCENT
52.927	57.53	57.53	0.00	0.00
56.456	67.35	67.37	0.02	0.03
59.616	77.36	77.32	-0.04	-0.05
62.592	87.77	87.78	0.01	0.01
66.587	103.68	103.64	-0.04	-0.04
71.163	124.57	124.61	0.04	0.03
75.820	149.40	149.42	0.02	0.01
80.134	175.82	175.86	0.04	0.02
85.916	217.16	217.15	-0.01	-0.01
91.230	261.73	261.72	-0.01	-0.01
97.635	324.97	324.95	-0.02	-0.01
104.233	402.38	402.36	-0.02	0.00
111.277	500.61	500.57	-0.04	-0.01
118.924	627.77	627.81	0.04	0.01
124.319	731.94	732.00	0.06	0.01
124.899	743.92	743.96	0.04	0.01
125.433	755.03	755.11	0.08	0.01
126.035	767.96	767.84	-0.12	-0.02
126.570	779.32	779.28	-0.04	0.00

STANDARD DEVIATION = 0.05

2,2,3,3-TETRAMETHYLBUTANE C8H18

SCOTT D.W., DOUSLIN D.R., GROSS M.E., OLIVER G.D., HUFMAN H.M.: J.AM.CHEM.SOC. 74,883(1952).

A = 7.86673 B = 1674.807 C = 230.779

B.P.(100) = 54.696

T	P EXPTL.	P CALCD.	DEV.	PERCENT
0.000	4.10	4.07	-0.03	-0.74
15.000	11.25	11.28	0.03	0.30
20.000	15.40	15.43	0.03	0.18
25.000	20.85	20.84	-0.01	-0.05
30.000	27.85	27.82	-0.03	-0.09
35.000	36.74	36.75	0.01	0.02
40.000	48.06	48.04	-0.02	-0.04
45.000	62.18	62.19	0.01	0.02
50.000	79.81	79.78	-0.03	-0.04
55.000	101.41	101.45	0.04	0.04
60.000	127.91	127.94	0.03	0.03
65.000	160.13	160.10	-0.03	-0.02

STANDARD DEVIATION = 0.03

2,2,4-TRIMETHYLPENTANE C8H18

WILLINGHAM C.J.,TAYLOR W.J.,PIGNOCCO J.M.,ROSSINI F.D.:
J.RESEARCH NATL.BUR.STANDARDS 35,219(1945).

A = 6.80304 B = 1252.590 C = 220.119

B.P.(760) = 99.238

	T	P EXPTL.	P CALCD.	DEV.	PERCENT
	24.358	47.79	47.81	0.02	0.04
	28.249	57.49	57.51	0.02	0.04
	31.668	67.33	67.34	0.01	0.01
	34.746	77.34	77.33	-0.01	-0.02
	37.628	87.76	87.75	-0.01	-0.01
	41.517	103.68	103.64	-0.04	-0.04
*	45.975	124.54	124.66	0.12	0.09
	50.496	149.41	149.40	-0.01	0.00
	54.698	175.84	175.85	0.01	0.00
	60.342	217.15	217.20	0.05	0.02
	65.523	261.75	261.74	-0.01	-0.01
	71.778	324.98	324.96	-0.02	-0.01
	78.232	402.39	402.40	0.01	0.00
	85.131	500.65	500.67	0.02	0.00
	92.624	627.84	627.86	0.02	0.00
	97.917	731.99	732.02	0.03	0.00
	98.487	743.97	743.99	0.02	0.00
	99.014	755.22	755.20	-0.02	0.00
	99.607	768.02	767.96	-0.06	-0.01
	100.130	779.37	779.36	-0.01	0.00

STANDARD DEVIATION = 0.03

2,2,4-TRIMETHYLPENTANE C8H18

MILAZZO G.:
ANNALI DI CHIMICA 46,1105(1956).

A = 6.71684 B = 1219.557 C = 217.697

B.P.(10) = -4.370

T	P EXPTL.	P CALCD.	DEV.	PERCENT
-78.510	0.01	0.01	0.00	-25.51
-70.110	0.04	0.03	-0.01	-18.81
-60.200	0.11	0.09	-0.01	-12.07
-50.060	0.30	0.28	-0.03	-9.01
-40.370	0.70	0.69	-0.01	-1.30
-30.160	1.70	1.64	-0.06	-3.75
-19.920	3.50	3.55	0.05	1.50
-14.950	5.00	5.03	0.03	0.63
0.170	13.15	13.16	0.01	0.05
9.990	23.10	22.94	-0.16	-0.70
19.270	37.00	37.18	0.18	0.49
21.280	41.00	41.08	0.08	0.19
25.290	50.00	49.87	-0.13	-0.26

STANDARD DEVIATION = 0.10

2,3,4-TRIMETHYLPENTANE C8H18

WILLINGHAM C.J.,TAYLOR W.J.,PIGNOCCO J.M.,ROSSINI F.D.: J.RESEARCH NATL.BUR.STANDARDS 35,219(1945).

A = 6.85569 B = 1316.133 C = 217.646

B.P.(760) = 113.467

T	P EXPTL.	P CALCD.	DEV.	PERCENT
36.568	47.71	47.69	-0.02	-0.04
40.606	57.47	57.46	-0.01	-0.02
44.127	67.29	67.29	-0.01	-0.01
47.305	77.33	77.31	-0.02	-0.03
50.280	87.75	87.78	0.03	0.03
54.290	103.69	103.71	0.02	0.02
58.865	124.70	124.71	0.01	0.01
63.517	149.48	149.50	0.02	0.02
67.835	175.97	175.98	0.01	0.00
73.616	217.23	217.25	0.02	0.01
78.935	261.82	261.82	0.00	0.00
85.341	325.01	324.96	-0.05	-0.02
91.960	402.46	402.43	-0.03	-0.01
99.028	500.70	500.69	-0.01	0.00
106.702	627.93	627.92	-0.01	0.00
112.121	732.12	732.12	0.00	0.00
112.703	744.05	744.07	0.02	0.00
113.241	755.29	755.25	-0.04	-0.01
113.852	768.11	768.11	0.00	0.00
114.381	779.31	779.38	0.07	0.01

STANDARD DEVIATION = 0.03

BUTYL TERT-BUTYL ETHER C8H18O

SHEEHAN R.J.,LANGER S.H.: J.CHEM.ENG.DATA 14,248(1969).

A = 6.95556 B = 1348.702 C = 206.303

B.P.(760) = 124.687

T	P EXPTL.	P CALCD.	DEV.	PERCENT
83.300	199.10	198.83	-0.27	-0.13
101.400	372.30	373.62	1.32	0.35
112.000	524.50	522.87	-1.63	-0.31
123.500	734.20	734.75	0.55	0.08

STANDARD DEVIATION = 2.18

BUTYL CARBITOL C8H18O

GARDNER G.S.,BREWER J.E.: IND.ENG.CHEM.29,179(1937).

A = 7.74114 B = 2056.904 C = 195.655

B.P.(10) = 109.472

T	P EXPTL.	P CALCD.	DEV.	PERCENT
50.200	0.20	0.24	0.04	18.52
53.700	0.30	0.31	0.01	3.55
60.500	0.40	0.51	0.11	28.58
62.900	0.50	0.61	0.11	22.12
66.900	0.70	0.81	0.11	15.31
76.600	1.40	1.54	0.14	9.64
78.700	1.70	1.75	0.05	3.15
82.400	2.00	2.21	0.21	10.32
88.000	3.10	3.09	-0.01	-0.38
92.300	4.20	3.96	-0.24	-5.65
100.400	6.40	6.22	-0.19	-2.89
109.100	9.90	9.81	-0.09	-0.89
117.600	15.00	14.96	-0.04	-0.27
126.800	22.90	23.03	0.13	0.56
144.900	50.10	50.27	0.17	0.33
152.900	69.30	69.17	-0.13	-0.19

STANDARD DEVIATION = 0.14

DIBUTYL ETHER C8H18O

CIDLINSKY J.,POLAK J.:
COLL.CZECH.CHEM.COMM. 34,1317(1969).

A = 6.79629 B = 1297.291 C = 191.029

B.P.(760) = 140.295

T	P EXPTL.	P CALCD.	DEV.	PERCENT
89.140	146.48	146.52	0.04	0.03
94.570	179.69	179.44	-0.25	-0.14
106.430	272.17	272.30	0.13	0.05
110.540	311.88	312.23	0.35	0.11
113.280	341.35	341.36	0.01	0.00
114.850	358.39	359.00	0.61	0.17
118.200	399.43	399.06	-0.37	-0.09
121.050	435.92	435.86	-0.06	-0.01
123.180	465.42	465.09	-0.34	-0.07
123.410	468.86	468.33	-0.53	-0.11
125.270	495.43	495.24	-0.19	-0.04
127.450	528.25	528.31	0.06	0.01
129.670	563.71	563.75	0.04	0.01
132.040	603.27	603.61	0.35	0.06
134.240	642.51	642.57	0.06	0.01
136.130	677.29	677.58	0.29	0.04
137.720	708.48	708.17	-0.31	-0.04
140.060	755.06	755.14	0.08	0.01

STANDARD DEVIATION = 0.31

DI-TERT.-BUTYL ETHER C8H18O

SMUTNY E.J.,BONDI A.:
J.PHYS.CHEM. 65,546(1961).

A = 6.93294 B = 1348.533 C = 223.790

B.P.(760) = 109.006

T	P EXPTL.	P CALCD.	DEV.	PERCENT
4.000	9.80	10.30	0.50	5.11
14.000	18.40	18.27	-0.13	-0.69
18.000	22.40	22.68	0.28	1.25
22.000	28.30	27.95	-0.35	-1.23
26.000	34.50	34.22	-0.28	-0.81
30.000	42.30	41.63	-0.67	-1.59
34.000	50.80	50.33	-0.47	-0.93
38.000	59.20	60.50	1.30	2.19
109.000	760.00	759.86	-0.14	-0.02

STANDARD DEVIATION = 0.70

2-ETHYL-1-HEXANOL C8H18O

DYKYJ J.,SEPRAKOVA M.,PAULECH J.:
CHEM.ZVESTI 15,465(1962).

A = 6.91473 B = 1339.704 C = 147.814

B.P.(760) = 184.296

T	P EXPTL.	P CALCD.	DEV.	PERCENT
74.070	7.40	7.53	0.13	1.78
80.880	11.50	11.39	-0.11	-0.92
83.140	13.15	13.00	-0.15	-1.13
88.130	17.33	17.25	-0.08	-0.49
99.740	31.80	31.84	0.04	0.13
110.720	54.17	54.05	-0.12	-0.22
117.570	74.00	73.55	-0.45	-0.61
121.490	87.12	87.10	-0.02	-0.02
127.240	110.60	110.67	0.07	0.06
132.100	133.87	134.46	0.59	0.44
138.220	170.20	170.22	0.02	0.01
144.200	211.14	212.29	1.15	0.54
159.440	358.60	358.49	-0.11	-0.03
162.490	397.54	395.67	-1.87	-0.47
170.120	503.50	502.29	-1.22	-0.24
177.140	618.77	619.41	0.64	0.10
183.550	742.98	744.27	1.29	0.17

STANDARD DEVIATION = 0.80

2-ETHYL-4-METHYL-1-PENTANOL C8H18O

DYKYJ J.,SEPRAKOVA M.,PAULECH J.:
CHEM.ZVESTI 15,465(1962).

A = 6.58263 B = 1134.599 C = 129.195

B.P.(760) = 177.304

T	P EXPTL.	P CALCD.	DEV.	PERCENT
70.030	8.40	7.72	-0.68	-8.11
77.410	12.90	12.33	-0.57	-4.41
87.190	22.30	21.84	-0.46	-2.08
93.770	31.30	31.18	-0.12	-0.37
97.900	38.60	38.59	-0.01	-0.03
107.830	62.30	62.49	0.19	0.30
108.440	63.90	64.28	0.38	0.59
116.000	89.70	90.22	0.52	0.58
122.430	117.80	118.45	0.65	0.55
128.790	152.70	153.00	0.30	0.20
137.750	215.50	214.94	-0.57	-0.26
146.960	297.60	297.89	0.29	0.10
156.380	409.30	406.99	-2.31	-0.56
162.770	498.80	497.21	-1.59	-0.32
167.310	570.30	570.22	-0.08	-0.01
169.970	615.80	616.69	0.89	0.14
176.390	738.60	740.87	2.27	0.31

STANDARD DEVIATION = 1.08

1-OCTANOL C8H18O

AMBROSE D.,SPRAKE C.H.S.: J.CHEM.THERMODYNAMICS 2,631(1970).

A = 6.74980 B = 1257.560 C = 129.877

B.P.(760) = 195.160

T	P EXPTL.	P CALCD.	DEV.	PERCENT
113.274	37.99	37.83	-0.16	-0.42
117.718	46.96	46.85	-0.11	-0.24
122.526	58.58	58.54	-0.04	-0.07
127.503	73.07	73.08	0.01	0.01
133.164	93.01	93.10	0.09	0.09
139.162	118.85	118.99	0.14	0.12
147.224	162.61	162.75	0.14	0.08
151.974	193.96	194.09	0.13	0.06
157.372	235.32	235.42	0.10	0.04
161.868	274.93	274.99	0.06	0.02
167.897	336.27	336.19	-0.08	-0.02
173.981	408.60	408.45	-0.15	-0.04
179.523	484.82	484.48	-0.34	-0.07
184.300	558.87	558.56	-0.31	-0.05
190.026	659.05	658.75	-0.30	-0.05
195.114	759.23	759.04	-0.19	-0.02
195.504	767.05	767.19	0.14	0.02
201.198	893.81	894.08	0.27	0.03
206.123	1015.82	1016.36	0.54	0.05

STANDARD DEVIATION = 0.23

1-OCTANOL C8H18O

GEISELER G.,FRUWERT J.,HUETTING R.: CHEM.BER. 99,1594(1966).

A = 9.26440 B = 2553.435 C = 221.596

B.P.(10) = 87.372

T	P EXPTL.	P CALCD.	DEV.	PERCENT
20.000	0.05	0.05	0.00	-0.82
30.000	0.13	0.13	0.00	0.35
40.000	0.32	0.32	0.00	-0.40
50.000	0.73	0.73	0.00	-0.12
60.000	1.57	1.57	0.00	0.18
70.000	3.22	3.22	0.00	-0.05
80.000	6.28	6.28	0.00	0.01

STANDARD DEVIATION = 0.00

2-OCTANOL C8H18O

GEISELER G.,FRUWERT J.,HUETTING R.:
CHEM.BER. 99,1594(1966).

A = 6.89511 B = 1335.886 C = 151.798

B.P.(10) = 74.811

T	P EXPTL.	P CALCD.	DEV.	PERCENT
10.000	0.06	0.04	-0.02	-27.48
20.000	0.15	0.13	-0.02	-12.28
30.000	0.36	0.35	-0.01	-2.14
40.000	0.84	0.85	0.01	1.34
50.000	1.87	1.88	0.01	0.78
60.000	3.88	3.87	-0.01	-0.25
70.000	7.45	7.45	0.00	-0.01
80.000	13.55	13.55	0.00	0.01

STANDARD DEVIATION = 0.01

3-OCTANOL C8H18O

GEISELER G.,FRUWERT J.,HUETTING R.:
CHEM.BER. 99,1594(1966).

A = 6.60563 B = 1184.563 C = 139.587

B.P.(10) = 71.729

T	P EXPTL.	P CALCD.	DEV.	PERCENT
10.000	0.06	0.05	-0.01	-18.98
20.000	0.16	0.15	-0.01	-4.75
30.000	0.43	0.42	-0.01	-2.91
40.000	1.01	1.02	0.01	1.22
50.000	2.27	2.28	0.01	0.35
60.000	4.69	4.68	-0.01	-0.14
70.000	8.99	8.99	0.00	0.00
80.000	16.26	16.26	0.00	0.00

STANDARD DEVIATION = 0.01

4-OCTANOL C8H18O

GEISELER G.,FRUWERT J.,HUETTING R.: CHEM.BER. 99,1594(1966).

A = 7.18141 B = 1445.565 C = 163.001

B.P.(10) = 70.856

T	P EXPTL.	P CALCD.	DEV.	PERCENT
10.000	0.08	0.07	-0.01	-16.34
20.000	0.19	0.19	0.00	0.80
30.000	0.52	0.49	-0.03	-5.49
40.000	1.14	1.15	0.01	0.82
50.000	2.48	2.48	0.00	0.07
60.000	4.98	5.00	0.02	0.43
70.000	9.51	9.49	-0.02	-0.20
80.000	17.08	17.08	0.00	0.03

STANDARD DEVIATION = 0.02

VINYLTRIETHOXYSILANE C8H18O3SI

JENKINS A.C.,CHAMBERS G.F.: IND.ENG.CHEM. 46,2367(1954).

A = 7.51294 B = 1785.331 C = 224.846

B.P.(100) = 98.997

T	P EXPTL.	P CALCD.	DEV.	PERCENT
61.100	18.40	18.59	0.19	1.05
70.000	28.70	28.70	0.00	-0.02
81.000	47.00	47.38	0.38	0.81
95.500	87.40	87.06	-0.34	-0.39
105.900	132.50	130.33	-2.17	-1.64
128.000	279.00	283.88	4.88	1.75
148.100	535.20	531.91	-3.29	-0.61

STANDARD DEVIATION = 3.15

DIBUTYL SULFIDE C8H18S

BAUER H.,BURSCHKIES K.:

BER. 68,1243(1935).

A = 6.76926 B = 1208.804 C = 217.508

B.P.(100) = 35.949

T	P EXPTL.	P CALCD.	DEV.	PERCENT
10.000	29.00	28.58	-0.42	-1.46
12.000	32.00	31.79	-0.21	-0.65
14.000	35.00	35.30	0.30	0.87
16.000	39.10	39.13	0.03	0.09
20.000	48.00	47.83	-0.17	-0.35
22.000	52.60	52.75	0.15	0.29
24.000	58.00	58.08	0.08	0.14
26.000	63.60	63.85	0.25	0.39
28.000	70.00	70.08	0.08	0.11
30.000	77.00	76.80	-0.20	-0.26
32.000	84.50	84.05	-0.45	-0.54
34.000	91.00	91.84	0.84	0.92
37.000	104.60	104.64	0.04	0.04
38.000	109.00	109.21	0.21	0.20
40.000	119.50	118.86	-0.64	-0.54

STANDARD DEVIATION = 0.39

ETHYLTRIETHOXYSILANE C8H20O3SI

JENKINS A.C.,CHAMBERS G.F.:

IND.ENG.CHEM. 46,2367(1954).

A = 6.88682 B = 1377.931 C = 182.951

B.P.(100) = 99.018

T	P EXPTL.	P CALCD.	DEV.	PERCENT
64.500	21.20	20.81	-0.39	-1.83
81.000	46.20	46.39	0.19	0.41
90.500	70.20	70.43	0.23	0.33
121.700	231.40	231.12	-0.28	-0.12
153.400	616.60	616.75	0.15	0.02

STANDARD DEVIATION = 0.41

OCTAMETHYLCYCLOTETRASILOXANE C8H24O4SI4

OSTHOFF R.C.,GRUBB W.T.: J.AM.CHEM.SOC. 76,399(1954).

A = 6.62601 B = 1280.000 C = 166.086

B.P.(100) = 110.611

T	P EXPTL.	P CALCD.	DEV.	PERCENT
30.400	1.50	1.29	-0.21	-13.81
42.400	3.20	3.07	-0.13	-4.20
49.200	4.90	4.79	-0.11	-2.22
53.600	6.30	6.30	0.00	0.04
62.800	10.90	10.81	-0.09	-0.85
72.800	18.60	18.53	-0.07	-0.39
83.700	31.90	31.74	-0.16	-0.50
93.500	49.70	49.55	-0.15	-0.29
99.900	65.40	65.12	-0.28	-0.43
104.800	79.00	79.57	0.57	0.73
116.700	124.50	125.78	1.28	1.03
123.200	159.60	158.97	-0.63	-0.39
134.800	236.10	235.45	-0.65	-0.27
139.600	275.60	274.60	-1.00	-0.36
149.700	372.80	373.79	0.99	0.27
155.300	439.70	439.80	0.10	0.02

STANDARD DEVIATION = 0.62

M-TOLYL TRIFLUOROACETATE C9H7F3O2

SHEEHAN R.J.,LANGER S.H.: J.CHEM.ENG.DATA 14,248(1969).

A = 7.68100 B = 1874.843 C = 223.480

B.P.(760) = 167.096

T	P EXPTL.	P CALCD.	DEV.	PERCENT
90.600	51.30	51.49	0.19	0.36
106.500	100.20	99.84	-0.36	-0.36
125.000	199.70	199.96	0.26	0.13
166.000	736.80	736.71	-0.09	-0.01

STANDARD DEVIATION = 0.49

P-TOLYL TRIFLUOROACETATE — $C_9H_7F_3O_2$

SHEEHAN R.J., LANGER S.H.: J.CHEM.ENG.DATA 14,248(1969).

A = 7.91381 B = 2055.410 C = 238.990

B.P.(760) = 169.397

T	P EXPTL.	P CALCD.	DEV.	PERCENT
92.400	51.30	51.45	0.15	0.30
108.600	100.40	100.11	-0.29	-0.29
127.200	199.70	199.91	0.21	0.11
168.300	736.70	736.63	-0.07	-0.01

STANDARD DEVIATION = 0.40

QUINOLINE — C_9H_7N

MALANOWSKI S.: BULL.ACAD.POLON.SCI.SER.SCI.CHIM.9, 71(1961).

A = 6.81759 B = 1668.731 C = 186.256

B.P.(760) = 237.626

T	P EXPTL.	P CALCD.	DEV.	PERCENT
164.670	115.50	115.44	-0.06	-0.05
165.370	117.90	117.99	0.09	0.07
178.150	173.10	173.09	-0.01	-0.01
187.510	225.40	225.40	0.00	0.00
192.950	261.30	261.21	-0.09	-0.03
199.210	307.90	307.94	0.04	0.01
199.180	307.90	307.70	-0.20	-0.07
204.020	348.10	348.19	0.09	0.03
208.420	388.40	388.59	0.19	0.05
212.640	430.60	430.74	0.14	0.03
216.530	472.70	472.73	0.03	0.01
219.750	510.00	509.89	-0.11	-0.02
224.460	568.48	568.34	-0.14	-0.02
227.980	615.44	615.37	-0.07	-0.01
229.430	635.44	635.61	0.17	0.03
230.260	647.57	647.42	-0.15	-0.02
231.590	666.58	666.71	0.13	0.02
232.230	676.32	676.15	-0.17	-0.02
232.920	686.54	686.45	-0.09	-0.01
233.290	692.05	692.02	-0.03	0.00
234.280	707.01	707.10	0.09	0.01
234.800	714.98	715.13	0.15	0.02
235.270	722.23	722.44	0.21	0.03
235.650	728.56	728.40	-0.16	-0.02
236.900	748.13	748.26	0.13	0.02
237.250	753.81	753.90	0.09	0.01
237.750	762.18	762.00	-0.18	-0.02
237.940	765.23	765.11	-0.12	-0.02

STANDARD DEVIATION = 0.13

ISOQUINOLINE C9H7N

MALANOWSKI S.: BULL.ACAD.POLON.SCI.SER.SCI.CHIM.9, 71(1961).

A = 6.91217 B = 1723.443 C = 184.265

B.P.(760) = 243.244

	T	P EXPTL.	P CALCD.	DEV.	PERCENT
	166.780	100.60	100.63	0.03	0.03
	174.030	126.60	126.49	-0.11	-0.09
	177.210	139.40	139.43	0.03	0.02
*	182.430	168.00	163.02	-4.98	-2.96
	184.950	175.60	175.52	-0.08	-0.05
	189.320	198.90	199.03	0.13	0.07
	193.140	221.60	221.62	0.02	0.01
	196.610	243.80	243.90	0.10	0.04
*	199.250	212.10	262.04	49.94	23.54
	201.710	279.90	279.90	0.00	0.00
	201.690	279.90	279.75	-0.15	-0.05
	203.490	293.50	293.43	-0.07	-0.02
	204.820	303.70	303.88	0.18	0.06
	207.140	322.70	322.81	0.11	0.04
	209.150	340.20	339.98	-0.22	-0.07
	211.070	357.20	357.05	-0.15	-0.04
	212.570	371.00	370.85	-0.15	-0.04
	215.790	401.90	401.94	0.04	0.01
	216.700	411.20	411.09	-0.11	-0.03
	218.660	430.40	431.36	0.96	0.22
	220.500	451.30	451.11	-0.19	-0.04
	221.930	466.80	466.96	0.16	0.03
	224.300	494.50	494.18	-0.32	-0.06
	226.990	526.60	526.60	0.00	0.00
	230.040	565.08	565.37	0.29	0.05
	225.770	512.00	511.69	-0.31	-0.06
	230.620	573.40	572.99	-0.41	-0.07
	232.950	604.36	604.43	0.07	0.01
	234.090	620.47	620.30	-0.17	-0.03
	235.530	640.94	640.81	-0.13	-0.02
	236.990	661.96	662.16	0.20	0.03
	238.140	679.43	679.36	-0.07	-0.01
	239.420	699.11	698.92	-0.19	-0.03
	240.470	713.92	715.29	1.37	0.19
	241.190	726.41	726.69	0.28	0.04
	242.070	741.18	740.82	-0.37	-0.05
	242.740	751.92	751.71	-0.21	-0.03
	243.290	761.13	760.75	-0.38	-0.05
	243.520	765.14	764.56	-0.58	-0.08
	243.700	767.22	767.55	0.33	0.04

STANDARD DEVIATION = 0.36

O-METHYLSTYRENE C9H10

CLEMENTS H.E.,WISE K.V.,JOHNSEN S.E.J.:
J.AM.CHEM.SOC. 75,1593(1953).

A = 7.21287 B = 1664.083 C = 214.585

B.P.(100) = 104.641

T	P EXPTL.	P CALCD.	DEV.	PERCENT
32.010	2.90	2.91	0.02	0.51
40.850	4.97	4.99	0.02	0.42
58.510	13.15	13.17	0.02	0.12
58.540	13.24	13.19	-0.05	-0.41
70.200	23.35	23.42	0.07	0.30
82.800	41.47	41.41	-0.06	-0.14
82.830	41.51	41.47	-0.04	-0.10
100.350	84.83	84.91	0.08	0.10
112.350	132.75	132.71	-0.04	-0.03

STANDARD DEVIATION = 0.06

M-METHYLSTYRENE C9H10

BUCK F.R.,COLES K.F.,KENNEDY G.T.,MORTON F.:
J.CHEM.SOC. 1949,2377.

A = 7.06423 B = 1564.740 C = 204.083

B.P.(760) = 169.952

T	P EXPTL.	P CALCD.	DEV.	PERCENT
67.600	20.20	20.17	-0.03	-0.13
72.300	25.80	25.28	-0.52	-2.03
80.700	37.00	37.13	0.13	0.35
87.000	49.00	48.82	-0.18	-0.36
93.800	62.80	64.77	1.97	3.13
96.500	73.00	72.20	-0.80	-1.10
106.900	108.00	107.80	-0.20	-0.18
121.000	180.00	178.18	-1.82	-1.01
128.300	228.00	227.29	-0.71	-0.31
138.500	312.00	313.86	1.86	0.60
147.600	411.00	412.03	1.03	0.25
159.200	568.00	571.47	3.47	0.61
169.100	748.00	743.47	-4.53	-0.61

STANDARD DEVIATION = 2.14

M-METHYLSTYRENE C9H10

CLEMENTS H.E.,WISE K.V.,JOHNSEN S.E.J.: J.AM.CHEM.SOC. 75,1593(1953).

A = 7.11224 B = 1615.081 C = 210.809

B.P.(100) = 105.115

T	P EXPTL.	P CALCD.	DEV.	PERCENT
41.780	5.15	5.23	0.08	1.47
43.220	5.58	5.68	0.10	1.80
49.900	8.28	8.27	-0.01	-0.17
52.120	9.46	9.32	-0.14	-1.44
55.730	11.24	11.29	0.05	0.47
57.800	12.55	12.57	0.02	0.19
65.200	18.34	18.23	-0.11	-0.62
70.950	23.94	23.99	0.05	0.23
71.010	24.02	24.06	0.04	0.17
71.020	24.02	24.07	0.05	0.22
71.100	24.36	24.16	-0.20	-0.81
77.120	31.76	31.84	0.08	0.24
99.330	80.23	80.29	0.06	0.07
111.800	127.66	127.62	-0.04	-0.03

STANDARD DEVIATION = 0.10

P-METHYLSTYRENE C9H10

BUCK F.R.,COLES K.F.,KENNEDY G.T.,MORTON F.: J.CHEM.SOC. 1949,2377.

A = 7.01119 B = 1535.073 C = 200.732

B.P.(760) = 170.923

T	P EXPTL.	P CALCD.	DEV.	PERCENT
68.600	20.80	20.49	-0.31	-1.47
70.600	22.40	22.58	0.18	0.78
75.100	28.00	27.92	-0.08	-0.28
78.800	33.00	33.08	0.08	0.25
80.600	35.40	35.87	0.47	1.33
82.700	39.00	39.37	0.37	0.95
84.200	41.80	42.04	0.24	0.57
87.900	49.20	49.29	0.09	0.18
94.100	64.20	63.76	-0.44	-0.68
104.400	96.20	95.57	-0.63	-0.65
115.100	142.20	141.50	-0.70	-0.49
131.900	249.00	249.03	0.03	0.01
144.000	360.00	361.60	1.60	0.44
154.700	493.00	492.36	-0.64	-0.13
163.700	624.00	629.42	5.42	0.87
170.000	748.00	742.20	-5.80	-0.78

STANDARD DEVIATION = 2.28

P-METHYLSTYRENE C9H10

CLEMENTS H.E.,WISE K.V.,JOHNSEN S.E.J.:
J.AM.CHEM.SOC. 75,1593(1953).

A = 7.04425 B = 1591.082 C = 209.441

B.P.(10) = 53.798

T	P EXPTL.	P CALCD.	DEV.	PERCENT
31.820	2.82	2.81	-0.01	-0.20
41.760	5.17	5.13	-0.04	-0.72
41.830	5.20	5.15	-0.05	-0.89
52.170	9.05	9.17	0.12	1.33
53.940	9.98	10.08	0.10	0.95
59.580	13.48	13.49	0.01	0.05
59.910	13.91	13.71	-0.20	-1.42
66.620	19.04	19.09	0.05	0.24
75.400	28.94	28.73	-0.21	-0.72
76.190	29.81	29.77	-0.04	-0.12
80.200	35.39	35.56	0.17	0.48
82.440	39.06	39.18	0.12	0.32
90.900	55.80	55.80	0.00	0.00
96.930	71.00	70.94	-0.06	-0.08

STANDARD DEVIATION = 0.12

PROPIOPHENONE C9H10O

DREISBACH R.R.,SHRADER S.A.:
IND.ENG.CHEM. 41,2879(1949).

A = 7.36975 B = 1894.146 C = 204.830

B.P.(100) = 147.914

T	P EXPTL.	P CALCD.	DEV.	PERCENT
132.210	57.04	56.21	-0.83	-1.46
136.410	66.39	65.91	-0.48	-0.72
140.020	75.86	75.35	-0.51	-0.67
154.730	123.76	126.41	2.65	2.14
183.840	315.52	313.57	-1.95	-0.62
201.320	507.50	508.25	0.75	0.15

STANDARD DEVIATION = 2.05

BENZYL ACETATE C9H10O2

GARDNER G.S.,BREWER J.E.: IND.ENG.CHEM.29,179(1937).

A = 8.45705 B = 2623.206 C = 259.067

B.P.(100) = 147.188

T	P EXPTL.	P CALCD.	DEV.	PERCENT
46.000	0.40	0.72	0.32	80.39
49.000	0.70	0.88	0.18	25.00
51.600	0.90	1.03	0.13	14.56
54.200	1.00	1.21	0.21	21.16
64.500	2.50	2.24	-0.26	-10.47
75.800	4.50	4.20	-0.30	-6.62
89.000	8.60	8.33	-0.27	-3.16
102.000	15.40	15.56	0.16	1.01
115.600	28.50	28.55	0.05	0.18
129.700	50.90	51.23	0.33	0.65
144.100	89.60	89.24	-0.36	-0.40
156.000	137.00	137.12	0.12	0.09

STANDARD DEVIATION = 0.28

ETHYL TRANS-BETA-(2-FURAN)-ACRYLATE C9H10O3

FROMM F.,LOEFFLER M.C.: J.PHYS.CHEM. 60,252(1956).

A = 6.04137 B = 984.896 C = 75.127

B.P.(100) = 168.576

T	P EXPTL.	P CALCD.	DEV.	PERCENT
155.000	62.00	57.75	-4.25	-6.85
168.000	98.00	97.82	-0.18	-0.19
168.000	99.00	97.82	-1.18	-1.19
177.000	133.00	136.47	3.47	2.61
194.000	234.00	240.87	6.87	2.93
205.000	341.00	335.34	-5.66	-1.66
215.000	437.00	443.27	6.27	1.43
220.000	502.00	506.03	4.03	0.80
222.000	546.00	532.89	-13.11	-2.40
227.000	603.00	604.64	1.64	0.27

STANDARD DEVIATION = 6.98

O-ISOPROPYLBROMOBENZENE $C_9H_{11}BR$

DREISBACH R.R.,SHRADER S.A.:
IND.ENG.CHEM. 41,2879(1949).

A = 6.71783 B = 1462.680 C = 170.918

B.P.(760) = 210.284

T	P EXPTL.	P CALCD.	DEV.	PERCENT
131.660	75.86	76.52	0.66	0.87
145.080	123.76	122.76	-1.00	-0.80
175.830	315.52	315.90	0.38	0.12
193.680	507.50	508.25	0.75	0.15
210.240	760.00	759.22	-0.78	-0.10

STANDARD DEVIATION = 1.17

CYCLOHEXYL PENTAFLUOROPROPIONATE $C_9H_{11}F_5O_2$

SHEEHAN R.J.,LANGER S.H.:
J.CHEM.ENG.DATA 14,248(1969).

A = 7.72546 B = 1844.726 C = 224.891

B.P.(760) = 155.886

T	P EXPTL.	P CALCD.	DEV.	PERCENT
82.000	52.30	51.81	-0.49	-0.93
103.300	126.30	127.22	0.92	0.73
126.500	299.90	299.00	-0.90	-0.30
155.000	740.10	740.49	0.39	0.05

STANDARD DEVIATION = 1.43

1-METHYL-2-ETHYLBENZENE C9H12

FORZIATI A.F.,NORRIS W.R.,ROSSINI F.D.: J.RESEARCH NATL.BUR.STANDARDS 43,555(1949).

A = 6.99453 B = 1529.313 C = 206.607

B.P.(760) = 165.153

	T	P EXPTL.	P CALCD.	DEV.	PERCENT
*	81.146	48.13	47.85	-0.28	-0.59
	85.618	57.73	57.70	-0.03	-0.05
	89.448	67.45	67.43	-0.02	-0.02
	92.949	77.43	77.49	0.06	0.08
	96.200	87.91	87.91	0.00	0.01
	100.584	103.81	103.79	-0.03	-0.02
	105.598	124.82	124.76	-0.06	-0.04
	110.711	149.59	149.63	0.04	0.03
	115.436	176.06	176.09	0.03	0.02
	121.762	217.35	217.38	0.03	0.01
	127.574	261.92	261.95	0.03	0.01
	134.570	325.16	325.13	-0.03	-0.01
	141.792	402.73	402.69	-0.04	-0.01
	149.482	500.98	500.92	-0.06	-0.01
	157.825	628.17	628.19	0.02	0.00
	163.706	732.39	732.37	-0.02	0.00
	164.337	744.38	744.32	-0.06	-0.01
	164.925	755.53	755.58	0.05	0.01
	165.591	768.46	768.51	0.05	0.01
	166.174	779.94	779.96	0.02	0.00

STANDARD DEVIATION = 0.04

1-METHYL-3-ETHYLBENZENE C9H12

FORZIATI A.F.,NORRIS W.R.,ROSSINI F.D.:
J.RESEARCH NATL.BUR.STANDARDS 43,555(1949).

A = 7.00944 B = 1524.727 C = 208.001

B.P.(760) = 161.305

	T	P EXPTL.	P CALCD.	DEV.	PERCENT
*	78.105	48.13	47.88	-0.25	-0.51
	82.525	57.73	57.71	-0.02	-0.03
	86.293	67.45	67.37	-0.08	-0.12
	89.793	77.47	77.51	0.04	0.05
	93.022	87.91	87.96	0.05	0.06
	97.368	103.81	103.84	0.03	0.03
	102.326	124.82	124.78	-0.04	-0.03
	107.383	149.59	149.60	0.01	0.00
	112.074	176.06	176.11	0.05	0.03
	118.338	217.35	217.38	0.03	0.01
	124.082	261.92	261.84	-0.08	-0.03
	131.027	325.16	325.15	-0.01	0.00
	138.178	402.73	402.71	-0.02	-0.01
	145.795	500.98	500.97	-0.01	0.00
	154.053	628.17	628.22	0.05	0.01
	159.871	732.39	732.34	-0.05	-0.01
	160.498	744.38	744.33	-0.05	-0.01
	161.080	755.54	755.60	0.06	0.01
	161.735	768.46	768.44	-0.02	0.00
	162.316	779.94	779.97	0.03	0.00

STANDARD DEVIATION = 0.05

1-METHYL-4-ETHYLBENZENE C9H12

FORZIATI A.F.,NORRIS W.R.,ROSSINI F.D.:
J.RESEARCH NATL.BUR.STANDARDS 43,555(1949).

A = 7.00236 B = 1530.089 C = 209.252

B.P.(760) = 161.989

	T	P EXPTL.	P CALCD.	DEV.	PERCENT
	78.306	48.13	48.02	-0.11	-0.24
*	82.701	57.43	57.74	0.31	0.54
	86.523	67.45	67.49	0.04	0.05
	89.988	77.47	77.47	0.00	-0.01
	93.252	87.91	87.96	0.05	0.06
	97.630	103.81	103.86	0.05	0.05
	102.619	124.82	124.80	-0.02	-0.02
	107.710	149.59	149.62	0.03	0.02
	112.422	176.07	176.08	0.01	0.01
	118.727	217.35	217.35	0.00	0.00
	131.499	325.16	325.10	-0.06	-0.02
	138.701	402.73	402.68	-0.05	-0.01
	146.368	500.98	500.92	-0.06	-0.01
	154.684	628.18	628.18	0.00	0.00
	160.548	732.39	732.40	0.01	0.00
	161.179	744.39	744.39	0.00	0.00
	161.761	755.55	755.57	0.02	0.00
	162.424	768.46	768.48	0.02	0.00
	163.008	779.95	780.00	0.05	0.01

STANDARD DEVIATION = 0.05

PROPYLBENZENE C9H12

FORZIATI A.F.,NORRIS W.R.,ROSSINI F.D.:
J.RESEARCH NATL.BUR.STANDARDS 43,555(1949).

A = 6.95398 B = 1492.978 C = 207.323

B.P.(760) = 159.217

T	P EXPTL.	P CALCD.	DEV.	PERCENT
75.818	48.02	47.98	-0.04	-0.08
80.181	57.71	57.69	-0.02	-0.04
83.993	67.45	67.46	0.01	0.01
87.457	77.50	77.49	-0.01	-0.01
90.688	87.93	87.94	0.01	0.01
95.049	103.84	103.85	0.01	0.01
100.020	124.78	124.82	0.04	0.03
105.085	149.61	149.64	0.03	0.02
109.781	176.09	176.12	0.03	0.02
116.060	217.38	217.38	0.00	0.00
128.794	325.26	325.19	-0.07	-0.02
135.972	402.80	402.73	-0.07	-0.02
143.625	501.03	501.01	-0.02	0.00
151.921	628.20	628.19	-0.01	0.00
157.779	732.43	732.43	0.00	0.00
158.408	744.35	744.40	0.05	0.01
158.991	755.54	755.61	0.07	0.01
159.654	768.55	768.53	-0.02	0.00
160.239	780.09	780.08	-0.01	0.00

STANDARD DEVIATION = 0.04

ISOPROPYLBENZENE C9H12

WILLINGHAM C.J.,TAYLOR W.J.,PIGNOCCO J.M.,ROSSINI F.D.:
J.RESEARCH NATL.BUR.STANDARDS 35,219(1945).

A = 6.93160 B = 1457.318 C = 207.370

B.P.(760) = 152.392

T	P EXPTL.	P CALCD.	DEV.	PERCENT
70.020	47.65	47.63	-0.02	-0.03
74.365	57.41	57.40	-0.01	-0.01
78.155	67.24	67.24	0.00	0.00
81.579	77.29	77.28	-0.01	-0.01
84.768	87.73	87.73	0.00	0.00
89.077	103.66	103.67	0.01	0.01
93.991	124.67	124.68	0.01	0.01
98.975	149.44	149.45	0.01	0.00
103.604	175.89	175.91	0.02	0.01
109.802	217.18	217.21	0.03	0.01
115.495	261.76	261.75	-0.01	0.00
122.353	324.95	324.92	-0.03	-0.01
129.433	402.44	402.42	-0.02	0.00
136.983	500.72	500.67	-0.05	-0.01
145.176	627.94	627.92	-0.02	0.00
150.956	732.11	732.12	0.01	0.00
151.576	744.08	744.06	-0.02	0.00
152.152	755.27	755.28	0.01	0.00
152.798	767.96	768.03	0.07	0.01
153.367	779.39	779.40	0.01	0.00

STANDARD DEVIATION = 0.03

1,2,3-TRIMETHYLBENZENE C9H12

FORZIATI A.F.,NORRIS W.R.,ROSSINI F.D.:
J.RESEARCH NATL.BUR.STANDARDS 43,555(1949).

A = 7.04507 B = 1596.959 C = 207.409

B.P.(760) = 176.083

T	P EXPTL.	P CALCD.	DEV.	PERCENT
90.332	48.13	48.03	-0.10	-0.22
94.826	57.73	57.71	-0.02	-0.04
98.770	67.45	67.50	0.05	0.07
102.336	77.47	77.51	0.04	0.05
105.663	87.91	87.93	0.02	0.02
110.157	103.81	103.83	0.02	0.02
115.287	124.81	124.81	0.00	0.00
120.504	149.59	149.62	0.03	0.02
125.333	176.07	176.07	0.00	0.00
131.800	217.35	217.36	0.01	0.00
137.737	261.91	261.91	0.00	0.00
144.882	325.16	325.09	-0.07	-0.02
152.260	402.72	402.70	-0.02	0.00
160.106	500.98	500.93	-0.05	-0.01
168.614	628.18	628.19	0.01	0.00
174.606	732.39	732.33	-0.06	-0.01
175.252	744.34	744.33	-0.01	0.00
175.852	755.55	755.61	0.06	0.01
176.527	768.46	768.47	0.01	0.00
177.126	779.95	780.02	0.07	0.01

STANDARD DEVIATION = 0.05

1,2,4-TRIMETHYLBENZENE C9H12

FORZIATI A.F.,NORRIS W.R.,ROSSINI F.D.:
J.RESEARCH NATL.BUR.STANDARDS 43,555(1949).

A = 7.04732 B = 1575.673 C = 208.826

B.P.(760) = 169.350

T	P EXPTL.	P CALCD.	DEV.	PERCENT
84.804	48.13	47.99	-0.14	-0.30
89.259	57.73	57.72	-0.01	-0.02
93.155	67.44	67.53	0.09	0.14
96.650	77.47	77.49	0.02	0.02
99.940	87.91	87.94	0.03	0.03
104.369	103.81	103.84	0.03	0.03
109.418	124.81	124.79	-0.02	-0.02
114.572	149.59	149.65	0.06	0.04
119.328	176.07	176.07	0.00	0.00
125.694	217.35	217.30	-0.05	-0.02
131.556	261.91	261.93	0.02	0.01
138.599	325.16	325.11	-0.05	-0.02
145.867	402.72	402.68	-0.04	-0.01
153.603	500.98	500.93	-0.05	-0.01
161.991	628.18	628.23	0.05	0.01
167.896	732.39	732.36	-0.03	0.00
168.534	744.39	744.38	-0.01	0.00
169.121	755.55	755.59	0.04	0.00
169.788	768.47	768.47	0.00	0.00
170.377	779.96	779.99	0.03	0.00

STANDARD DEVIATION = 0.05

1,3,5-TRIMETHYLBENZENE C9H12

FORZIATI A.F.,NORRIS W.R.,ROSSINI F.D.:
J.RESEARCH NATL.BUR.STANDARDS 43,555(1949).

A = 7.07638 B = 1571.005 C = 209.728

B.P.(760) = 164.716

T	P EXPTL.	P CALCD.	DEV.	PERCENT
81.488	48.12	48.06	-0.06	-0.13
85.857	57.73	57.74	0.01	0.02
89.662	67.44	67.46	0.02	0.03
93.131	77.47	77.47	0.00	0.00
96.386	87.92	87.96	0.04	0.05
100.747	103.81	103.84	0.03	0.03
105.716	124.82	124.76	-0.06	-0.05
110.789	149.59	149.59	0.00	0.00
115.489	176.07	176.10	0.03	0.01
121.765	217.36	217.37	0.01	0.01
134.464	325.17	325.13	-0.04	-0.01
141.618	402.72	402.71	-0.01	0.00
149.228	500.98	500.95	-0.04	-0.01
157.477	628.19	628.21	0.02	0.00
163.289	732.39	732.43	0.04	0.01
163.911	744.39	744.35	-0.04	-0.01
164.489	755.56	755.57	0.01	0.00
165.146	768.47	768.47	0.00	0.00
165.725	779.97	779.99	0.02	0.00

STANDARD DEVIATION = 0.03

3-ETHYL-5-MFTHYLPHENOL C9H12O

HANDLEY R.,ET AL.: J.CHEM.SOC. 1964,4404.

A = 7.04083 B = 1615.438 C = 152.601

B.P.(760) = 235.724

T	P EXPTL.	P CALCD.	DEV.	PERCENT
195.082	248.05	248.05	0.00	0.00
201.372	299.99	299.98	-0.01	0.00
206.593	349.47	349.49	0.02	0.01
211.388	400.55	400.57	0.02	0.01
215.581	450.03	450.01	-0.02	0.00
219.456	499.96	499.94	-0.02	0.00
222.985	549.16	549.18	0.02	0.00
226.374	600.03	600.04	0.01	0.00
229.550	651.04	651.04	0.00	0.00
232.279	697.57	697.55	-0.02	0.00
235.080	748.03	748.00	-0.03	0.00
237.858	800.86	800.85	-0.01	0.00
240.440	852.57	852.57	0.00	0.00
242.656	899.02	899.03	0.01	0.00
244.794	945.71	945.71	0.00	0.00
247.169	999.77	999.79	0.02	0.00

STANDARD DEVIATION = 0.02

HEMELLITENOL C9H12O

VONTERRES E.,ET AL.:
BRENNSTOFF CHEM. 36,272(1955).

A = 6.97207 B = 1563.460 C = 134.167

B.P.(760) = 247.979

T	P EXPTL.	P CALCD.	DEV.	PERCENT
123.500	10.00	8.02	-1.98	-19.77
143.200	25.00	21.64	-3.36	-13.44
161.300	50.00	47.93	-2.07	-4.15
172.000	75.00	73.37	-1.63	-2.18
180.500	100.00	100.80	0.80	0.80
191.600	150.00	148.85	-1.15	-0.77
201.000	200.00	202.93	2.93	1.47
208.000	250.00	252.80	2.80	1.12
215.000	300.00	312.16	12.16	4.05
216.000	325.00	321.49	-3.51	-1.08
219.000	350.00	350.83	0.83	0.24
224.500	400.00	410.19	10.19	2.55
227.100	450.00	440.91	-9.09	-2.02
232.000	500.00	503.81	3.81	0.76
234.000	550.00	531.45	-18.55	-3.37
239.000	600.00	605.85	5.85	0.97
241.000	650.00	637.82	-12.18	-1.87
245.000	700.00	705.77	5.77	0.82
248.000	760.00	760.38	0.38	0.05

STANDARD DEVIATION = 7.76

MESITOL C9H12O

VONTERRES E.,ET AL.: BRENNSTOFF CHEM. 36,272(1955).

A = 6.65919 B = 1392.299 C = 147.969

B.P.(760) = 220.522

	T	P EXPTL.	P CALCD.	DEV.	PERCENT
	94.000	10.00	8.04	-1.96	-19.62
	114.700	25.00	22.84	-2.17	-8.66
	132.000	50.00	48.54	-1.46	-2.91
	143.000	75.00	74.84	-0.16	-0.21
*	156.100	100.00	120.51	20.51	20.51
	163.000	150.00	152.02	2.02	1.35
	171.500	200.00	200.00	0.00	0.00
	179.000	250.00	251.76	1.76	0.70
	185.000	300.00	300.41	0.41	0.14
	188.000	325.00	327.38	2.38	0.73
	190.300	350.00	349.33	-0.67	-0.19
	195.500	400.00	403.23	3.23	0.81
	199.900	450.00	453.75	3.75	0.83
	204.000	500.00	505.18	5.18	1.04
	206.500	550.00	538.70	-11.30	-2.06
	210.500	600.00	595.90	-4.10	-0.68
	212.800	650.00	630.86	-19.14	-2.94
	218.000	700.00	715.77	15.77	2.25
	220.600	760.00	761.40	1.40	0.18

STANDARD DEVIATION = 7.47

3-METHYL-5-ETHYLPHENOL C9H12O

VONTERRES E.,ET AL.:

BRENNSTOFF CHEM. 36,272(1955).

A = 7.95799 B = 2236.830 C = 207.972

B.P.(760) = 232.593

T	P EXPTL.	P CALCD.	DEV.	PERCENT
111.500	10.00	9.04	-0.96	-9.56
131.600	25.00	23.49	-1.52	-6.06
148.400	50.00	48.01	-1.99	-3.98
158.900	75.00	72.60	-2.40	-3.19
167.200	100.00	99.05	-0.95	-0.95
179.000	150.00	150.54	0.54	0.36
188.000	200.00	203.72	3.72	1.86
194.500	250.00	251.34	1.34	0.54
201.000	300.00	308.03	8.03	2.68
202.700	325.00	324.52	-0.48	-0.15
205.000	350.00	348.00	-2.00	-0.57
210.000	400.00	403.99	3.99	1.00
213.000	450.00	441.07	-8.93	-1.98
218.000	500.00	509.18	9.18	1.84
221.200	550.00	557.22	7.22	1.31
224.000	600.00	602.30	2.30	0.38
225.700	650.00	631.12	-18.88	-2.91
228.800	700.00	686.62	-13.38	-1.91
233.000	760.00	768.24	8.24	1.08

STANDARD DEVIATION = 7.66

O-PROPYL PHENOL C9H12O

VONTERRES E.,ET AL.:
BRENNSTOFF CHEM. 36,272(1955).

A = 9.21533 B = 3253.991 C = 291.861

B.P.(760) = 221.831

	T	P EXPTL.	P CALCD.	DEV.	PERCENT
	104.300	10.00	10.03	0.03	0.35
*	113.000	25.00	15.07	-9.93	-39.73
	141.000	50.00	49.88	-0.12	-0.24
	150.800	75.00	73.17	-1.83	-2.44
	159.000	100.00	99.55	-0.45	-0.45
	170.000	150.00	147.89	-2.11	-1.41
	179.600	200.00	205.77	5.77	2.89
	185.200	250.00	247.98	-2.02	-0.81
	190.800	300.00	297.54	-2.46	-0.82
	193.600	325.00	325.41	0.41	0.13
	196.300	350.00	354.41	4.41	1.26
	200.400	400.00	402.74	2.74	0.69
	204.000	450.00	449.80	-0.20	-0.04
	207.600	500.00	501.55	1.55	0.31
	210.600	550.00	548.55	-1.45	-0.26
	214.000	600.00	606.38	6.38	1.06
	216.000	650.00	642.80	-7.20	-1.11
	219.000	700.00	700.98	0.98	0.14
	221.500	760.00	752.87	-7.13	-0.94

STANDARD DEVIATION = 3.89

P-PROPYL PHENOL C9H12O

VONTERRES E.,ET AL.:

BRENNSTOFF CHEM. 36,272(1955).

A = 8.32960 B = 2661.185 C = 253.702

B.P.(760) = 234.697

	T	P EXPTL.	P CALCD.	DEV.	PERCENT
*	0.000	10.00	0.01	-9.99	-99.93
	119.000	25.00	15.47	-9.53	-38.14
	148.300	50.00	51.26	1.26	2.52
	155.200	75.00	66.29	-8.71	-11.61
	167.300	100.00	101.98	1.98	1.98
	179.200	150.00	152.15	2.15	1.44
	188.500	200.00	204.91	4.91	2.46
	196.000	250.00	258.19	8.19	3.28
	201.300	300.00	302.60	2.60	0.87
	204.000	325.00	327.62	2.62	0.81
	207.000	350.00	357.46	7.46	2.13
	211.200	400.00	403.10	3.10	0.78
	215.100	450.00	449.82	-0.18	-0.04
	219.000	500.00	501.04	1.04	0.21
	222.100	550.00	545.19	-4.81	-0.87
	225.300	600.00	594.17	-5.83	-0.97
	228.400	650.00	645.12	-4.88	-0.75
	230.800	700.00	687.04	-12.96	-1.85
	234.500	760.00	756.15	-3.85	-0.51

STANDARD DEVIATION = 6.39

O-ISOPROPYL PHENOL C9H12O

VONTERRES E.,ET AL.: BRENNSTOFF CHEM. 36,272(1955).

A = 8.16684 B = 2342.842 C = 229.108

B.P.(760) = 214.106

T	P EXPTL.	P CALCD.	DEV.	PERCENT
96.700	10.00	9.46	-0.54	-5.38
116.500	25.00	24.43	-0.57	-2.28
133.200	50.00	50.16	0.16	0.33
143.100	75.00	74.54	-0.46	-0.61
150.600	100.00	99.25	-0.75	-0.75
162.000	150.00	150.16	0.16	0.11
170.200	200.00	199.33	-0.67	-0.33
177.000	250.00	249.93	-0.07	-0.03
183.000	300.00	303.26	3.26	1.09
185.300	325.00	326.11	1.11	0.34
187.600	350.00	350.41	0.41	0.12
191.700	400.00	397.51	-2.49	-0.62
196.000	450.00	452.55	2.55	0.57
200.000	500.00	509.38	9.38	1.88
202.500	550.00	547.86	-2.15	-0.39
205.400	600.00	595.52	-4.48	-0.75
208.000	650.00	641.16	-8.84	-1.36
210.800	700.00	693.56	-6.44	-0.92
214.500	760.00	768.25	8.25	1.09

STANDARD DEVIATION = 4.52

P-ISOPROPYL PHENOL C9H12O

VONTERRES E.,ET AL.:
BRENNSTOFF CHEM. 36,272(1955).

A = 8.66648 B = 2810.255 C = 257.945

B.P.(760) = 227.782

	T	P EXPTL.	P CALCD.	DEV.	PERCENT
	108.200	10.00	9.80	-0.20	-2.00
	128.400	25.00	24.69	-0.31	-1.24
	145.000	50.00	49.22	-0.78	-1.55
	156.000	75.00	75.42	0.42	0.57
	163.100	100.00	98.17	-1.83	-1.83
	175.000	150.00	149.78	-0.22	-0.15
	183.700	200.00	201.06	1.06	0.53
	190.300	250.00	249.46	-0.54	-0.21
	196.400	300.00	302.82	2.82	0.94
	199.000	325.00	328.38	3.38	1.04
	201.100	350.00	350.36	0.36	0.10
	205.300	400.00	398.12	-1.88	-0.47
	209.900	450.00	456.73	6.73	1.50
*	210.900	500.00	470.41	-29.59	-5.92
	216.000	550.00	545.73	-4.27	-0.78
	219.100	600.00	596.36	-3.64	-0.61
	221.600	650.00	640.06	-9.94	-1.53
	224.700	700.00	698.01	-1.99	-0.29
	228.200	760.00	768.74	8.74	1.15

STANDARD DEVIATION = 4.36

PROPYL PHENYL ETHER C9H12O

DREISBACH R.R.,SHRADER S.A.:
IND.ENG.CHEM. 41,2879(1949).

A = 7.73435 B = 2146.196 C = 252.344

B.P.(760) = 189.849

T	P EXPTL.	P CALCD.	DEV.	PERCENT
101.000	47.16	45.75	-1.41	-2.99
106.390	57.04	56.45	-0.59	-1.04
110.810	66.39	66.75	0.36	0.55
114.830	75.86	77.48	1.62	2.13
157.640	315.52	315.87	0.35	0.11
174.310	507.50	505.88	-1.62	-0.32
189.900	760.00	760.98	0.98	0.13

STANDARD DEVIATION = 1.48

ISOPROPYL PHENYL ETHER C9H12O

HEINRICH J.,SUROVY J.,DOJCANSKY J.: CHEM.ZVESTI 19,462(1965).

A = 6.51759 B = 1238.025 C = 162.970

B.P.(760) = 177.448

T	P EXPTL.	P CALCD.	DEV.	PERCENT
72.100	17.50	17.82	0.32	1.84
81.200	28.40	28.01	-0.39	-1.39
87.100	37.20	36.89	-0.31	-0.84
92.600	47.40	47.14	-0.26	-0.54
98.100	59.00	59.63	0.63	1.07
101.600	69.70	68.90	-0.80	-1.15
105.100	78.70	79.30	0.60	0.77
108.300	89.40	89.90	0.50	0.56
110.400	98.80	97.46	-1.34	-1.36
121.800	146.50	147.95	1.45	0.99
130.700	199.90	200.39	0.49	0.25
137.500	249.40	249.63	0.23	0.09
142.400	294.30	290.67	-3.63	-1.23
153.200	396.70	399.84	3.14	0.79
157.800	454.80	455.03	0.23	0.05
161.900	509.10	509.04	-0.06	-0.01
165.500	561.60	560.43	-1.17	-0.21
168.500	602.60	606.22	3.62	0.60
172.300	669.60	668.29	-1.31	-0.20
174.800	714.00	711.70	-2.30	-0.32

STANDARD DEVIATION = 1.73

PSEUDOCUMENOL C9H12O

VONTERRES E.,ET AL.:

BRENNSTOFF CHEM. 36,272(1955).

A = 6.91446 B = 1546.809 C = 151.548

B.P.(760) = 231.929

T	P EXPTL.	P CALCD.	DEV.	PERCENT
106.500	10.00	8.32	-1.68	-16.79
126.200	25.00	22.15	-2.85	-11.41
143.600	50.00	47.17	-2.83	-5.66
154.500	75.00	72.50	-2.50	-3.34
162.000	100.00	95.77	-4.23	-4.23
177.000	150.00	160.86	10.86	7.24
183.500	200.00	198.51	-1.49	-0.75
191.600	250.00	255.13	5.13	2.05
196.000	300.00	290.95	-9.05	-3.02
200.200	325.00	328.83	3.83	1.18
203.000	350.00	356.20	6.20	1.77
207.500	400.00	403.99	3.99	1.00
211.800	450.00	454.31	4.31	0.96
216.000	500.00	508.16	8.16	1.63
218.100	550.00	536.92	-13.08	-2.38
223.000	600.00	609.05	9.05	1.51
225.000	650.00	640.60	-9.40	-1.45
227.700	700.00	685.23	-14.77	-2.11
232.000	760.00	761.31	1.31	0.17

STANDARD DEVIATION = 7.85

ISOPSEUDOCUMENOL C9H12O

VONTERRES E.,ET AL.:
BRENNSTOFF CHEM. 36,272(1955).

A = 5.60196 B = 768.340 C = 48.528

B.P.(760) = 233.831

T	P EXPTL.	P CALCD.	DEV.	PERCENT
106.000	10.00	4.26	-5.74	-57.36
127.300	25.00	17.07	-7.93	-31.74
145.000	50.00	42.83	-7.17	-14.33
155.800	75.00	69.44	-5.56	-7.41
164.500	100.00	98.90	-1.10	-1.10
176.100	150.00	151.87	1.87	1.24
185.000	200.00	205.03	5.03	2.52
192.200	250.00	257.17	7.17	2.87
199.000	300.00	314.71	14.71	4.90
204.400	325.00	366.59	41.59	12.80
200.000	350.00	323.89	-26.11	-7.46
208.500	400.00	409.86	9.86	2.47
212.400	450.00	454.27	4.27	0.95
216.500	500.00	504.51	4.51	0.90
220.000	550.00	550.37	0.37	0.07
223.000	600.00	591.93	-8.07	-1.35
225.900	650.00	634.12	-15.88	-2.44
229.000	700.00	681.47	-18.53	-2.65
233.000	760.00	746.07	-13.93	-1.83

STANDARD DEVIATION =15.57

2,3,5-TRIMETHYLPHENOL C9H12O

HANDLEY R.,ET AL.:

J.CHEM.SOC. 1964,4404.

A = 7.08012 B = 1685.896 C = 166.141

B.P.(760) = 235.329

T	P EXPTL.	P CALCD.	DEV.	PERCENT
186.482	199.12	199.11	-0.01	0.00
193.614	247.68	247.68	0.00	0.00
200.127	300.05	300.06	0.01	0.01
205.448	349.23	349.24	0.01	0.00
210.422	400.93	400.92	-0.01	0.00
214.675	449.83	449.84	0.01	0.00
218.644	499.72	499.71	-0.01	0.00
222.382	550.67	550.65	-0.02	0.00
225.707	599.37	599.37	0.00	0.00
229.041	651.63	651.62	-0.01	0.00
231.852	698.43	698.43	0.00	0.00
234.759	749.61	749.62	0.01	0.00
237.500	800.58	800.56	-0.02	0.00
240.130	851.97	851.99	0.02	0.00
242.511	900.75	900.77	0.02	0.00
244.811	949.92	949.96	0.04	0.00
247.062	1000.16	1000.12	-0.04	0.00

STANDARD DEVIATION = 0.02

1,4-DIMETHYL-BICYCLO(2,2,1)-HEPTANE C9H16

BARUSHCHENKO R.M.,BELIKOVA N.A.,SKURATOV S.M.,PLATE A.F.: ZH.PHYS.KHIM. 44,3022(1970).

A = 6.76196 B = 1312.656 C = 213.532

B.P.(100) = 62.122

T	P EXPTL.	P CALCD.	DEV.	PERCENT
55.565	76.56	76.55	-0.01	-0.01
66.050	116.65	116.65	0.00	0.00
71.358	142.65	142.68	0.03	0.02
76.249	170.67	170.66	-0.01	0.00
82.860	215.41	215.37	-0.05	-0.02
89.064	265.45	265.45	0.00	0.00
96.191	334.00	334.04	0.04	0.01
103.470	417.94	417.94	0.00	0.00
111.011	521.61	521.59	-0.02	0.00
116.834	614.63	614.64	0.01	0.00
117.458	625.39	625.34	-0.05	-0.01
118.113	636.73	636.72	-0.01	0.00
118.691	646.88	646.89	0.01	0.00
119.367	658.91	658.95	0.04	0.01

STANDARD DEVIATION = 0.03

TRANS-2,3-DIMETHYL-BICYCLO-(2,2,1)-HEPTANE C9H16

BARUSHCHENKO R.M.,BELIKOVA N.A.,SKURATOV S.M.,PLATE A.F.: ZH.PHYS.KHIM. 44,3022(1970).

A = 6.86815 B = 1420.319 C = 212.941

B.P.(100) = 78.816

T	P EXPTL.	P CALCD.	DEV.	PERCENT
72.054	76.67	76.65	-0.02	-0.03
82.892	116.71	116.70	-0.01	-0.01
88.380	142.71	142.73	0.02	0.01
93.437	170.68	170.73	0.05	0.03
100.257	215.41	215.41	0.00	0.00
106.644	265.40	265.40	0.00	0.00
113.984	333.96	333.95	-0.01	0.00
121.474	417.83	417.81	-0.02	0.00
129.221	521.44	521.36	-0.08	-0.02
135.202	614.51	614.40	-0.11	-0.02
135.847	625.16	625.17	0.01	0.00
136.526	636.55	636.66	0.11	0.02
137.118	646.68	646.82	0.14	0.02
137.811	658.94	658.87	-0.07	-0.01

STANDARD DEVIATION = 0.07

CIS-HEXAHYDROINDANE C9H16

CAMIN D.L.,ROSSINI F.D.:

J.PHYS.CHEM. 59,1173(1955).

A = 6.86822 B = 1497.332 C = 207.669

B.P.(760) = 167.846

T	P EXPTL.	P CALCD.	DEV.	PERCENT
77.497	41.49	41.45	-0.04	-0.11
82.973	52.03	52.05	0.02	0.04
87.511	62.48	62.46	-0.02	-0.03
90.961	71.43	71.49	0.06	0.08
94.583	82.08	82.09	0.01	0.01
99.401	98.16	98.18	0.02	0.02
110.605	145.76	145.77	0.01	0.01
115.466	171.62	171.57	-0.05	-0.03
122.308	214.15	214.06	-0.09	-0.04
128.708	261.15	261.14	-0.01	-0.01
135.917	323.76	323.79	0.03	0.01
143.611	403.32	403.38	0.06	0.02
151.698	503.06	503.08	0.02	0.00
160.350	630.29	630.37	0.08	0.01
166.527	735.68	735.79	0.11	0.02
167.170	747.63	747.52	-0.11	-0.02
167.718	757.63	757.62	-0.01	0.00
168.407	770.57	770.47	-0.10	-0.01

STANDARD DEVIATION = 0.07

TRANS-HEXAHYDROINDANE C9H16

CAMIN D.L.,ROSSINI F.D.: J.PHYS.CHEM. 59,1173(1955).

A = 6.86119 B = 1475.698 C = 209.659

B.P.(760) = 161.084

T	P EXPTL.	P CALCD.	DEV.	PERCENT
71.756	41.49	41.43	-0.06	-0.14
77.171	51.99	52.04	0.05	0.10
81.651	62.48	62.44	-0.04	-0.06
85.066	71.45	71.48	0.03	0.04
88.648	82.06	82.09	0.03	0.04
93.415	98.19	98.20	0.01	0.01
104.486	145.76	145.78	0.02	0.01
109.299	171.61	171.62	0.01	0.01
116.059	214.13	214.09	-0.04	-0.02
122.378	261.15	261.11	-0.05	-0.02
129.500	323.76	323.70	-0.06	-0.02
137.115	403.33	403.36	0.03	0.01
145.118	503.07	503.14	0.07	0.01
153.673	630.29	630.42	0.13	0.02
159.779	735.68	735.79	0.11	0.02
160.413	747.64	747.48	-0.16	-0.02
160.955	757.63	757.58	-0.05	-0.01
161.642	770.58	770.54	-0.04	0.00

STANDARD DEVIATION = 0.07

1-NONENE

C9H18

FORZIATI A.F.,CAMIN D.L.,ROSSINI F.D.:
J.RESEARCH NATL.BUR.STANDARDS 45,406(1950).

A = 6.95777 B = 1437.862 C = 205.814

B.P.(760) = 146.867

T	P EXPTL.	P CALCD.	DEV.	PERCENT
66.607	47.89	47.83	-0.06	-0.13
70.874	57.69	57.69	0.00	-0.01
74.517	67.46	67.39	-0.07	-0.10
77.861	77.49	77.46	-0.03	-0.04
81.001	87.92	88.02	0.10	0.11
85.202	103.85	103.98	0.13	0.12
89.942	124.84	124.77	-0.07	-0.06
94.829	149.60	149.67	0.07	0.05
99.341	176.13	176.14	0.01	0.01
110.935	262.03	262.02	-0.01	-0.01
117.622	325.26	325.22	-0.04	-0.01
124.521	402.84	402.74	-0.10	-0.03
131.881	501.09	501.06	-0.03	-0.01
139.859	628.33	628.29	-0.04	-0.01
145.488	732.50	732.51	0.01	0.00
146.091	744.38	744.43	0.05	0.01
146.653	755.63	755.69	0.06	0.01
147.289	768.62	768.58	-0.04	-0.01
147.860	780.22	780.30	0.08	0.01

STANDARD DEVIATION = 0.07

PROPYLCYCLOHEXANE C9H18

FORZIATI A.F.,NORRIS W.R.,ROSSINI F.D.:
J.RESEARCH NATL.BUR.STANDARDS 43,555(1949).

A = 6.89260 B = 1464.346 C = 208.288

B.P.(760) = 156.723

T	P EXPTL.	P CALCD.	DEV.	PERCENT
72.691	48.01	47.98	-0.03	-0.07
77.085	57.70	57.71	0.01	0.02
80.871	67.44	67.37	-0.07	-0.11
84.375	77.49	77.46	-0.03	-0.04
87.641	87.92	87.96	0.04	0.05
92.026	103.83	103.89	0.06	0.05
97.017	124.77	124.81	0.04	0.04
102.111	149.60	149.62	0.02	0.01
106.842	176.08	176.12	0.04	0.02
113.165	217.37	217.37	0.00	0.00
118.982	261.97	261.92	-0.05	-0.02
126.004	325.24	325.21	-0.03	-0.01
133.245	402.79	402.75	-0.04	-0.01
140.965	501.03	500.97	-0.06	-0.01
149.347	628.19	628.17	-0.03	0.00
155.269	732.42	732.44	0.02	0.00
155.904	744.34	744.38	0.04	0.01
156.494	755.53	755.61	0.08	0.01
157.166	768.55	768.56	0.01	0.00
157.756	780.09	780.08	-0.01	0.00

STANDARD DEVIATION = 0.04

ISOPROPYLCYCLOHEXANE C9H18

FORZIATI A.F.,NORRIS W.R.,ROSSINI F.D.:
J.RESEARCH NATL.BUR.STANDARDS 43,555(1949).

A = 6.87314 B = 1453.201 C = 209.435

B.P.(760) = 154.563

T	P EXPTL.	P CALCD.	DEV.	PERCENT
70.515	48.09	48.11	0.02	0.04
74.868	57.78	57.77	-0.01	-0.02
78.690	67.51	67.53	0.02	0.03
82.165	77.56	77.55	-0.01	-0.01
85.400	87.98	87.96	-0.02	-0.02
89.788	103.90	103.89	-0.01	-0.01
94.793	124.85	124.87	0.02	0.02
99.887	149.67	149.67	0.00	0.00
104.612	176.16	176.12	-0.04	-0.02
110.953	217.44	217.47	0.03	0.01
116.782	262.04	262.08	0.04	0.02
123.806	325.32	325.33	0.01	0.00
131.051	402.85	402.83	-0.02	-0.01
138.782	501.08	501.05	-0.03	-0.01
147.177	628.25	628.23	-0.02	0.00
153.109	732.47	732.48	0.01	0.00
153.747	744.41	744.45	0.04	0.01
154.334	755.63	755.60	-0.03	0.00
155.009	768.59	768.59	0.00	0.00
155.602	780.14	780.14	0.00	0.00

STANDARD DEVIATION = 0.03

1,1,3-TRIMETHYLCYCLOHEXANE C9H18

FORZIATI A.F.,NORRIS W.R.,ROSSINI F.D.:
J.RESEARCH NATL.BUR.STANDARDS 43,555(1949).

A = 6.83951 B = 1394.882 C = 215.733

B.P.(760) = 136.625

	T	P EXPTL.	P CALCD.	DEV.	PERCENT
	54.669	47.99	47.97	-0.02	-0.04
	58.950	57.68	57.72	0.04	0.08
	62.624	67.43	67.36	-0.07	-0.11
	66.058	77.48	77.53	0.05	0.06
	69.180	87.91	87.84	-0.07	-0.08
	73.481	103.82	103.87	0.05	0.05
	78.336	124.76	124.77	0.01	0.01
*	82.305	149.59	144.31	-5.28	-3.53
	87.915	176.07	176.10	0.03	0.02
	94.080	217.35	217.36	0.01	0.00
	99.762	261.96	261.98	0.02	0.01
	106.606	325.23	325.19	-0.04	-0.01
	113.678	402.78	402.75	-0.03	-0.01
	121.220	501.02	500.98	-0.04	-0.01
	129.411	628.18	628.15	-0.03	-0.01
	135.203	732.42	732.42	0.00	0.00
	135.824	744.34	744.36	0.02	0.00
	136.401	755.52	755.59	0.07	0.01
	137.058	768.55	768.54	-0.01	0.00
	137.636	780.06	780.07	0.01	0.00

STANDARD DEVIATION = 0.04

METHYL CAPRYLATE C9H18O2

ROSE A.,SUPINA W.R.: J.CHEM.ENG.DATA6,173(1961).

A = 6.91645 B = 1496.316 C = 176.469

B.P.(100) = 127.879

T	P EXPTL.	P CALCD.	DEV.	PERCENT
100.300	32.70	32.37	-0.33	-1.02
105.500	40.80	40.72	-0.08	-0.20
109.800	49.00	48.92	-0.08	-0.16
114.700	59.90	59.90	0.00	0.01
115.000	60.80	60.64	-0.16	-0.27
118.500	69.60	69.77	0.17	0.24
120.300	74.80	74.89	0.09	0.12
121.000	77.00	76.97	-0.03	-0.04
124.000	86.30	86.40	0.10	0.12
126.900	96.10	96.41	0.31	0.32
128.100	100.70	100.82	0.12	0.12
128.400	101.60	101.95	0.35	0.35
128.600	102.60	102.71	0.11	0.11
129.400	105.50	105.79	0.29	0.27
131.100	112.50	112.58	0.08	0.07
131.500	113.80	114.23	0.43	0.38
131.400	113.90	113.82	-0.08	-0.07
131.500	114.20	114.23	0.03	0.03
135.100	130.50	130.00	-0.50	-0.39
137.300	141.40	140.48	-0.92	-0.65
137.600	142.60	141.96	-0.64	-0.45
138.800	148.00	148.01	0.01	0.01
141.000	159.60	159.66	0.06	0.04
145.700	186.50	187.05	0.55	0.29

STANDARD DEVIATION = 0.35

PELARGONIC ACID C9H18O2

KAHLBAUM G.W.A.:

Z.PHYS.CHEM. 13,14(1894).

A = 3.23594 B = 143.970 C = -75.644

B.P.(10) = 140.033

T	P EXPTL.	P CALCD.	DEV.	PERCENT
136.800	10.10	7.62	-2.48	-24.58
137.600	10.60	8.17	-2.43	-22.93
140.100	12.10	10.05	-2.05	-16.91
141.100	12.70	10.88	-1.82	-14.36
141.200	12.80	10.96	-1.84	-14.37
141.700	13.20	11.39	-1.81	-13.73
142.400	13.70	12.00	-1.70	-12.39
143.400	14.40	12.92	-1.48	-10.31
145.600	16.20	15.06	-1.14	-7.01
146.000	16.50	15.48	-1.02	-6.21
147.200	17.60	16.75	-0.85	-4.84
148.200	18.50	17.85	-0.65	-3.50
148.900	19.00	18.65	-0.35	-1.85
149.200	19.40	19.00	-0.40	-2.08
150.200	20.40	20.18	-0.22	-1.08
151.400	21.50	21.65	0.15	0.71
152.600	22.70	23.18	0.48	2.12
153.600	23.80	24.50	0.70	2.93
154.400	24.90	25.58	0.68	2.73
155.400	26.90	26.97	0.07	0.25
156.200	27.10	28.10	1.00	3.70
157.100	28.00	29.41	1.41	5.03
158.100	29.10	30.90	1.80	6.18
159.200	30.60	32.58	1.98	6.46
160.300	32.20	34.30	2.10	6.52
161.800	34.60	36.72	2.12	6.13
162.800	35.70	38.38	2.68	7.50
163.800	36.90	40.07	3.17	8.59
165.100	39.30	42.32	3.02	7.69
166.000	41.00	43.91	2.91	7.10
167.100	43.40	45.89	2.49	5.74
168.200	45.90	47.91	2.01	4.39
169.100	48.00	49.59	1.59	3.32
170.400	51.20	52.07	0.87	1.69
171.500	54.10	54.20	0.10	0.19
172.600	56.90	56.37	-0.53	-0.93
173.200	58.50	57.57	-0.93	-1.59
175.000	64.10	61.22	-2.88	-4.49
176.300	68.10	63.92	-4.18	-6.14
177.000	69.60	65.39	-4.21	-6.05

STANDARD DEVIATION = 2.00

PROPYL CAPROATE C9H18O2

BONHORST C.W.,ALTHOUSE P.M.,TRIEBOLD H.O.:
IND.ENG.CHEM. 40,2379(1948).

A = 8.66706 B = 2556.033 C = 262.889

B.P.(100) = 120.493

T	P EXPTL.	P CALCD.	DEV.	PERCENT
42.900	2.00	2.03	0.03	1.68
54.200	4.00	4.04	0.04	0.94
57.700	5.00	4.94	-0.06	-1.11
60.900	6.00	5.93	-0.07	-1.20
65.900	8.00	7.82	-0.18	-2.30
70.200	10.00	9.85	-0.15	-1.52
84.200	20.00	20.08	0.08	0.42
98.900	40.00	40.00	0.00	0.00
103.800	50.00	49.71	-0.29	-0.57
108.200	60.00	60.14	0.14	0.23
115.300	80.00	80.99	0.99	1.24
120.300	100.00	99.23	-0.77	-0.77

STANDARD DEVIATION = 0.44

1-CHLORONONANE C9H19CL

KEMME R.H.,KREPS S.I.:
J.CHEM.ENG.DATA 14,98(1969).

A = 7.04654 B = 1655.565 C = 192.260

B.P.(760) = 205.165

T	P EXPTL.	P CALCD.	DEV.	PERCENT
69.100	5.10	5.15	0.05	1.05
70.800	5.50	5.66	0.16	2.97
80.600	9.60	9.53	-0.07	-0.73
89.900	15.10	15.10	0.00	0.02
95.400	19.60	19.56	-0.05	-0.23
104.600	29.50	29.49	-0.01	-0.05
111.500	39.60	39.47	-0.13	-0.32
119.800	55.00	55.11	0.11	0.21
129.500	79.70	79.65	-0.05	-0.06
141.300	121.20	121.12	-0.08	-0.06
156.700	200.50	200.57	0.07	0.03
170.300	301.60	302.15	0.55	0.18
188.600	501.80	500.76	-1.04	-0.21
204.800	752.80	753.32	0.52	0.07

STANDARD DEVIATION = 0.40

1-CYCLOHEXYLAMINO-2-PROPANOL $C_9H_{19}NO$

MCDONALD R.A.,SHRADER S.A.,STULL D.R.:
J.CHEM.ENG.DATA 4,311(1959).

A = 7.01156 B = 1655.019 C = 162.588

B.P.(760) = 238.070

T	P EXPTL.	P CALCD.	DEV.	PERCENT
150.400	53.13	52.94	-0.19	-0.36
155.260	63.86	63.77	-0.09	-0.14
159.230	73.84	73.93	0.09	0.13
165.770	93.34	93.60	0.26	0.28
177.790	141.07	141.02	-0.05	-0.04
223.150	526.37	526.07	-0.30	-0.06
235.800	719.65	719.91	0.26	0.04
238.070	760.00	760.00	0.00	0.00

STANDARD DEVIATION = 0.24

3,3-DIETHYLPENTANE C_9H_{20}

FORZIATI A.F.,NORRIS W.R.,ROSSINI F.D.:
J.RESEARCH NATL.BUR.STANDARDS 43,555(1949).

A = 6.89603 B = 1453.480 C = 215.825

B.P.(760) = 146.167

T	P EXPTL.	P CALCD.	DEV.	PERCENT
62.882	48.02	47.97	-0.05	-0.11
67.240	57.71	57.71	0.00	0.00
71.026	67.46	67.45	-0.01	-0.01
74.484	77.50	77.51	0.01	0.02
77.699	87.93	87.94	0.01	0.02
82.043	103.84	103.85	0.01	0.01
87.000	124.79	124.82	0.03	0.03
92.048	149.62	149.62	0.00	0.00
96.737	176.10	176.13	0.03	0.01
103.002	217.38	217.37	-0.01	-0.01
115.727	325.26	325.21	-0.05	-0.01
122.903	402.80	402.75	-0.05	-0.01
130.558	501.04	501.04	0.00	0.00
138.862	628.21	628.22	0.01	0.00
144.725	732.44	732.40	-0.04	-0.01
145.356	744.36	744.37	0.01	0.00
145.942	755.56	755.63	0.07	0.01
146.606	768.57	768.55	-0.02	0.00
147.194	780.10	780.13	0.03	0.00

STANDARD DEVIATION = 0.03

NONANE

C9H20

FORZIATI A.F.,NORRIS W.R.,ROSSINI F.D.:
J.RESEARCH NATL.BUR.STANDARDS 43,555(1949).

A = 6.93764 B = 1430.459 C = 201.808

B.P.(760) = 150.797

	T	P EXPTL.	P CALCD.	DEV.	PERCENT
	70.343	48.04	48.03	-0.01	-0.02
	74.546	57.73	57.74	0.01	0.01
*	78.219	67.47	67.51	0.04	0.05
	81.548	77.52	77.51	-0.01	-0.01
	84.658	87.94	87.94	0.00	0.00
	88.864	103.86	103.85	-0.01	-0.01
	93.661	124.80	124.83	0.03	0.02
	98.545	149.63	149.64	0.01	0.01
	103.072	176.11	176.10	-0.01	-0.01
	109.136	217.40	217.40	0.00	0.00
	114.712	262.00	262.00	0.00	0.00
	121.433	325.27	325.29	0.02	0.01
	128.357	402.81	402.79	-0.02	-0.01
	135.741	501.05	501.02	-0.03	-0.01
	143.751	628.22	628.18	-0.04	-0.01
	149.409	732.45	732.44	-0.01	0.00
	150.017	744.37	744.41	0.04	0.01
	150.579	755.57	755.61	0.04	0.00
	151.222	768.57	768.58	0.01	0.00
	151.786	780.11	780.10	-0.01	0.00

STANDARD DEVIATION = 0.02

2,2,3,3-TETRAMETHYLPENTANE C_9H_{20}

FORZIATI A.F.,NORRIS W.R.,ROSSINI F.D.:
J.RESEARCH NATL.BUR.STANDARDS 43,555(1949).

A = 6.83060 B = 1398.668 C = 213.839

B.P.(760) = 140.273

T	P EXPTL.	P CALCD.	DEV.	PERCENT
57.834	48.10	48.11	0.01	0.02
62.087	57.79	57.76	-0.03	-0.06
65.828	67.52	67.52	0.00	-0.01
69.236	77.57	77.56	-0.01	-0.02
72.414	87.99	88.00	0.01	0.01
76.705	103.91	103.90	-0.01	-0.01
81.609	124.86	124.89	0.03	0.03
86.602	149.68	149.70	0.02	0.01
91.242	176.17	176.20	0.03	0.02
97.450	217.46	217.49	0.03	0.02
103.154	262.06	262.00	-0.06	-0.02
110.054	325.33	325.31	-0.02	-0.01
117.168	402.86	402.81	-0.05	-0.01
124.767	501.10	501.11	0.01	0.00
133.015	628.26	628.28	0.02	0.00
138.844	732.49	732.49	0.00	0.00
139.471	744.43	744.46	0.03	0.00
140.051	755.65	755.66	0.01	0.00
140.713	768.61	768.61	0.00	0.00
141.295	780.16	780.14	-0.02	0.00

STANDARD DEVIATION = 0.03

2,2,3,4-TETRAMETHYLPENTANE C9H20

FORZIATI A.F.,NORRIS W.R.,ROSSINI F.D.:
J.RESEARCH NATL.BUR.STANDARDS 43,555(1949).

A = 6.83418 B = 1375.594 C = 214.940

B.P.(760) = 133.016

T	P EXPTL.	P CALCD.	DEV.	PERCENT
52.028	48.07	48.03	-0.04	-0.08
56.243	57.76	57.76	0.00	-0.01
59.913	67.50	67.50	0.00	0.01
63.262	77.55	77.55	0.00	0.00
66.381	87.97	87.98	0.01	0.02
70.587	103.89	103.86	-0.03	-0.03
75.418	124.84	124.91	0.07	0.05
80.311	149.66	149.66	0.00	0.00
84.871	176.15	176.18	0.03	0.02
90.965	217.43	217.45	0.02	0.01
96.574	262.04	262.02	-0.02	-0.01
103.343	325.31	325.27	-0.04	-0.01
110.331	402.85	402.81	-0.04	-0.01
117.787	501.08	501.04	-0.05	-0.01
125.886	628.25	628.22	-0.03	-0.01
131.611	732.48	732.46	-0.02	0.00
132.227	744.41	744.44	0.03	0.00
132.798	755.62	755.68	0.06	0.01
133.446	768.60	768.59	-0.01	0.00
134.021	780.14	780.19	0.05	0.01

STANDARD DEVIATION = 0.04

2,2,4,4-TETRAMETHYLPENTANE C9H20

FORZIATI A.F.,NORRIS W.R.,ROSSINI F.D.:
J.RESEARCH NATL.BUR.STANDARDS 43,555(1949).

A = 6.79620 B = 1324.588 C = 216.019

B.P.(760) = 122.284

	T	P EXPTL.	P CALCD.	DEV.	PERCENT
*	42.956	48.09	48.03	-0.07	-0.14
	47.079	57.78	57.76	-0.02	-0.04
	50.671	67.52	67.52	0.00	0.00
	53.944	77.56	77.56	0.00	0.00
	56.991	87.99	87.99	0.00	-0.01
	61.117	103.91	103.91	0.00	0.00
	65.828	124.86	124.89	0.03	0.02
	70.625	149.68	149.68	0.00	0.00
	75.085	176.17	176.19	0.02	0.01
	81.052	217.45	217.46	0.01	0.00
	86.549	262.06	262.05	-0.01	0.00
	93.183	325.33	325.32	-0.01	0.00
	100.031	402.86	402.84	-0.02	-0.01
	107.342	501.10	501.06	-0.04	-0.01
	115.289	628.27	628.27	0.00	0.00
	120.906	732.49	732.48	-0.01	0.00
	121.510	744.43	744.44	0.01	0.00
	122.072	755.65	755.71	0.06	0.01
	122.709	768.61	768.64	0.03	0.00
	123.267	780.16	780.11	-0.05	-0.01

STANDARD DEVIATION = 0.03

2,3,3,4-TETRAMETHYLPENTANE C9H20

FORZIATI A.F.,NORRIS W.R.,ROSSINI F.D.:
J.RESEARCH NATL.BUR.STANDARDS 43,555(1949).

A = 6.86311 B = 1419.742 C = 214.963

B.P.(760) = 141.551

T	P EXPTL.	P CALCD.	DEV.	PERCENT
59.010	48.02	47.98	-0.04	-0.08
63.320	57.71	57.72	0.01	0.01
67.037	67.45	67.38	-0.07	-0.10
70.491	77.50	77.53	0.03	0.04
73.670	87.93	87.95	0.02	0.03
77.969	103.84	103.86	0.02	0.02
82.872	124.78	124.80	0.02	0.02
87.881	149.61	149.65	0.04	0.03
98.732	217.38	217.39	0.01	0.00
111.345	325.26	325.22	-0.04	-0.01
118.461	402.80	402.75	-0.05	-0.01
126.053	501.04	501.00	-0.04	-0.01
134.294	628.20	628.16	-0.04	-0.01
140.120	732.44	732.43	-0.01	0.00
140.746	744.36	744.39	0.03	0.00
141.326	755.56	755.62	0.06	0.01
141.987	768.57	768.56	-0.01	0.00
142.571	780.10	780.14	0.04	0.01

STANDARD DEVIATION = 0.04

2,2,5-TRIMETHYLHEXANE C9H20

FORZIATI A.F.,NORRIS W.R.,ROSSINI F.D.:
J.RESEARCH NATL.BUR.STANDARDS 43,555(1949).

A = 6.83775 B = 1325.541 C = 210.908

B.P.(760) = 124.083

T	P EXPTL.	P CALCD.	DEV.	PERCENT
46.141	48.03	47.97	-0.06	-0.12
50.208	57.72	57.72	0.00	-0.01
53.743	67.46	67.47	0.01	0.02
56.968	77.51	77.52	0.01	0.01
59.973	87.94	87.96	0.02	0.03
68.655	124.80	124.82	0.02	0.01
73.381	149.62	149.66	0.04	0.02
77.756	176.11	176.10	-0.01	0.00
83.624	217.39	217.40	0.01	0.00
95.539	325.27	325.26	-0.01	0.00
102.256	402.81	402.73	-0.08	-0.02
109.434	501.05	501.02	-0.03	-0.01
117.225	628.22	628.21	-0.01	0.00
122.731	732.45	732.44	-0.01	0.00
123.324	744.37	744.42	0.05	0.01
123.868	755.57	755.55	-0.02	0.00
124.497	768.58	768.58	0.00	0.00
125.050	780.12	780.18	0.06	0.01

STANDARD DEVIATION = 0.04

2,4,4-TRIMETHYLHEXANE

C9H20

FORZIATI A.F.,NORRIS W.R.,ROSSINI F.D.:
J.RESEARCH NATL.BUR.STANDARDS 43,555(1949).

A = 6.85654 B = 1371.813 C = 214.400

B.P.(760) = 130.646

T	P EXPTL.	P CALCD.	DEV.	PERCENT
50.648	48.02	47.95	-0.07	-0.14
54.826	57.72	57.70	-0.02	-0.04
58.467	67.46	67.48	0.02	0.02
61.776	77.51	77.51	0.00	0.00
64.862	87.93	87.96	0.03	0.03
69.023	103.85	103.84	-0.01	-0.01
73.786	124.79	124.85	0.06	0.05
78.631	149.62	149.65	0.03	0.02
83.123	176.10	176.10	0.00	0.00
89.149	217.39	217.42	0.03	0.01
101.368	325.27	325.22	-0.05	-0.02
108.264	402.81	402.74	-0.07	-0.02
115.621	501.05	500.95	-0.10	-0.02
129.260	732.45	732.44	-0.01	0.00
129.869	744.37	744.44	0.07	0.01
130.429	755.57	755.62	0.05	0.01
131.070	768.57	768.57	0.00	0.00
131.636	780.11	780.15	0.04	0.01

STANDARD DEVIATION = 0.05

1-NONANOL C9H20O

KEMME R.H.,KREPS S.I.: J.CHEM.ENG.DATA 14,98(1969).

A = 6.67575 B = 1276.626 C = 123.101

B.P.(760) = 213.302

T	P EXPTL.	P CALCD.	DEV.	PERCENT
91.700	5.60	5.40	-0.20	-3.56
102.100	10.50	10.16	-0.34	-3.24
108.900	15.30	14.90	-0.40	-2.64
114.400	20.30	19.98	-0.32	-1.60
122.400	30.20	29.90	-0.30	-1.00
128.800	40.60	40.53	-0.07	-0.17
135.800	55.50	55.56	0.06	0.12
144.700	80.40	81.03	0.63	0.79
155.000	120.70	121.68	0.98	0.81
169.000	201.40	201.95	0.55	0.27
181.000	302.40	300.40	-2.00	-0.66
198.000	502.50	501.14	-1.36	-0.27
213.600	763.90	765.89	1.99	0.26

STANDARD DEVIATION = 1.10

CYCLOHEXYL TRIMETHYLSILYL ETHER C9H20SIO

SHEEHAN R.J.,LANGER S.H.: J.CHEM.ENG.DATA 14,248(1969).

A = 8.09052 B = 2276.617 C = 267.943

B.P.(760) = 169.053

T	P EXPTL.	P CALCD.	DEV.	PERCENT
90.700	55.10	55.29	0.19	0.34
104.900	96.80	96.47	-0.33	-0.34
125.100	198.50	198.71	0.21	0.11
168.100	740.40	740.33	-0.07	-0.01

STANDARD DEVIATION = 0.44

PROPYL BORATE C9H21BO3

CHRISTOPHER P.M.,SHILMAN A.:
J.CHEM.ENG.DATA 12,333(1967).

A = 7.39983 B = 1741.100 C = 206.412

B.P.(760) = 178.871

T	P EXPTL.	P CALCD.	DEV.	PERCENT
85.100	28.50	26.74	-1.76	-6.17
111.200	82.20	82.79	0.59	0.72
131.300	174.60	175.49	0.89	0.51
136.500	212.00	210.10	-1.90	-0.90
146.500	296.00	292.62	-3.38	-1.14
154.800	369.40	379.89	10.49	2.84
167.100	548.00	547.51	-0.49	-0.09
171.200	622.60	615.18	-7.42	-1.19
175.800	692.60	699.03	6.43	0.93
178.800	764.00	758.53	-5.47	-0.72

STANDARD DEVIATION = 6.05

ISOPROPYL BORATE C9H21BO3

CHRISTOPHER P.M.,SHILMAN A.:
J.CHEM.ENG.DATA 12,333(1967).

A = 8.06982 B = 2120.204 C = 269.125

B.P.(760) = 139.470

T	P EXPTL.	P CALCD.	DEV.	PERCENT
65.100	56.40	53.23	-3.17	-5.62
85.700	120.80	124.30	3.50	2.90
98.100	198.10	197.80	-0.30	-0.15
107.100	272.30	271.86	-0.44	-0.16
114.500	348.40	349.18	0.78	0.23
119.100	411.20	406.01	-5.19	-1.26
124.900	485.20	488.57	3.37	0.69
129.900	563.90	570.63	6.73	1.19
133.500	631.00	636.60	5.60	0.89
136.900	707.80	704.63	-3.17	-0.45
138.900	757.20	747.40	-9.80	-1.29

STANDARD DEVIATION = 5.52

M-TOLYL PENTAFLUOROPROPIONATE C10H7F5O2

SHEEHAN R.J.,LANGER S.H.:
J.CHEM.ENG.DATA 14,248(1969).

A = 7.42720 B = 1707.592 C = 201.695

B.P.(760) = 173.898

T	P EXPTL.	P CALCD.	DEV.	PERCENT
98.400	54.60	54.58	-0.02	-0.04
113.000	100.20	100.24	0.04	0.04
131.400	199.90	199.88	-0.02	-0.01
172.600	732.90	732.90	0.00	0.00

STANDARD DEVIATION = 0.05

P-TOLYL PENTAFLUOROPROPIONATE C10H7F5O2

SHEEHAN R.J.,LANGER S.H.:
J.CHEM.ENG.DATA 14,248(1969).

A = 8.07860 B = 2223.789 C = 252.110

B.P.(760) = 175.724

T	P EXPTL.	P CALCD.	DEV.	PERCENT
98.800	54.60	55.13	0.53	0.97
113.500	100.20	99.12	-1.08	-1.07
133.000	200.60	201.45	0.85	0.42
174.400	732.60	732.27	-0.33	-0.05

STANDARD DEVIATION = 1.51

NAPHTHALENE C10H8

CAMIN D.L.,ROSSINI F.D.:

J.PHYS.CHEM. 59,1173(1955).

A = 6.81812 B = 1585.857 C = 184.819

B.P.(760) = 217.958

T	P EXPTL.	P CALCD.	DEV.	PERCENT
126.325	52.12	52.63	0.51	0.98
130.836	62.54	62.24	-0.30	-0.47
134.548	71.53	71.20	-0.33	-0.46
138.677	82.12	82.39	0.27	0.33
143.930	98.26	98.67	0.41	0.42
155.766	145.78	145.16	-0.62	-0.43
161.104	171.60	171.28	-0.32	-0.19
168.540	214.14	213.88	-0.26	-0.12
175.526	261.16	261.32	0.16	0.06
183.336	323.72	324.00	0.28	0.09
191.702	403.34	403.88	0.54	0.13
200.471	503.08	503.63	0.55	0.11
216.539	736.00	736.02	0.02	0.00
217.237	748.03	747.73	-0.30	-0.04
217.848	758.40	758.11	-0.29	-0.04
218.638	772.00	771.70	-0.30	-0.04

STANDARD DEVIATION = 0.41

NAPHTHALENE C10H8

FOWLER L.,TRUMP N.W.,VOGLER C.E.:
J.CHEM.ENG.DATA13,209(1968).

A = 7.03358 B = 1756.328 C = 204.842

B.P.(100) = 144.080

	T	P EXPTL.	P CALCD.	DEV.	PERCENT
	80.330	7.51	7.49	-0.02	-0.21
	82.880	8.55	8.50	-0.05	-0.61
*	83.350	8.80	8.70	-0.10	-1.19
	85.690	9.71	9.74	0.03	0.26
	88.850	11.30	11.31	0.01	0.08
	92.740	13.58	13.54	-0.04	-0.30
	93.280	13.85	13.88	0.03	0.19
	94.310	14.64	14.54	-0.10	-0.68
	98.660	17.67	17.65	-0.02	-0.12
	102.930	21.25	21.23	-0.02	-0.08
	106.190	24.33	24.37	0.04	0.15
	109.230	27.73	27.64	-0.09	-0.34
	109.490	27.88	27.93	0.05	0.18
	109.800	28.28	28.29	0.01	0.03
	115.550	35.63	35.63	0.00	-0.01
	115.570	35.67	35.65	-0.02	-0.04
	119.220	41.03	41.10	0.07	0.17
	124.440	50.05	50.09	0.04	0.08
	124.910	50.98	50.98	0.00	-0.01
	125.910	52.95	52.90	-0.05	-0.09
	133.280	69.06	69.06	0.00	0.00
	133.690	69.97	70.07	0.10	0.14
	138.480	82.63	82.77	0.14	0.17
	139.610	86.08	86.04	-0.04	-0.05
	140.830	89.64	89.68	0.04	0.04
	149.640	120.00	119.94	-0.06	-0.05
	150.380	122.80	122.82	0.02	0.02
*	158.010	155.70	156.04	0.34	0.22
	159.080	161.30	161.24	-0.06	-0.04
	159.540	163.60	163.52	-0.08	-0.05
	168.700	214.80	214.66	-0.14	-0.06
	179.510	290.90	291.07	0.17	0.06

STANDARD DEVIATION = 0.07

ALPHA-NAPHTHOL

C10H8O

VONTERRES E.,ET AL.:

BRENNSTOFF CHEM. 36,272(1955).

A = 7.51563 B = 2227.353 C = 198.170

B.P.(760) = 282.399

	T	P EXPTL.	P CALCD.	DEV.	PERCENT
	141.500	10.00	9.08	-0.92	-9.17
	164.800	25.00	23.94	-1.06	-4.23
	184.200	50.00	49.04	-0.96	-1.93
	196.900	75.00	75.47	0.47	0.62
	206.000	100.00	101.09	1.09	1.09
	219.300	150.00	151.45	1.45	0.97
	229.000	200.00	200.18	0.18	0.09
	236.400	250.00	245.59	-4.41	-1.76
	244.000	300.00	300.82	0.82	0.27
	246.500	325.00	321.09	-3.91	-1.20
	250.200	350.00	353.16	3.16	0.90
	255.100	400.00	399.64	-0.36	-0.09
	260.300	450.00	454.36	4.36	0.97
	264.800	500.00	506.55	6.55	1.31
*	265.700	550.00	517.56	-32.44	-5.90
	272.000	600.00	600.20	0.20	0.03
	275.200	650.00	646.13	-3.87	-0.60
	278.300	700.00	693.32	-6.68	-0.95
	282.500	760.00	761.70	1.70	0.22

STANDARD DEVIATION = 3.43

BETA-NAPHTHOL C10H8O

VONTERRES E.,ET AL.: BRENNSTOFF CHEM. 36,272(1955).

A = 7.40789 B = 2162.342 C = 190.280

B.P.(760) = 287.366

	T	P EXPTL.	P CALCD.	DEV.	PERCENT
	144.000	10.00	8.69	-1.31	-13.06
	168.000	25.00	23.58	-1.42	-5.68
	188.000	50.00	49.16	-0.84	-1.67
	200.500	75.00	74.90	-0.10	-0.13
	210.000	100.00	101.35	1.35	1.35
	223.100	150.00	150.32	0.32	0.21
	233.000	200.00	199.23	-0.77	-0.39
*	246.000	250.00	282.86	32.86	13.14
	248.400	300.00	301.09	1.09	0.36
	251.000	325.00	321.91	-3.09	-0.95
	254.200	350.00	349.15	-0.85	-0.24
	260.200	400.00	405.33	5.33	1.33
	265.100	450.00	456.51	6.51	1.45
	269.000	500.00	500.93	0.93	0.19
	273.000	550.00	550.08	0.08	0.02
	276.600	600.00	597.61	-2.39	-0.40
	279.700	650.00	641.16	-8.84	-1.36
	283.000	700.00	690.32	-9.68	-1.38
	288.000	760.00	770.57	10.57	1.39

STANDARD DEVIATION = 5.03

QUINALDINE $C_{10}H_9N$

MALANOWSKI S.:
BULL.ACAD.POLON.SCI.SER.SCI.CHIM.9, 71(1961).

A = 7.17899 B = 1857.842 C = 184.496

B.P.(760) = 247.743

T	P EXPTL.	P CALCD.	DEV.	PERCENT
178.300	114.20	114.31	0.11	0.10
186.350	148.00	147.66	-0.34	-0.23
191.410	172.40	172.46	0.06	0.03
195.110	192.60	192.69	0.09	0.05
197.980	209.80	209.69	-0.11	-0.05
200.640	226.30	226.53	0.23	0.10
202.950	241.80	242.04	0.24	0.10
204.900	256.10	255.80	-0.30	-0.12
207.350	273.80	273.99	0.19	0.07
207.310	273.90	273.68	-0.22	-0.08
209.480	290.60	290.65	0.05	0.02
211.550	307.60	307.62	0.02	0.01
213.130	321.10	321.11	0.01	0.00
215.020	337.80	337.88	0.08	0.02
216.630	352.90	352.71	-0.19	-0.05
218.280	368.20	368.46	0.26	0.07
219.750	383.20	382.97	-0.23	-0.06
221.540	401.10	401.26	0.16	0.04
222.930	415.80	415.95	0.15	0.04
224.690	435.30	435.16	-0.14	-0.03
226.270	453.00	453.02	0.02	0.00
227.680	469.40	469.45	0.05	0.01
229.200	487.70	487.70	0.00	0.00
230.900	508.90	508.78	-0.12	-0.02
231.930	522.20	521.90	-0.30	-0.06
233.520	542.57	542.70	0.13	0.02
234.920	561.48	561.56	0.08	0.01
235.960	576.20	575.91	-0.29	-0.05
* 236.340	589.92	581.22	-8.70	-1.47
* 237.110	605.93	592.11	-13.82	-2.28
239.060	620.24	620.43	0.19	0.03
240.220	637.65	637.78	0.13	0.02
241.580	658.34	658.62	0.28	0.04
242.520	674.17	673.34	-0.83	-0.12
243.570	689.54	690.09	0.55	0.08
244.520	705.73	705.53	-0.20	-0.03
245.290	717.82	718.25	0.43	0.06
246.210	733.50	733.68	0.18	0.02
246.920	745.99	745.77	-0.22	-0.03
247.340	753.10	753.00	-0.10	-0.01
247.860	762.18	762.03	-0.15	-0.02

STANDARD DEVIATION = 0.26

3-METHYLISOQUINOLINE C10H9N

MALONOWSKA B.,WECSILE J.: BULL.ACAD.POL.SCI. 12,239(1964).

A = 6.96924 B = 1717.330 C = 166.892

B.P.(760) = 253.155

T	P EXPTL.	P CALCD.	DEV.	PERCENT
176.310	92.01	92.34	0.33	0.36
185.880	126.70	126.22	-0.48	-0.38
191.590	150.90	150.89	-0.01	0.00
194.920	167.01	167.02	0.01	0.01
202.410	208.38	208.46	0.08	0.04
205.480	227.63	227.70	0.07	0.03
208.410	247.29	247.39	0.10	0.04
210.580	262.74	262.83	0.09	0.04
213.280	283.01	283.13	0.12	0.04
216.310	308.02	307.40	-0.62	-0.20
219.840	337.60	337.77	0.17	0.05
223.620	373.27	372.91	-0.36	-0.10
227.070	407.34	407.48	0.14	0.04
231.180	451.82	451.98	0.16	0.04
236.910	519.98	520.40	0.42	0.08
238.800	544.59	544.69	0.10	0.02
241.510	580.93	581.08	0.15	0.03
243.740	612.43	612.45	0.02	0.00
245.600	639.66	639.63	-0.03	0.00
247.520	668.63	668.68	0.05	0.01
249.260	695.91	695.90	-0.01	0.00
250.530	716.31	716.31	0.00	0.00
252.340	746.27	746.22	-0.05	-0.01
253.300	762.71	762.47	-0.24	-0.03
254.590	785.02	784.75	-0.27	-0.03

STANDARD DEVIATION = 0.24

LEPIDINE C10H9N

MALONOWSKI S.:

BULL.ACAD.POL.SCI.CHIM. 9,71(1961

A = 7.27121 B = 1946.137 C = 177.643

B.P.(760) = 265.628

	T	P EXPTL.	P CALCD.	DEV.	PERCENT
	198.580	125.40	125.42	0.02	0.02
	202.460	141.70	141.64	-0.06	-0.04
	210.480	180.70	180.71	0.01	0.00
	216.660	216.60	216.55	-0.05	-0.02
	221.000	245.00	245.08	0.08	0.03
	224.560	270.90	270.71	-0.19	-0.07
	230.210	315.60	315.89	0.29	0.09
	232.800	338.30	338.57	0.27	0.08
	237.920	387.60	387.32	-0.28	-0.07
	242.470	435.50	435.30	-0.20	-0.05
	245.500	469.70	469.85	0.15	0.03
	248.930	511.60	511.62	0.02	0.00
	251.700	547.49	547.49	0.00	0.00
*	252.890	570.83	563.52	-7.31	-1.28
	255.940	606.16	606.32	0.16	0.03
	257.250	625.77	625.50	-0.27	-0.04
	257.250	625.77	625.50	-0.27	-0.04
	259.560	660.51	660.49	-0.02	0.00
	261.440	690.04	690.13	0.09	0.01
	264.960	748.13	748.48	0.35	0.05
	265.410	756.43	756.21	-0.22	-0.03
	265.820	763.21	763.32	0.11	0.01

STANDARD DEVIATION = 0.20

6-METHYLQUINOLINE C10H9N

MALONOWSKA B.,WECSILE J.: BULL.ACAD.POL.SCI. 12,239(1964).

A = 6.92718 B = 1746.076 C = 166.456

B.P.(760) = 265.061

	T	P EXPTL.	P CALCD.	DEV.	PERCENT
	186.490	95.50	95.51	0.01	0.01
	193.030	117.49	117.50	0.01	0.01
	199.450	142.99	142.97	-0.02	-0.01
	204.670	166.86	166.87	0.01	0.01
	208.580	186.81	186.82	0.01	0.01
	212.310	207.65	207.62	-0.03	-0.01
	216.550	233.52	233.51	-0.01	0.00
	221.060	263.87	263.86	-0.01	-0.01
	224.700	290.58	290.60	0.02	0.01
*	228.340	320.21	319.49	-0.72	-0.23
	231.400	345.53	345.52	-0.01	0.00
	235.060	378.84	378.86	0.02	0.00
	237.630	403.79	403.77	-0.02	-0.01
	239.940	427.25	427.26	0.01	0.00
	242.590	455.53	455.54	0.01	0.00
	245.650	490.06	490.03	-0.03	-0.01
	248.320	521.76	521.79	0.03	0.01
	251.070	556.20	556.19	-0.01	0.00
	253.570	588.99	589.00	0.01	0.00
	255.340	613.10	613.14	0.04	0.01
	257.520	643.90	643.93	0.03	0.01
	259.800	677.46	677.44	-0.02	0.00
	261.480	703.00	703.00	0.00	0.00
	263.340	732.21	732.17	-0.04	-0.01
*	264.290	763.80	747.43	-16.37	-2.14
	266.140	777.89	777.87	-0.02	0.00

STANDARD DEVIATION = 0.02

7-METHYLQUINOLINE $C_{10}H_9N$

MALONOWSKI S.:

BULL.ACAD.POL.SCI.CHIM. 9,71(1961).

A = 7.59766 B = 2229.359 C = 214.932

B.P.(760) = 257.706

T	P EXPTL.	P CALCD.	DEV.	PERCENT
238.030	473.80	474.15	0.35	0.07
239.890	496.60	496.65	0.05	0.01
241.950	522.40	522.57	0.17	0.03
243.780	546.87	546.53	-0.34	-0.06
245.510	570.19	570.00	-0.19	-0.03
247.080	592.34	592.01	-0.33	-0.06
248.850	617.45	617.65	0.20	0.03
250.500	642.12	642.37	0.25	0.04
251.950	664.64	664.75	0.11	0.02
252.980	681.59	681.04	-0.55	-0.08
252.990	681.59	681.20	-0.39	-0.06
254.070	698.62	698.62	0.00	0.00
254.950	713.64	713.09	-0.55	-0.08
256.080	731.66	732.03	0.37	0.05
256.960	746.79	747.06	0.27	0.04
257.740	760.87	760.59	-0.28	-0.04
258.000	764.28	765.15	0.87	0.11

STANDARD DEVIATION = 0.41

8-METHYLQUINOLINE C10H9N

MALONOWSKI S.: BULL.ACAD.POL.SCI.CHIM. 9,71(1961).

A = 7.58069 B = 2206.046 C = 221.497

B.P.(760) = 247.887

T	P EXPTL.	P CALCD.	DEV.	PERCENT
227.330	462.90	462.96	0.06	0.01
229.230	485.60	485.59	-0.01	0.00
230.540	501.90	501.71	-0.19	-0.04
232.150	522.40	522.12	-0.28	-0.05
233.690	542.15	542.28	0.13	0.02
235.130	560.85	561.70	0.85	0.15
236.730	584.22	583.95	-0.27	-0.05
237.970	602.11	601.69	-0.42	-0.07
239.360	622.08	622.09	0.01	0.00
240.610	640.71	640.91	0.20	0.03
241.950	661.72	661.61	-0.11	-0.02
243.910	692.81	692.87	0.06	0.01
246.100	729.03	729.20	0.17	0.02
247.320	750.81	750.11	-0.70	-0.09
247.710	756.93	756.90	-0.03	0.00
248.150	764.12	764.61	0.49	0.06

STANDARD DEVIATION = 0.39

M-DIACETYLBENZENE C10H10O2

KHOREVSKAYA A.S.,BYK S.SH.: ZH.PRIKL.KHIM. 40,464(1967).

A = 0.05624 B = 64.188 C = -196.965

B.P.(10) = 128.952

T	P EXPTL.	P CALCD.	DEV.	PERCENT
50.140	1.48	3.11	1.63	110.45
96.740	3.04	4.97	1.93	63.61
109.900	5.96	6.22	0.26	4.29
121.900	8.84	8.15	-0.69	-7.77
134.400	12.96	12.08	-0.88	-6.77
144.640	18.60	19.18	0.58	3.14

STANDARD DEVIATION = 1.64

P-DIACETYLBENZENE C10H10O2

KHOREVSKAYA A.S.,BYK S.SH.: ZH.PRIKL.KHIM. 40,464(1967).

A = 2.80371 B = 177.251 C = -46.430

B.P.(10) = 144.700

T	P EXPTL.	P CALCD.	DEV.	PERCENT
115.700	1.54	1.76	0.22	14.12
117.100	2.10	1.98	-0.13	-5.95
121.500	2.84	2.77	-0.07	-2.44
121.650	2.94	2.80	-0.14	-4.74
129.950	5.14	4.80	-0.34	-6.57
139.100	7.40	7.78	0.38	5.14
149.000	11.56	11.90	0.34	2.96
157.200	16.24	15.98	-0.26	-1.61

STANDARD DEVIATION = 0.33

DIMETHYL PHTHALATE C10H10O4

GARDNER G.S.,BREWER J.E.: IND.ENG.CHEM.29,179(1937).

A = 4.52232 B = 700.305 C = 51.424

B.P.(10) = 147.395

T	P EXPTL.	P CALCD.	DEV.	PERCENT
82.000	0.30	0.19	-0.11	-37.41
85.000	0.40	0.25	-0.16	-38.77
90.000	0.50	0.37	-0.13	-25.61
99.000	0.60	0.74	0.14	22.64
103.000	0.90	0.97	0.07	7.93
105.300	1.20	1.13	-0.07	-5.65
107.400	1.60	1.30	-0.30	-18.92
110.600	1.70	1.59	-0.11	-6.75
115.000	2.00	2.06	0.06	3.12
118.000	2.40	2.45	0.05	2.02
118.800	2.50	2.56	0.06	2.41
123.800	3.30	3.36	0.06	1.67
129.500	4.40	4.48	0.08	1.90
136.900	6.10	6.36	0.26	4.32
146.500	9.70	9.64	-0.06	-0.62
146.900	9.70	9.80	0.10	1.03
150.600	11.50	11.37	-0.13	-1.10
151.000	11.60	11.55	-0.05	-0.40

STANDARD DEVIATION = 0.14

M-ETHYLSTYRENE C10H12

DREISBACH R.R.,SHRADER S.A.: IND.ENG.CHEM. 41,2879(1949).

A = 4.50728 B = 414.145 C = 45.724

B.P.(100) = 119.454

T	P EXPTL.	P CALCD.	DEV.	PERCENT
99.670	45.41	45.59	0.18	0.39
104.430	56.80	56.12	-0.68	-1.19
108.600	66.41	66.63	0.22	0.33
112.090	76.00	76.39	0.39	0.51
127.880	132.45	132.34	-0.11	-0.08

STANDARD DEVIATION = 0.59

TETRALIN C10H12

HERZ W.,SCHUFTAN P.: Z.PHYS.CHEM.101,269(1922).

A = 7.07055 B = 1741.304 C = 208.259

B.P.(760) = 207.352

T	P EXPTL.	P CALCD.	DEV.	PERCENT
93.800	20.00	20.22	0.22	1.10
103.000	30.00	29.93	-0.07	-0.22
140.000	118.00	117.63	-0.37	-0.31
150.000	162.00	162.21	0.21	0.13
167.500	273.00	273.18	0.18	0.07
206.200	740.00	739.88	-0.12	-0.02

STANDARD DEVIATION = 0.31

BETA-PHENYL ETHYL ACETATE $C_{10}H_{12}O_2$

DREISBACH R.R.,SHRADER S.A.:
IND.ENG.CHEM. 41,2879(1949).

A = 6.83428 B = 1555.150 C = 160.798

B.P.(760) = 232.565

T	P EXPTL.	P CALCD.	DEV.	PERCENT
149.410	66.39	66.23	-0.16	-0.24
153.100	75.86	75.85	-0.01	-0.01
167.240	123.76	124.03	0.27	0.22
197.900	315.52	315.31	-0.21	-0.07
232.570	760.00	760.08	0.08	0.01

STANDARD DEVIATION = 0.27

BUTYLBENZENE $C_{10}H_{14}$

FORZIATI A.F.,NORRIS W.R.,ROSSINI F.D.:
J.RESEARCH NATL.BUR.STANDARDS 43,555(1949).

A = 6.98922 B = 1582.219 C = 201.849

B.P.(760) = 183.269

T	P EXPTL.	P CALCD.	DEV.	PERCENT
96.233	48.12	48.00	-0.12	-0.25
100.814	57.73	57.75	0.02	0.04
104.778	67.44	67.47	0.03	0.05
108.403	77.47	77.52	0.05	0.07
111.762	87.92	87.91	-0.01	-0.01
116.322	103.81	103.84	0.03	0.03
121.506	124.82	124.76	-0.06	-0.05
126.797	149.60	149.58	-0.02	-0.02
138.300	217.36	217.60	0.24	0.11
151.541	325.17	325.05	-0.12	-0.04
159.032	402.73	402.61	-0.12	-0.03
167.011	500.99	500.88	-0.11	-0.02
175.666	628.20	628.16	-0.04	-0.01
181.767	732.40	732.37	-0.03	0.00
182.429	744.40	744.45	0.05	0.01
183.036	755.57	755.66	0.09	0.01
183.725	768.48	768.55	0.07	0.01
184.329	779.98	780.00	0.02	0.00

STANDARD DEVIATION = 0.10

ISOBUTYLBENZENE C10H14

FORZIATI A.F.,NORRIS W.R.,ROSSINI F.D.:
J.RESEARCH NATL.BUR.STANDARDS 43,555(1949).

A = 6.93556 B = 1530.057 C = 204.591

B.P.(760) = 172.759

T	P EXPTL.	P CALCD.	DEV.	PERCENT
86.638	48.12	48.06	-0.06	-0.13
91.118	57.73	57.72	-0.01	-0.01
95.026	67.44	67.43	-0.01	-0.02
98.620	77.47	77.51	0.04	0.06
101.946	87.92	87.93	0.01	0.01
106.450	103.81	103.85	0.04	0.04
111.582	124.82	124.81	-0.01	-0.01
116.808	149.60	149.60	0.00	0.00
121.659	176.07	176.09	0.02	0.01
128.149	217.36	217.37	0.01	0.01
134.112	261.91	261.91	0.00	0.00
141.301	325.17	325.12	-0.05	-0.02
148.724	402.73	402.70	-0.03	-0.01
156.632	500.99	500.94	-0.05	-0.01
165.217	628.21	628.23	0.02	0.00
171.270	732.41	732.40	-0.01	0.00
171.920	744.41	744.36	-0.05	-0.01
172.526	755.59	755.63	0.04	0.01
173.209	768.49	768.50	0.01	0.00
173.814	779.99	780.04	0.05	0.01

STANDARD DEVIATION = 0.04

SEC.-BUTYLBENZENE C10H14

FORZIATI A.F.,NORRIS W.R.,ROSSINI F.D.:
J.RESEARCH NATL.BUR.STANDARDS 43,555(1949).

A = 6.94219 B = 1533.947 C = 204.386

B.P.(760) = 173.305

	T	P EXPTL.	P CALCD.	DEV.	PERCENT
*	87.118	48.12	47.86	-0.26	-0.53
	91.684	57.72	57.70	-0.02	-0.04
	95.620	67.43	67.47	0.04	0.07
	99.179	77.47	77.46	-0.01	-0.01
	102.523	87.92	87.93	0.01	0.01
	107.009	103.81	103.79	-0.02	-0.02
	112.151	124.82	124.79	-0.03	-0.03
	117.387	149.60	149.63	0.03	0.02
	122.232	176.08	176.09	0.01	0.01
	128.715	217.36	217.34	-0.02	-0.01
	134.683	261.90	261.94	0.04	0.01
	141.867	325.17	325.13	-0.04	-0.01
	149.288	402.73	402.73	0.00	0.00
	157.194	500.99	501.01	0.02	0.00
	165.768	628.21	628.22	0.01	0.00
	171.820	732.41	732.45	0.04	0.01
	172.468	744.40	744.37	-0.03	0.00
	173.068	755.59	755.54	-0.05	-0.01
	173.754	768.49	768.48	-0.01	0.00
	174.358	779.99	780.01	0.02	0.00

STANDARD DEVIATION = 0.03

TERT.-BUTYLBENZENE C10H14

FORZIATI A.F.,NORRIS W.R.,ROSSINI F.D.:
J.RESEARCH NATL.BUR.STANDARDS 43,555(1949).

A = 6.92255 B = 1505.987 C = 203.490

B.P.(760) = 169.119

T	P EXPTL.	P CALCD.	DEV.	PERCENT
83.877	48.20	48.07	-0.13	-0.26
88.312	57.72	57.75	0.03	0.05
92.194	67.43	67.50	0.07	0.11
95.715	77.50	77.49	-0.01	-0.01
99.017	87.92	87.94	0.02	0.02
103.471	103.81	103.85	0.04	0.04
108.546	124.82	124.80	-0.02	-0.01
113.720	149.60	149.60	0.00	0.00
118.524	176.08	176.10	0.02	0.01
124.936	217.36	217.31	-0.05	-0.02
137.968	325.18	325.15	-0.03	-0.01
145.315	402.72	402.70	-0.02	0.00
153.149	500.99	500.98	-0.01	0.00
161.649	628.21	628.24	0.03	0.00
167.646	732.41	732.44	0.03	0.00
168.287	744.41	744.33	-0.08	-0.01
168.886	755.59	755.58	-0.01	0.00
169.565	768.50	768.50	0.00	0.00
170.165	780.00	780.05	0.05	0.01

STANDARD DEVIATION = 0.05

O-CYMENE C10H14

MCDONALD R.A.,SHRADER S.A.,STULL D.R.: J.CHEM.ENG.DATA 4,311(1959).

A = 7.26610 B = 1768.448 C = 224.946

B.P.(760) = 178.323

T	P EXPTL.	P CALCD.	DEV.	PERCENT
81.170	31.28	30.84	-0.45	-1.42
93.160	50.84	50.91	0.07	0.14
108.550	91.55	91.90	0.35	0.39
129.480	188.84	189.01	0.17	0.09
153.230	389.53	388.90	-0.63	-0.16
176.700	730.41	729.62	-0.79	-0.11
178.660	766.60	766.44	-0.16	-0.02
179.710	785.45	786.77	1.32	0.17

STANDARD DEVIATION = 0.79

M-CYMENE C10H14

MCDONALD R.A.,SHRADER S.A.,STULL D.R.: J.CHEM.ENG.DATA 4,311(1959).

A = 7.12374 B = 1644.950 C = 212.756

B.P.(760) = 174.936

T	P EXPTL.	P CALCD.	DEV.	PERCENT
78.760	30.88	30.27	-0.61	-1.98
90.680	50.34	50.43	0.09	0.17
105.780	90.51	91.13	0.62	0.68
126.630	189.20	189.19	-0.01	-0.01
150.060	389.70	388.95	-0.75	-0.19
173.340	729.60	729.91	0.31	0.04
174.520	752.20	752.05	-0.15	-0.02
176.330	786.70	787.06	0.36	0.05

STANDARD DEVIATION = 0.56

P-CYMENE C10H14

MCDONALD R.A.,SHRADER S.A.,STULL D.R.: J.CHEM.ENG.DATA 4,311(1959).

A = 7.05074 B = 1608.909 C = 208.721

B.P.(760) = 177.115

T	P EXPTL.	P CALCD.	DEV.	PERCENT
107.040	90.20	90.24	0.04	0.05
128.240	188.93	188.79	-0.14	-0.07
151.900	388.12	388.37	0.25	0.07
175.350	727.30	727.19	-0.11	-0.01
176.460	748.50	747.69	-0.81	-0.11
177.360	764.68	764.64	-0.04	0.00
178.420	784.21	785.00	0.79	0.10

STANDARD DEVIATION = 0.59

1,2-DIETHYLBENZENE C10H14

FORZIATI A.F.,NORRIS W.R.,ROSSINI F.D.: J.RESEARCH NATL.BUR.STANDARDS 43,555(1949).

A = 6.99506 B = 1581.389 C = 200.946

B.P.(760) = 183.423

T	P EXPTL.	P CALCD.	DEV.	PERCENT
96.729	48.19	48.15	-0.04	-0.09
101.263	57.88	57.85	-0.03	-0.06
105.223	67.59	67.60	0.01	0.02
108.822	77.64	77.62	-0.02	-0.02
112.191	88.05	88.09	0.04	0.04
116.728	104.00	104.00	0.00	0.00
121.906	124.95	124.99	0.04	0.03
127.171	149.76	149.79	0.03	0.02
132.059	176.26	176.28	0.02	0.01
138.590	217.54	217.55	0.01	0.00
144.596	262.13	262.13	0.00	0.00
151.832	325.42	325.37	-0.05	-0.01
159.290	402.93	402.89	-0.04	-0.01
167.235	501.16	501.09	-0.07	-0.01
175.853	628.32	628.28	-0.04	-0.01
181.936	732.54	732.54	0.00	0.00
182.590	744.50	744.52	0.02	0.00
183.197	755.75	755.77	0.02	0.00
183.885	768.64	768.69	0.05	0.01
184.493	780.21	780.25	0.04	0.00

STANDARD DEVIATION = 0.04

1,3-DIETHYLBENZENE C10H14

FORZIATI A.F.,NORRIS W.R.,ROSSINI F.D.:
J.RESEARCH NATL.BUR.STANDARDS 43,555(1949).

A = 7.00940 B = 1578.620 C = 201.262

B.P.(760) = 181.101

T	P EXPTL.	P CALCD.	DEV.	PERCENT
95.092	48.18	48.15	-0.03	-0.06
99.573	57.86	57.80	-0.06	-0.10
103.524	67.58	67.60	0.02	0.03
107.096	77.63	77.62	-0.01	-0.02
110.436	88.04	88.07	0.03	0.03
114.946	103.98	104.00	0.02	0.02
120.082	124.94	124.98	0.04	0.03
125.303	149.75	149.75	0.00	0.00
130.157	176.25	176.27	0.02	0.01
136.638	217.53	217.54	0.01	0.01
142.597	262.12	262.12	0.00	0.00
149.777	325.41	325.39	-0.02	-0.01
157.169	402.92	402.85	-0.07	-0.02
165.050	501.15	501.07	-0.08	-0.02
173.595	628.31	628.26	-0.05	-0.01
179.628	732.53	732.56	0.03	0.00
180.275	744.49	744.51	0.02	0.00
180.877	755.74	755.77	0.03	0.00
181.558	768.64	768.66	0.02	0.00
182.162	780.20	780.25	0.05	0.01

STANDARD DEVIATION = 0.04

1,4-DIETHYLBENZENE C10H14

FORZIATI A.F.,NORRIS W.R.,ROSSINI F.D.:
J.RESEARCH NATL.BUR.STANDARDS 43,555(1949).

A = 7.00250 B = 1590.654 C = 202.171

B.P.(760) = 183.751

T	P EXPTL.	P CALCD.	DEV.	PERCENT
96.817	48.17	48.13	-0.04	-0.09
101.370	57.86	57.83	-0.03	-0.05
105.353	67.58	67.62	0.04	0.05
108.962	77.63	77.63	0.00	0.00
112.339	88.04	88.09	0.05	0.06
116.893	103.98	104.02	0.04	0.04
122.043	124.94	124.83	-0.11	-0.09
127.360	149.75	149.79	0.04	0.03
132.256	176.24	176.26	0.02	0.01
138.811	217.53	217.56	0.03	0.01
144.823	262.12	262.06	-0.06	-0.02
152.086	325.40	325.38	-0.02	-0.01
159.566	402.92	402.94	0.02	0.00
167.530	501.14	501.14	0.00	0.00
176.164	628.31	628.28	-0.03	-0.01
182.260	732.53	732.52	-0.01	0.00
182.916	744.49	744.50	0.01	0.00
183.524	755.73	755.75	0.02	0.00
184.212	768.63	768.64	0.01	0.00
184.821	780.20	780.19	-0.01	0.00

STANDARD DEVIATION = 0.04

NICOTINE C10H14N

YOUNG H.D.,NELSON O.A.: IND.ENG.CHEM.21,321(1929).

A = 6.78897 B = 1650.347 C = 176.371

B.P.(760) = 245.912

T	P EXPTL.	P CALCD.	DEV.	PERCENT
133.800	28.40	29.39	0.99	3.49
140.100	37.40	37.51	0.11	0.29
146.900	48.30	48.29	-0.01	-0.03
155.000	64.70	64.36	-0.34	-0.53
161.000	83.70	78.92	-4.78	-5.71
170.000	104.10	105.75	1.65	1.59
178.400	137.50	137.12	-0.38	-0.28
184.500	164.30	164.34	0.04	0.02
189.300	186.40	188.70	2.30	1.23
194.800	220.10	220.11	0.01	0.01
198.900	248.00	246.16	-1.84	-0.74
199.900	254.00	252.88	-1.12	-0.44
212.300	343.00	349.01	6.01	1.75
218.900	406.70	410.90	4.20	1.03
225.100	473.30	476.67	3.37	0.71
229.400	530.50	526.96	-3.54	-0.67
236.100	623.20	613.54	-9.66	-1.55
246.200	763.00	764.67	1.67	0.22

STANDARD DEVIATION = 3.75

O-SEC BUTYLPHENOL

$C_{10}H_{14}O$

HANDLEY R.,ET AL.:

J.CHEM.SOC. 1964,4404.

A = 6.95193 B = 1593.737 C = 163.794

B.P.(760) = 227.681

T	P EXPTL.	P CALCD.	DEV.	PERCENT
178.735	199.09	199.10	0.01	0.00
185.796	247.21	247.20	-0.01	0.00
192.314	299.57	299.56	-0.01	0.00
197.770	349.95	349.97	0.02	0.00
202.698	401.15	401.14	-0.01	0.00
206.860	448.86	448.87	0.01	0.00
210.886	499.28	499.26	-0.02	0.00
214.558	549.07	549.05	-0.02	0.00
218.023	599.58	599.56	-0.02	0.00
221.317	650.92	650.94	0.02	0.00
224.159	698.02	698.00	-0.02	0.00
227.163	750.59	750.62	0.03	0.00
229.890	801.03	801.05	0.02	0.00
232.465	851.05	851.07	0.02	0.00
234.936	901.32	901.34	0.02	0.00
237.136	948.05	948.03	-0.02	0.00
239.512	1000.60	1000.56	-0.04	0.00

STANDARD DEVIATION = 0.02

O-TERT BUTYL PHENOL

$C_{10}H_{14}O$

MCDONALD R.A.,SHRADER S.A.,STULL D.R.:

J.CHEM.ENG.DATA 4,311(1959).

A = 7.21756 B = 1822.809 C = 196.229

B.P.(760) = 224.088

T	P EXPTL.	P CALCD.	DEV.	PERCENT
134.650	51.64	51.12	-0.52	-1.01
150.710	91.32	91.96	0.64	0.70
172.820	189.87	189.82	-0.05	-0.03
197.670	389.38	389.00	-0.38	-0.10
222.340	728.85	728.95	0.10	0.01
223.510	749.84	749.61	-0.23	-0.03
224.350	764.40	764.73	0.33	0.04
225.430	784.49	784.53	0.04	0.01

STANDARD DEVIATION = 0.45

P-TERT BUTYL PHENOL $C_{10}H_{14}O$

HANDLEY R.,ET AL.:

J.CHEM.SOC. 1964,4404.

A = 7.00038 B = 1627.506 C = 155.240

B.P.(760) = 239.827

T	P EXPTL.	P CALCD.	DEV.	PERCENT
198.121	247.96	248.08	0.12	0.05
204.527	299.69	299.64	-0.05	-0.02
209.901	349.47	349.28	-0.19	-0.05
214.892	401.20	401.12	-0.08	-0.02
219.142	449.92	449.98	0.06	0.01
223.088	499.38	499.50	0.12	0.02
226.818	550.17	550.22	0.05	0.01
230.203	599.77	599.71	-0.06	-0.01
233.457	650.48	650.57	0.09	0.01
236.342	698.48	698.46	-0.02	0.00
239.305	750.45	750.51	0.06	0.01
242.018	800.75	800.81	0.06	0.01
244.560	850.34	850.31	-0.03	0.00
247.015	900.44	900.37	-0.07	-0.01
249.336	949.83	949.80	-0.03	0.00
251.608	1000.24	1000.22	-0.02	0.00

STANDARD DEVIATION = 0.09

BUTYL PHENYL ETHER $C_{10}H_{14}O$

DREISBACH R.R.,SHRADER S.A.:

IND.ENG.CHEM. 41,2879(1949).

A = 7.29966 B = 1882.700 C = 215.820

B.P.(760) = 210.241

T	P EXPTL.	P CALCD.	DEV.	PERCENT
118.880	47.16	47.27	0.11	0.24
123.530	57.04	56.45	-0.59	-1.03
127.730	66.39	66.00	-0.39	-0.59
132.190	75.86	77.58	1.72	2.27
145.380	123.76	122.27	-1.49	-1.21
176.510	315.52	316.88	1.36	0.43
193.970	507.50	507.40	-0.10	-0.02
210.200	760.00	759.24	-0.76	-0.10

STANDARD DEVIATION = 1.28

3,5-DIETHYLPHENOL $C_{10}H_{14}O$

VONTERRES E.,ET AL.: BRENNSTOFF CHEM. 36,272(1955).

A = 7.65133 B = 2227.854 C = 218.584

B.P.(760) = 248.420

T	P EXPTL.	P CALCD.	DEV.	PERCENT
114.500	10.00	9.18	-0.82	-8.21
136.500	25.00	23.83	-1.17	-4.67
155.000	50.00	48.74	-1.26	-2.52
166.200	75.00	72.69	-2.31	-3.08
176.000	100.00	101.22	1.22	1.22
188.400	150.00	150.41	0.41	0.27
198.600	200.00	204.70	4.70	2.35
205.300	250.00	248.61	-1.39	-0.55
212.000	300.00	300.13	0.13	0.04
215.100	325.00	326.81	1.81	0.56
217.200	350.00	345.98	-4.02	-1.15
223.000	400.00	403.83	3.83	0.96
226.700	450.00	444.75	-5.25	-1.17
231.000	500.00	496.56	-3.44	-0.69
236.000	550.00	562.95	12.95	2.36
239.000	600.00	606.18	6.18	1.03
241.300	650.00	641.14	-8.86	-1.36
245.000	700.00	700.84	0.84	0.12
248.000	760.00	752.52	-7.48	-0.98

STANDARD DEVIATION = 5.30

DURENOL

C10H14O

VONTERRES E.,ET AL.:

BRENNSTOFF CHEM. 36,272(1955).

A = 7.75835 B = 2432.625 C = 250.094

B.P.(760) = 248.647

T	P EXPTL.	P CALCD.	DEV.	PERCENT
108.600	10.00	9.47	-0.53	-5.28
131.800	25.00	24.46	-0.54	-2.16
150.700	50.00	48.85	-1.15	-2.31
163.000	75.00	74.05	-0.95	-1.26
172.100	100.00	99.19	-0.81	-0.81
186.000	150.00	151.40	1.40	0.93
196.000	200.00	201.92	1.92	0.96
204.000	250.00	251.91	1.91	0.76
210.900	300.00	302.99	2.99	1.00
213.200	325.00	321.83	-3.17	-0.98
216.700	350.00	352.36	2.36	0.68
221.500	400.00	398.14	-1.86	-0.47
226.600	450.00	452.09	2.09	0.46
230.200	500.00	493.71	-6.29	-1.26
234.900	550.00	552.79	2.79	0.51
238.000	600.00	594.86	-5.14	-0.86
241.400	650.00	644.01	-5.99	-0.92
245.200	700.00	702.85	2.85	0.41
249.000	760.00	766.05	6.05	0.80

STANDARD DEVIATION = 3.53

N,N-DIETHYLANILINE $C_{10}H_{15}N$

NELSON O.A.,WALES H.: J.AM.CHEM.SOC. 47,867 (1927).

A = 7.46601 B = 1993.572 C = 218.516

B.P.(760) = 216.268

	T	P EXPTL.	P CALCD.	DEV.	PERCENT
	50.000	1.60	1.10	-0.50	-31.22
	60.000	2.70	2.03	-0.67	-24.70
	70.000	4.20	3.60	-0.60	-14.29
	79.000	5.90	5.82	-0.08	-1.27
*	81.600	6.40	6.66	0.26	4.03
	86.930	9.70	8.69	-1.01	-10.37
	96.490	13.80	13.72	-0.08	-0.59
	100.930	17.40	16.80	-0.60	-3.45
	112.920	28.80	28.25	-0.55	-1.90
	123.960	43.20	44.15	0.95	2.20
	134.160	64.70	65.06	0.36	0.56
	140.400	81.10	81.58	0.48	0.59
	146.760	101.50	101.93	0.43	0.42
	154.530	133.80	132.43	-1.37	-1.03
	162.600	167.80	171.84	4.04	2.41
	164.080	183.10	180.04	-3.06	-1.67
	175.400	254.40	254.16	-0.24	-0.09
	182.100	308.00	308.85	0.85	0.28
	184.230	328.90	328.14	-0.76	-0.23
	189.400	381.20	379.14	-2.06	-0.54
	199.300	493.40	494.99	1.59	0.32
	205.190	576.10	576.67	0.57	0.10
	207.100	604.40	605.40	1.00	0.17
	212.200	686.00	687.87	1.87	0.27
	213.460	710.80	709.58	-1.22	-0.17
	215.620	747.20	748.11	0.91	0.12
	216.300	761.20	760.58	-0.62	-0.08
	217.220	778.60	777.73	-0.87	-0.11
	217.520	786.00	783.38	-2.62	-0.33
	218.400	799.20	800.18	0.98	0.12

STANDARD DEVIATION = 1.47

DIPENTENE C10H16

PICKETT O.A.,PETERSON J.M.:
IND.ENG.CHEM. 21,325(1929).

A = 7.11163 B = 1613.415 C = 207.840

B.P.(760) = 173.508

	T	P EXPTL.	P CALCD.	DEV.	PERCENT
	21.100	1.50	1.16	-0.34	-22.69
	21.200	1.60	1.17	-0.43	-27.01
	55.700	11.50	9.76	-1.74	-15.11
*	56.200	12.30	10.03	-2.27	-18.49
	78.500	29.80	29.99	0.19	0.65
	78.900	30.10	30.54	0.44	1.46
	79.000	30.30	30.68	0.38	1.25
	79.700	31.70	31.66	-0.04	-0.12
	80.200	32.20	32.38	0.18	0.56
	80.500	32.50	32.82	0.32	0.98
	81.800	34.80	34.77	-0.03	-0.08
	81.900	35.00	34.93	-0.07	-0.21
	82.000	36.50	35.08	-1.42	-3.89
	82.600	36.70	36.02	-0.68	-1.85
	82.900	38.50	36.50	-2.00	-5.19
	90.000	51.40	49.50	-1.90	-3.70
	91.200	52.10	52.04	-0.06	-0.12
	92.100	54.20	54.01	-0.19	-0.34
	93.500	57.90	57.21	-0.69	-1.19
	98.900	70.50	71.08	0.58	0.82
	101.600	78.50	79.00	0.50	0.64
	102.100	79.50	80.55	1.05	1.32
	102.500	81.10	81.80	0.70	0.87
	102.700	82.40	82.44	0.04	0.04
	103.200	83.10	84.04	0.94	1.13
	103.600	84.70	85.34	0.64	0.75
	105.700	91.80	92.43	0.63	0.69
	108.000	99.50	100.76	1.26	1.27
	109.000	103.50	104.58	1.08	1.04
	109.500	105.40	106.53	1.13	1.07
	130.100	218.50	217.46	-1.04	-0.48
	131.500	228.00	227.55	-0.46	-0.20
	132.500	235.00	234.98	-0.02	-0.01
	133.300	240.00	241.08	1.08	0.45
	134.000	247.00	246.51	-0.49	-0.20
	134.100	250.30	247.30	-3.00	-1.20
	161.900	560.00	559.74	-0.26	-0.05
	166.400	631.50	631.62	0.12	0.02
	166.900	643.50	640.04	-3.46	-0.54
	167.200	648.50	645.14	-3.36	-0.52
	167.500	650.50	650.27	-0.23	-0.04
	167.900	655.50	657.15	1.65	0.25
	168.800	667.50	672.87	5.37	0.80
*	169.100	670.50	678.17	7.67	1.14
	169.200	683.50	679.94	-3.56	-0.52
	170.300	695.50	699.71	4.21	0.61

STANDARD DEVIATION = 1.71

ALPHA-PINENE C10H16

HAWKINS J.E.,ARMSTRONG G.T.: J.AM.CHEM.SOC. 76,3756(1954).

A = 6.85253 B = 1446.380 C = 208.027

B.P.(760) = 156.143

	T	P EXPTL.	P CALCD.	DEV.	PERCENT
	19.440	3.06	3.12	0.06	1.90
	21.290	3.43	3.51	0.08	2.30
*	27.240	4.91	5.07	0.16	3.19
*	29.710	5.93	5.87	-0.06	-1.03
	37.060	9.01	8.93	-0.08	-0.85
	46.910	15.11	15.10	-0.01	-0.05
	53.680	21.24	21.17	-0.07	-0.31
	54.170	21.58	21.68	0.10	0.48
	56.560	24.26	24.32	0.06	0.25
	57.080	24.82	24.93	0.11	0.44
	57.520	25.45	25.45	0.00	0.01
	62.270	31.66	31.73	0.07	0.22
	65.520	36.44	36.73	0.29	0.80
	66.530	38.41	38.41	0.00	0.01
	68.260	41.66	41.44	-0.22	-0.52
	70.440	45.55	45.55	0.00	-0.01
	75.090	55.73	55.43	-0.30	-0.54
	76.030	57.74	57.63	-0.11	-0.19
	77.110	60.25	60.25	0.00	0.00
	79.740	66.79	67.04	0.25	0.37
	84.700	81.45	81.56	0.11	0.13
	86.220	86.53	86.50	-0.03	-0.04
	88.410	93.85	94.04	0.19	0.20
	92.000	107.25	107.57	0.32	0.30
	95.260	121.24	121.20	-0.04	-0.03
	102.240	155.39	155.17	-0.22	-0.14
	105.890	175.74	175.79	0.05	0.03
	106.750	181.64	180.96	-0.68	-0.37
	110.520	205.80	205.10	-0.70	-0.34
	113.490	226.03	225.90	-0.13	-0.06
	115.060	238.48	237.56	-0.92	-0.39
	122.100	296.11	295.96	-0.15	-0.05
	125.010	322.25	323.24	0.99	0.31
	129.860	371.35	373.13	1.78	0.48
	135.710	440.12	441.28	1.16	0.26
*	147.470	605.53	608.01	2.48	0.41
	155.760	754.57	752.71	-1.86	-0.25
*	155.750	756.01	752.52	-3.49	-0.46

STANDARD DEVIATION = 0.60

BETA-PINENE C10H16

HAWKINS J.E.,ARMSTRONG G.T.:
J.AM.CHEM.SOC. 76,3756(1954).

A = 6.89837 B = 1511.744 C = 210.243

B.P.(760) = 166.041

	T	P EXPTL.	P CALCD.	DEV.	PERCENT
*	18.710	1.89	1.97	0.08	4.48
	20.040	2.13	2.16	0.03	1.22
	23.080	2.59	2.63	0.04	1.36
	26.810	3.35	3.32	-0.03	-0.90
	29.440	3.88	3.90	0.02	0.52
	31.670	4.44	4.46	0.02	0.43
	32.020	4.44	4.55	0.11	2.54
	36.770	5.95	6.00	0.05	0.87
	37.080	6.03	6.11	0.08	1.30
	39.400	6.97	6.96	-0.01	-0.11
	41.330	7.78	7.75	-0.03	-0.41
	45.510	9.67	9.71	0.04	0.46
	49.400	11.85	11.91	0.06	0.52
	50.100	12.33	12.35	0.02	0.15
	52.310	13.59	13.82	0.23	1.69
	56.560	17.01	17.07	0.06	0.35
	56.930	17.56	17.38	-0.18	-1.02
	58.850	19.13	19.07	-0.06	-0.29
	59.470	19.65	19.65	0.00	0.00
	62.610	22.92	22.80	-0.12	-0.54
	62.970	23.13	23.18	0.05	0.23
	65.910	26.51	26.55	0.04	0.15
	68.740	30.11	30.17	0.06	0.20
	72.220	35.08	35.19	0.11	0.30
	75.420	40.38	40.39	0.01	0.03
	78.580	46.06	46.15	0.09	0.20
	80.870	51.08	50.74	-0.34	-0.66
	81.220	52.12	51.48	-0.64	-1.23
	85.850	62.00	62.05	0.05	0.08
	90.790	74.88	75.25	0.37	0.50
	94.110	85.10	85.37	0.27	0.31
	97.320	96.22	96.19	-0.03	-0.03
	103.400	119.62	119.79	0.17	0.14
	108.800	144.67	144.54	-0.13	-0.09
	115.490	180.89	180.85	-0.04	-0.02
	120.050	209.81	209.60	-0.21	-0.10
	125.320	248.11	247.33	-0.78	-0.32
	131.380	298.24	297.29	-0.95	-0.32
	136.010	338.32	340.69	2.37	0.70
*	149.540	488.06	497.22	9.16	1.88
	161.210	672.31	673.84	1.53	0.23
	165.790	755.97	755.31	-0.66	-0.09
	165.910	759.02	757.54	-1.48	-0.20

STANDARD DEVIATION = 0.59

TERPINOLENE C10H16

PICKETT O.A.,PETERSON J.M.: IND.ENG.CHEM. 21,325(1929).

A = 7.16996 B = 1706.420 C = 210.766

B.P.(100) = 119.298

	T	P EXPTL.	P CALCD.	DEV.	PERCENT
	40.500	1.20	2.39	1.19	99.30
*	45.500	4.20	3.24	-0.96	-22.74
	58.200	6.50	6.69	0.19	2.96
	59.700	7.50	7.26	-0.24	-3.23
	63.300	8.40	8.78	0.38	4.56
	83.000	22.50	22.97	0.47	2.10
	83.500	23.60	23.50	-0.10	-0.42
	84.000	24.80	24.04	-0.76	-3.07
	84.100	25.30	24.15	-1.15	-4.56
	107.600	63.40	64.57	1.17	1.85
	107.700	63.70	64.82	1.12	1.76
	108.500	68.60	66.86	-1.74	-2.54
	108.600	69.00	67.11	-1.89	-2.73
	108.700	74.10	67.37	-6.73	-9.08
	108.800	74.50	67.63	-6.87	-9.22
	131.200	148.20	151.33	3.13	2.11
	131.400	149.20	152.35	3.15	2.11
	131.800	151.00	154.41	3.41	2.26
	132.100	152.00	155.97	3.97	2.61
	133.500	161.40	163.41	2.01	1.24
	133.600	161.60	163.95	2.35	1.45
	133.900	162.80	165.59	2.79	1.71
	157.800	343.80	346.80	3.00	0.87
	158.200	357.00	350.83	-6.17	-1.73
	164.100	414.30	414.84	0.54	0.13
	175.900	559.60	571.21	11.61	2.08
	176.000	571.30	572.72	1.42	0.25
	176.400	579.20	578.76	-0.44	-0.08
	178.000	602.20	603.45	1.25	0.21
	178.200	609.00	606.59	-2.41	-0.40
	178.500	619.00	611.33	-7.67	-1.24
	178.800	622.30	616.10	-6.20	-1.00
	178.900	623.90	617.70	-6.20	-0.99

STANDARD DEVIATION = 4.14

CAMPHOR

C10H16O

DE WILDE J.H.:

Z.ANORG.ALLGEM.CHEM. 233,411(1937).

A = 6.10572 B = 1043.585 C = 116.391

B.P.(760) = 207.211

T	P EXPTL.	P CALCD.	DEV.	PERCENT
178.050	368.10	364.27	-3.83	-1.04
179.150	379.30	375.50	-3.80	-1.00
180.150	390.00	385.94	-4.06	-1.04
181.150	398.40	396.59	-1.81	-0.45
182.250	408.30	408.57	0.27	0.07
183.250	419.50	419.69	0.19	0.04
186.350	456.10	455.61	-0.49	-0.11
188.550	481.60	482.46	0.86	0.18
191.550	514.10	520.95	6.85	1.33
191.650	517.00	522.27	5.27	1.02
193.550	544.10	547.86	3.76	0.69
195.750	587.60	578.63	-8.97	-1.53
198.750	619.50	622.62	3.12	0.50
201.750	664.80	669.04	4.24	0.64
203.950	699.20	704.66	5.46	0.78
206.750	749.10	751.98	2.88	0.38
211.050	817.80	829.12	11.32	1.38
215.650	920.30	917.85	-2.45	-0.27
220.850	1027.90	1026.20	-1.70	-0.17
225.450	1143.40	1129.47	-13.93	-1.22
225.850	1150.40	1138.79	-11.61	-1.01
232.350	1290.80	1298.00	7.21	0.56

STANDARD DEVIATION = 6.50

CIS-DECALIN C10H18

CAMIN D.L., ROSSINI F.D.: J.PHYS.CHEM. 59, 1173(1955).

A = 6.87540 B = 1594.561 C = 203.405

B.P.(760) = 195.776

T	P EXPTL.	P CALCD.	DEV.	PERCENT
99.883	41.47	41.48	0.01	0.02
105.685	52.06	52.06	0.00	0.00
110.490	62.49	62.44	-0.05	-0.08
114.152	71.43	71.46	0.03	0.04
118.004	82.08	82.08	0.00	0.00
123.132	98.17	98.21	0.04	0.04
128.731	118.71	118.70	-0.01	-0.01
135.021	145.76	145.78	0.02	0.01
140.176	171.62	171.55	-0.07	-0.04
147.456	214.11	214.13	0.02	0.01
154.245	261.15	261.19	0.04	0.01
161.885	323.76	323.74	-0.02	-0.01
170.056	403.33	403.37	0.04	0.01
178.629	503.05	502.94	-0.11	-0.02
187.823	630.29	630.39	0.10	0.02
194.370	735.66	735.68	0.02	0.00
195.055	747.50	747.45	-0.05	-0.01
195.635	757.57	757.53	-0.04	-0.01
196.376	770.52	770.56	0.04	0.00

STANDARD DEVIATION = 0.05

TRANS-DECALIN C10H18

CAMIN D.L.,ROSSINI F.D.:

J.PHYS.CHEM. 59, 1173(1955).

A = 6.86137 B = 1568.089 C = 206.663

B.P.(760) = 187.274

T	P EXPTL.	P CALCD.	DEV.	PERCENT
92.360	41.48	41.43	-0.05	-0.12
98.129	52.07	52.07	0.00	0.00
102.891	62.49	62.48	-0.01	-0.02
106.500	71.42	71.47	0.05	0.07
110.316	82.10	82.11	0.01	0.01
115.358	98.15	98.14	-0.01	-0.01
120.918	118.72	118.71	-0.01	-0.01
127.140	145.77	145.79	0.02	0.01
132.255	171.62	171.64	0.02	0.01
139.441	214.13	214.13	0.00	0.00
146.156	261.15	261.16	0.01	0.01
153.719	323.75	323.73	-0.02	-0.01
161.801	403.33	403.30	-0.03	-0.01
170.297	503.06	502.97	-0.09	-0.02
179.395	630.29	630.34	0.05	0.01
185.885	735.67	735.74	0.07	0.01
186.563	747.55	747.50	-0.05	-0.01
187.140	757.60	757.63	0.03	0.00
187.867	770.55	770.54	-0.01	0.00

STANDARD DEVIATION = 0.04

ISOBORNEOL C10H180

GREKHNEV M.A.:

LESOKHIM.PROM. 5,11(1936).

A = -1.22645 B = 860.273 C = -413.400

B.P.(100) = 146.768

T	P EXPTL.	P CALCD.	DEV.	PERCENT
78.000	23.80	21.80	-2.00	-8.40
100.000	30.20	33.00	2.80	9.27
109.000	38.10	39.78	1.68	4.41
138.000	76.20	78.94	2.74	3.59
159.000	149.10	142.93	-6.17	-4.14
184.000	328.60	333.93	5.33	1.62

STANDARD DEVIATION = 5.43

FENCHYL ALCOHOL $C_{10}H_{18}O$

PICKETT O.A.,PETERSON J.M.:
IND.ENG.CHEM. 21,325(1929).

A = 5.69274 B = 797.553 C = 84.584

B.P.(760) = 199.049

T	P EXPTL.	P CALCD.	DEV.	PERCENT
59.100	2.50	1.39	-1.11	-44.53
63.600	3.50	2.04	-1.46	-41.59
64.400	3.90	2.19	-1.72	-43.98
71.100	5.30	3.71	-1.59	-29.93
72.100	5.60	4.00	-1.60	-28.49
73.000	5.90	4.28	-1.62	-27.43
74.000	6.50	4.61	-1.89	-29.11
77.100	7.90	5.75	-2.15	-27.17
78.100	8.80	6.17	-2.63	-29.89
108.500	36.60	36.49	-0.11	-0.31
108.900	38.10	37.21	-0.89	-2.33
109.000	37.70	37.39	-0.31	-0.81
137.600	119.60	126.80	7.20	6.02
138.000	122.50	128.70	6.20	5.06
138.200	128.00	129.66	1.66	1.29
167.100	339.30	334.09	-5.21	-1.54
168.100	344.90	343.88	-1.02	-0.30
168.500	355.20	347.85	-7.35	-2.07
169.100	358.90	353.87	-5.03	-1.40
195.500	704.00	700.14	-3.86	-0.55
196.900	724.30	723.35	-0.95	-0.13
198.600	747.00	752.24	5.24	0.70
198.800	751.30	755.69	4.39	0.58
199.200	759.30	762.62	3.32	0.44

STANDARD DEVIATION = 3.80

ALPHA-TERPINEOL C10H18O

PICKETT O.A.,PETERSON J.M.:

IND.ENG.CHEM. 21,325(1929).

A = 8.14120 B = 2479.443 C = 253.662

B.P.(760) = 217.680

	T	P EXPTL.	P CALCD.	DEV.	PERCENT
	83.800	5.90	6.22	0.32	5.44
	89.900	8.10	8.40	0.30	3.71
	90.600	8.60	8.69	0.09	1.04
	94.700	9.60	10.56	0.96	10.02
	100.000	12.60	13.50	0.90	7.17
	100.700	12.70	13.94	1.24	9.77
	105.800	18.60	17.52	-1.08	-5.80
	106.600	19.20	18.15	-1.05	-5.47
	107.300	20.10	18.72	-1.38	-6.89
	107.900	20.60	19.21	-1.39	-6.73
	108.200	20.70	19.47	-1.23	-5.96
	108.600	21.00	19.81	-1.19	-5.67
	108.700	21.20	19.90	-1.30	-6.16
	119.100	29.10	30.88	1.78	6.11
	119.900	29.70	31.91	2.21	7.43
	140.900	72.60	71.97	-0.63	-0.87
	141.500	74.60	73.57	-1.03	-1.38
	142.300	73.70	75.75	2.05	2.78
	172.000	207.20	207.15	-0.05	-0.02
	173.300	216.50	215.79	-0.71	-0.33
	174.000	220.40	220.56	0.16	0.07
	174.500	223.00	224.03	1.03	0.46
	177.600	246.00	246.56	0.56	0.23
	178.000	249.50	249.60	0.10	0.04
	178.300	251.90	251.91	0.01	0.00
	178.800	255.80	255.79	-0.01	-0.01
	208.400	597.10	595.88	-1.22	-0.20
	209.500	612.00	613.62	1.62	0.27
	209.600	614.50	615.26	0.76	0.12
	209.600	614.50	615.26	0.76	0.12
	209.700	617.80	616.90	-0.90	-0.15
	216.900	747.50	744.88	-2.62	-0.35
*	216.500	749.50	737.23	-12.27	-1.64

STANDARD DEVIATION = 1.22

ISOBUTYLCYCLOHEXANE C10H20

FORZIATI A.F.,NORRIS W.R.,ROSSINI F.D.:
J.RES.NATL.BUR.STANDARDS 43,555(1949).

A = 6.86797 B = 1493.101 C = 203.157

B.P.(760) = 171.321

T	P EXPTL.	P CALCD.	DEV.	PERCENT
84.752	48.09	48.08	-0.01	-0.03
89.248	57.78	57.77	-0.01	-0.02
93.184	67.51	67.53	0.02	0.03
96.767	77.56	77.57	0.01	0.01
100.094	87.98	87.97	-0.01	-0.02
104.610	103.91	103.89	-0.02	-0.02
109.766	124.86	124.88	0.02	0.02
115.008	149.68	149.66	-0.02	-0.01
119.887	176.17	176.19	0.02	0.01
126.405	217.45	217.47	0.02	0.01
132.404	262.05	262.06	0.01	0.00
139.640	325.33	325.33	0.00	0.00
147.103	402.86	402.84	-0.02	-0.01
155.065	501.09	501.05	-0.04	-0.01
163.711	628.26	628.21	-0.05	-0.01
169.822	732.49	732.47	-0.02	0.00
170.480	744.43	744.46	0.03	0.00
171.089	755.65	755.69	0.04	0.00
171.780	768.60	768.59	-0.01	0.00
172.394	780.16	780.20	0.04	0.00

STANDARD DEVIATION = 0.03

SEC.-BUTYLCYCLOHEXANE C10H20

FORZIATI A.F.,NORRIS W.R.,ROSSINI F.D.:
J.RESEARCH NATL.BUR.STANDARDS 43,555(1949).

A = 6.89096 B = 1530.701 C = 202.373

B.P.(760) = 179.334

T	P EXPTL.	P CALCD.	DEV.	PERCENT
91.458	48.09	48.03	-0.06	-0.13
96.842	57.78	57.75	-0.03	-0.06
100.048	67.51	67.52	0.01	0.02
103.688	77.56	77.56	0.00	0.01
107.082	87.98	88.01	0.03	0.03
111.660	103.90	103.90	0.00	0.00
116.900	124.85	124.92	0.07	0.05
122.224	149.67	149.71	0.04	0.03
127.163	176.16	176.17	0.01	0.01
133.776	217.45	217.43	-0.02	-0.01
139.861	262.05	261.99	-0.06	-0.02
147.212	325.32	325.34	0.02	0.00
154.776	402.86	402.78	-0.08	-0.02
162.856	501.09	501.05	-0.04	-0.01
171.626	628.26	628.29	0.03	0.00
177.817	732.48	732.50	0.02	0.00
178.481	744.42	744.44	0.02	0.00
179.098	755.64	755.66	0.02	0.00
179.799	768.60	768.57	-0.03	0.00
180.421	780.15	780.18	0.03	0.00

STANDARD DEVIATION = 0.04

TERT.-BUTYLCYCLOHEXANE C10H20

FORZIATI A.F.,NORRIS W.R.,ROSSINI F.D.:
J.RESEARCH NATL.BUR.STANDARDS 43,555(1949).

A = 6.85680 B = 1501.724 C = 206.108

B.P.(760) = 171.590

T	P EXPTL.	P CALCD.	DEV.	PERCENT
84.033	48.00	47.97	-0.03	-0.07
88.600	57.70	57.70	0.00	0.00
92.550	67.44	67.38	-0.06	-0.08
96.200	77.49	77.49	0.00	0.01
99.582	87.92	87.95	0.03	0.03
104.146	103.83	103.87	0.04	0.04
109.341	124.77	124.80	0.03	0.02
114.649	149.60	149.62	0.02	0.01
119.573	176.08	176.10	0.02	0.01
126.159	217.37	217.35	-0.02	-0.01
132.228	261.98	261.96	-0.02	-0.01
139.542	325.25	325.21	-0.04	-0.01
147.093	402.79	402.76	-0.03	-0.01
155.147	501.03	501.00	-0.03	-0.01
163.894	628.20	628.21	0.01	0.00
170.071	732.43	732.41	-0.02	0.00
170.735	744.35	744.36	0.01	0.00
171.351	755.54	755.60	0.06	0.01
172.054	768.56	768.57	0.01	0.00
172.670	780.10	780.09	-0.01	0.00

STANDARD DEVIATION = 0.03

1-DECENE

C10H20

FORZIATI A.F.,CAMIN D.L.,ROSSINI F.D.:
J.RESEARCH NATL.BUR.STANDARDS 45,406(1950).

A = 6.95433 B = 1497.527 C = 197.056

B.P.(760) = 170.569

	T	P EXPTL.	P CALCD.	DEV.	PERCENT
*	86.774	47.98	47.66	-0.32	-0.66
	91.308	57.71	57.70	-0.01	-0.03
	95.134	67.48	67.47	-0.01	-0.01
	98.604	77.51	77.50	-0.01	-0.02
	101.844	87.93	87.94	0.01	0.01
	106.223	103.87	103.88	0.01	0.01
	111.213	124.86	124.87	0.01	0.01
	116.283	149.64	149.65	0.01	0.00
	120.995	176.14	176.15	0.01	0.00
*	127.265	217.44	217.22	-0.22	-0.10
	140.063	325.22	325.23	0.01	0.00
	147.265	402.86	402.81	-0.05	-0.01
	154.939	501.12	501.10	-0.02	0.00
	169.134	732.53	732.56	0.03	0.00
	169.762	744.42	744.46	0.04	0.01
	170.345	755.67	755.65	-0.02	0.00
	171.012	768.65	768.61	-0.04	-0.01
	171.605	780.26	780.28	0.02	0.00

STANDARD DEVIATION = 0.03

CAPRIC ACID $C_{10}H_{20}O_2$

KAHLBAUM G.W.A.: Z.PHYS.CHEM. 13,14(1894).

A = 6.25530 B = 1106.290 C = 57.961

B.P.(10) = 152.548

T	P EXPTL.	P CALCD.	DEV.	PERCENT
152.800	10.00	10.15	0.15	1.46
152.800	10.20	10.15	-0.05	-0.53
153.700	10.60	10.68	0.08	0.76
154.000	10.80	10.86	0.06	0.59
154.200	11.00	10.99	-0.01	-0.11
154.600	11.30	11.24	-0.06	-0.54
155.000	11.60	11.49	-0.11	-0.91
155.400	11.80	11.76	-0.04	-0.38
156.200	12.40	12.29	-0.11	-0.87
157.000	12.80	12.85	0.05	0.38
157.200	13.20	12.99	-0.21	-1.59
158.200	13.60	13.72	0.12	0.89
158.100	13.80	13.65	-0.15	-1.11
158.400	14.20	13.87	-0.33	-2.31
158.600	14.40	14.02	-0.38	-2.61
159.500	15.10	14.72	-0.38	-2.50
160.100	15.50	15.21	-0.29	-1.90
160.700	16.00	15.70	-0.30	-1.87
161.200	16.40	16.12	-0.28	-1.69
162.700	17.50	17.45	-0.05	-0.29
164.000	18.60	18.67	0.07	0.38
165.200	19.80	19.86	0.06	0.29
166.300	21.00	21.00	0.00	0.01
167.200	21.80	21.98	0.18	0.81
168.200	22.70	23.10	0.40	1.78
169.300	23.90	24.40	0.50	2.09
170.300	25.10	25.63	0.53	2.10
171.200	26.30	26.78	0.48	1.81
171.700	27.50	27.43	-0.07	-0.25
172.700	28.80	28.78	-0.02	-0.06
173.600	30.20	30.05	-0.16	-0.51
174.600	31.60	31.50	-0.10	-0.32
175.900	33.30	33.48	0.18	0.53
176.900	34.80	35.07	0.27	0.77
178.100	36.90	37.06	0.16	0.42
179.100	38.80	38.78	-0.02	-0.05
180.500	41.00	41.31	0.31	0.75
182.100	44.10	44.35	0.25	0.58
183.000	46.10	46.15	0.05	0.10
183.900	48.30	48.00	-0.30	-0.62
184.900	50.40	50.13	-0.27	-0.54
185.800	52.40	52.11	-0.29	-0.56
187.100	55.30	55.08	-0.22	-0.41

STANDARD DEVIATION = 0.24

1-CHLORODECANE C10H21CL

KEMME R.H.,KREPS S.I.:

J.CHEM.ENG.DATA 14,98(1969).

A = 6.93986 B = 1639.055 C = 177.940

B.P.(760) = 225.863

T	P EXPTL.	P CALCD.	DEV.	PERCENT
86.200	5.50	5.43	-0.07	-1.32
98.900	10.50	10.45	-0.05	-0.44
106.200	15.00	14.84	-0.16	-1.08
112.800	20.20	20.06	-0.14	-0.69
122.700	30.80	30.76	-0.04	-0.13
129.200	40.10	40.12	0.02	0.05
137.100	54.70	54.60	-0.10	-0.19
147.700	80.00	80.63	0.63	0.79
159.300	120.50	120.13	-0.37	-0.31
175.700	201.60	201.86	0.26	0.13
189.500	301.40	301.38	-0.02	-0.01
208.600	501.50	500.65	-0.85	-0.17
225.600	754.70	755.38	0.68	0.09

STANDARD DEVIATION = 0.43

DECANE C10H22

WILLINGHAM C.J.,TAYLOR W.J.,PIGNOCCO J.M.,ROSSINI F.D.:
J.RESEARCH NATL.BUR.STANDARDS 35,219(1945).

A = 6.95707 B = 1503.568 C = 194.738

B.P.(760) = 174.122

T	P EXPTL.	P CALCD.	DEV.	PERCENT
94.481	57.37	57.32	-0.05	-0.08
98.352	67.16	67.14	-0.02	-0.02
101.859	77.20	77.21	0.01	0.01
105.118	87.65	87.65	0.00	0.00
109.526	103.60	103.61	0.01	0.01
114.540	124.58	124.60	0.02	0.02
119.640	149.36	149.41	0.05	0.04
124.372	175.90	175.92	0.02	0.01
130.690	217.15	217.16	0.01	0.01
136.499	261.71	261.70	-0.01	0.00
143.495	324.91	324.86	-0.05	-0.01
150.718	402.44	402.39	-0.05	-0.01
158.419	500.72	500.67	-0.05	-0.01
166.772	627.97	627.96	-0.01	0.00
172.661	732.13	732.15	0.02	0.00
173.295	744.11	744.13	0.02	0.00
173.882	755.32	755.36	0.04	0.01
174.538	768.07	768.07	0.00	0.00
175.121	779.47	779.50	0.03	0.00

STANDARD DEVIATION = 0.03

DIAMYL ETHER C10H22O

DREISBACH R.R.,SHRADER S.A.:
IND.ENG.CHEM. 41,2879(1949).

A = 7.06710 B = 1604.769 C = 196.584

B.P.(760) = 186.755

T	P EXPTL.	P CALCD.	DEV.	PERCENT
105.460	57.04	56.76	-0.28	-0.48
112.860	75.86	76.06	0.20	0.26
126.060	123.76	123.97	0.21	0.17
154.670	315.52	315.08	-0.44	-0.14
171.370	507.50	507.89	0.39	0.08
186.750	760.00	759.90	-0.10	-0.01

STANDARD DEVIATION = 0.41

1-DECANOL C10H22O

AMBROSE D.,SPRAKE C.H.S.: J.CHEM.THERMODYNAMICS 2,631(1970).

A = 6.72145 B = 1366.062 C = 124.641

B.P.(760) = 231.046

T	P EXPTL.	P CALCD.	DEV.	PERCENT
127.261	19.92	19.88	-0.04	-0.19
129.526	22.26	22.22	-0.04	-0.17
133.573	27.00	26.98	-0.02	-0.08
138.245	33.51	33.50	-0.01	-0.03
141.817	39.32	39.33	0.01	0.02
145.863	46.88	46.92	0.04	0.09
149.414	54.50	54.55	0.05	0.10
154.602	67.47	67.52	0.05	0.07
159.206	81.00	81.05	0.05	0.07
163.693	96.27	96.31	0.04	0.04
168.653	115.80	115.82	0.02	0.02
174.653	143.61	143.60	-0.01	-0.01
177.568	158.96	158.92	-0.04	-0.02
182.740	189.39	189.34	-0.05	-0.03
189.377	235.21	235.06	-0.15	-0.06
194.484	276.08	275.93	-0.15	-0.06
201.039	336.59	336.47	-0.12	-0.03
207.352	404.28	404.31	0.03	0.01
213.119	475.27	475.30	0.03	0.01
218.923	556.13	556.28	0.15	0.03
225.489	660.01	660.48	0.47	0.07
230.471	749.21	749.19	-0.02	0.00
237.598	891.55	891.83	0.28	0.03
243.055	1014.43	1014.49	0.06	0.01
249.366	1172.68	1172.03	-0.65	-0.06
* 255.172	1334.72	1332.82	-1.90	-0.14

STANDARD DEVIATION = 0.19

DIISOAMYL SULFIDE C10H22S

BAUER H.,BURSCHKIES K.: BER. 68,1243(1935).

A = -1.95981 B = 390.608 C = -219.334

B.P.(10) = 87.363

T	P EXPTL.	P CALCD.	DEV.	PERCENT
10.000	0.70	0.81	0.11	15.10
15.000	0.80	0.90	0.10	11.87
20.000	0.90	1.00	0.10	11.05
25.000	1.00	1.12	0.12	12.25
30.000	1.20	1.27	0.07	5.70
35.000	1.40	1.44	0.04	3.06
40.000	1.70	1.65	-0.05	-2.76
45.000	2.00	1.91	-0.09	-4.56
50.000	2.30	2.22	-0.08	-3.35
55.000	2.70	2.61	-0.09	-3.23
60.000	3.20	3.10	-0.10	-3.05
65.000	3.80	3.73	-0.08	-1.97
70.000	4.40	4.53	0.13	2.90
75.000	5.50	5.58	0.08	1.42
80.000	7.00	6.98	-0.02	-0.35

STANDARD DEVIATION = 0.10

BIS-PENTAMETHYLDISILANYL ETHER C10H30OSI4

CRAIG A.D.,ET AL.: J.CHEM.SOC. 1962,548.

A = 8.16144 B = 2575.250 C = 273.323

B.P.(100) = 144.639

T	P EXPTL.	P CALCD.	DEV.	PERCENT
87.700	10.70	10.67	-0.03	-0.26
100.300	18.60	18.57	-0.03	-0.16
109.000	26.60	26.65	0.05	0.18
113.900	32.40	32.43	0.03	0.08
119.400	40.20	40.18	-0.02	-0.05
122.700	45.60	45.57	-0.03	-0.07
126.500	52.50	52.54	0.04	0.07
129.700	59.10	59.10	0.00	0.01
132.400	65.20	65.18	-0.02	-0.03
136.000	74.10	74.12	0.02	0.03
139.800	84.70	84.69	-0.01	-0.01
145.400	102.60	102.61	0.01	0.01
150.400	121.30	121.28	-0.03	-0.02
168.400	214.50	214.50	0.00	0.00
182.800	327.70	327.71	0.01	0.00

STANDARD DEVIATION = 0.03

1-METHYLNAPHTHALENE C11H10

CAMIN D.L.,ROSSINI F.D.:
J.PHYS.CHEM. 59,1173(1955).

A = 7.03409 B = 1825.445 C = 194.833

B.P.(760) = 244.686

T	P EXPTL.	P CALCD.	DEV.	PERCENT
142.140	41.43	41.39	-0.04	-0.09
153.600	62.36	62.38	0.02	0.04
157.539	71.33	71.39	0.06	0.09
161.689	82.07	82.02	-0.05	-0.06
167.212	98.17	98.19	0.02	0.02
179.971	145.83	145.78	-0.05	-0.04
185.505	171.62	171.62	0.00	0.00
193.280	214.08	214.14	0.06	0.03
200.536	261.26	261.23	-0.03	-0.01
208.677	323.74	323.72	-0.02	-0.01
217.375	403.25	403.30	0.05	0.01
226.498	502.97	502.95	-0.02	0.00
236.243	630.25	630.18	-0.07	-0.01
243.177	735.67	735.36	-0.31	-0.04
243.949	747.48	747.88	0.40	0.05
244.555	757.79	757.83	0.04	0.01
245.326	770.74	770.64	-0.10	-0.01

STANDARD DEVIATION = 0.14

2-METHYLNAPHTHALENE C11H10

CAMIN D.L.,ROSSINI F.D.: J.PHYS.CHEM. 59,1173(1955).

A = 7.07084 B = 1842.309 C = 198.636

B.P.(760) = 241.054

T	P EXPTL.	P CALCD.	DEV.	PERCENT
139.193	41.52	41.44	-0.08	-0.19
145.431	52.14	52.04	-0.10	-0.20
150.665	62.55	62.60	0.05	0.08
154.576	71.55	71.61	0.06	0.08
158.689	82.07	82.22	0.15	0.19
164.155	98.26	98.33	0.07	0.07
176.722	145.76	145.44	-0.32	-0.22
182.322	171.63	171.73	0.10	0.06
190.033	214.15	214.18	0.03	0.01
197.234	261.17	261.22	0.05	0.02
205.329	323.75	323.79	0.04	0.01
213.963	403.40	403.36	-0.04	-0.01
223.026	503.21	503.12	-0.09	-0.02
239.613	736.00	736.27	0.27	0.04
240.336	748.07	748.10	0.03	0.00
240.957	758.42	758.38	-0.04	-0.01
241.760	772.03	771.84	-0.19	-0.02

STANDARD DEVIATION = 0.14

2,4-DIMETHYLQUINOLINE C11H11N

MALONOWSKA B.,WECSILE J.:
BULL.ACAD.POL.SCI. 12,239(1964).

A = 7.02536 B = 1830.292 C = 174.443

B.P.(760) = 267.172

	T	P EXPTL.	P CALCD.	DEV.	PERCENT
	185.350	88.44	86.75	-1.69	-1.91
	189.690	99.47	99.75	0.28	0.28
	195.310	118.74	118.94	0.20	0.17
	202.250	144.86	146.73	1.87	1.29
	211.440	191.54	191.53	-0.01	-0.01
	218.100	230.61	230.52	-0.09	-0.04
	225.530	281.60	281.39	-0.21	-0.07
	231.470	328.60	328.31	-0.29	-0.09
	237.400	381.58	381.25	-0.33	-0.09
	241.960	426.58	426.45	-0.13	-0.03
	246.430	475.01	474.85	-0.16	-0.03
	250.090	517.81	517.66	-0.15	-0.03
	253.160	556.00	555.91	-0.09	-0.02
	256.560	600.73	600.85	0.12	0.02
	259.570	642.73	643.00	0.28	0.04
	262.190	681.18	681.59	0.41	0.06
*	265.080	715.67	726.24	10.57	1.48
	267.530	765.18	765.89	0.71	0.09
	269.450	799.04	798.14	-0.90	-0.11

STANDARD DEVIATION = 0.75

2,6-DIMETHYLQUINOLINE C11H11N

MALONOWSKA B.,WECSILE J.: BULL.ACAD.POL.SCI. 12,239(1964).

A = 6.93112 B = 1748.726 C = 166.367

B.P.(760) = 265.384

	T	P EXPTL.	P CALCD.	DEV.	PERCENT
	188.460	100.63	100.63	0.00	0.00
	191.160	109.60	109.63	0.03	0.03
	196.680	130.11	130.11	0.00	0.00
	201.370	149.87	149.88	0.01	0.01
	206.620	174.85	174.86	0.01	0.00
	210.800	197.10	197.08	-0.02	-0.01
	214.370	217.82	217.83	0.01	0.00
	217.550	237.82	237.77	-0.05	-0.02
	221.220	262.64	262.60	-0.04	-0.02
*	224.750	289.13	288.41	-0.72	-0.25
	229.030	322.46	322.41	-0.05	-0.02
	233.180	358.38	358.38	0.00	0.00
	237.050	394.76	394.76	0.00	0.00
	240.520	429.84	429.83	-0.01	0.00
	243.810	465.38	465.34	-0.04	-0.01
	246.470	495.36	495.73	0.37	0.07
	249.900	537.23	537.21	-0.02	0.00
	251.060	551.83	551.85	0.02	0.00
	253.860	588.56	588.48	-0.08	-0.01
	256.120	619.49	619.43	-0.06	-0.01
*	258.170	650.08	648.60	-1.48	-0.23
	260.110	677.12	677.20	0.08	0.01
	262.450	712.99	713.00	0.01	0.00
	263.450	728.82	728.75	-0.07	-0.01
	264.380	743.60	743.64	0.04	0.01
	265.630	764.09	764.03	-0.06	-0.01
	266.290	775.04	774.97	-0.07	-0.01

STANDARD DEVIATION = 0.09

2,4,5-TRIMETHYLSTYRENE

$C_{11}H_{14}$

BUCK F.R., COLES K.F., KENNEDY G.T., MORTON F.:
J.CHEM.SOC. 1949, 2377.

A = 7.33150 B = 1880.658 C = 205.680

B.P.(760) = 216.875

T	P EXPTL.	P CALCD.	DEV.	PERCENT
79.500	5.00	5.46	0.46	9.12
97.900	13.00	13.70	0.70	5.35
106.800	21.00	20.56	-0.44	-2.10
113.600	28.00	27.62	-0.38	-1.37
142.000	83.00	83.62	0.62	0.75
165.800	188.00	185.73	-2.27	-1.21
177.000	261.00	261.25	0.25	0.09
198.400	470.00	475.69	5.69	1.21
216.300	754.00	749.46	-4.54	-0.60

STANDARD DEVIATION = 3.15

2,4,6-TRIMETHYLSTYRENE

$C_{11}H_{14}$

BUCK F.R., COLES K.F., KENNEDY G.T., MORTON F.:
J.CHEM.SOC. 1949, 2377.

A = 7.08905 B = 1702.607 C = 195.931

B.P.(760) = 208.658

T	P EXPTL.	P CALCD.	DEV.	PERCENT
89.900	14.00	13.56	-0.44	-3.12
97.100	19.00	19.00	0.00	-0.01
102.200	25.00	23.88	-1.12	-4.46
109.900	32.00	33.26	1.26	3.93
114.500	40.00	40.22	0.22	0.54
117.700	46.00	45.75	-0.25	-0.55
123.700	58.00	57.84	-0.16	-0.27
134.500	88.00	86.37	-1.63	-1.85
139.200	101.00	102.01	1.01	1.00
141.700	111.00	111.24	0.24	0.21
147.500	134.00	135.33	1.33	1.00
151.800	155.00	155.85	0.85	0.55
155.800	178.00	177.17	-0.83	-0.46
165.200	238.00	236.81	-1.19	-0.50
172.800	297.00	296.19	-0.81	-0.27
176.400	328.00	328.26	0.26	0.08
179.100	354.00	354.11	0.11	0.03
182.400	388.00	387.92	-0.09	-0.02
186.400	432.00	432.33	0.33	0.08
* 192.900	500.00	513.17	13.17	2.63
197.100	570.00	571.55	1.55	0.27
199.800	611.00	611.81	0.81	0.13
205.000	694.00	695.69	1.69	0.24
207.900	750.00	746.29	-3.71	-0.49

STANDARD DEVIATION = 1.26

1,3,5-TRIMETHYL-2-ETHYLBENZENE C11H16

BUCK F.R.,COLES K.F.,KENNEDY G.T.,MORTON F.: J.CHEM.SOC. 1949,2377.

A = 6.79084 B = 1505.818 C = 174.685

B.P.(760) = 210.432

T	P EXPTL.	P CALCD.	DEV.	PERCENT
88.500	11.80	11.73	-0.07	-0.59
91.300	13.80	13.48	-0.32	-2.35
96.300	17.80	17.14	-0.66	-3.71
102.000	22.80	22.31	-0.49	-2.15
107.000	28.00	27.87	-0.13	-0.48
110.100	32.00	31.86	-0.14	-0.43
115.400	39.00	39.80	0.80	2.05
119.000	46.60	46.08	-0.52	-1.12
122.200	53.00	52.33	-0.67	-1.26
125.100	58.00	58.59	0.59	1.02
130.000	70.00	70.57	0.57	0.82
133.900	80.50	81.49	0.99	1.22
140.500	103.00	103.10	0.10	0.10
151.900	151.00	151.37	0.37	0.24
161.200	204.00	203.10	-0.90	-0.44
174.000	300.00	296.67	-3.33	-1.11
185.000	401.00	402.11	1.11	0.28
195.900	528.00	533.93	5.93	1.12
210.200	760.00	755.88	-4.12	-0.54

STANDARD DEVIATION = 2.08

1,4,5-TRIMETHYL-2-ETHYLBENZENE C11H16

BUCK F.R.,COLES K.F.,KENNEDY G.T.,MORTON F.: J.CHEM.SOC. 1949,2377.

A = 3.02926 B = 116.352 C = -34.562

B.P.(10) = 91.899

T	P EXPTL.	P CALCD.	DEV.	PERCENT
87.300	11.00	6.65	-4.35	-39.51
98.300	16.00	15.99	-0.01	-0.08
100.200	19.00	18.06	-0.94	-4.97
101.700	20.00	19.78	-0.22	-1.10
103.600	21.00	22.08	1.08	5.13
109.400	27.00	29.82	2.82	10.45
109.100	28.00	29.40	1.40	4.99
117.400	39.00	42.14	3.14	8.05
122.200	50.00	50.31	0.31	0.61
126.300	58.00	57.67	-0.33	-0.57
132.100	71.00	68.61	-2.39	-3.37

STANDARD DEVIATION = 2.42

1-UNDECENE C11H22

FORZIATI A.F.,CAMIN D.L.,ROSSINI F.D.:
J.RESEARCH NATL.BUR.STANDARDS 45,406(1950).

A = 6.97002 B = 1564.785 C = 189.993

B.P.(760) = 192.669

	T	P EXPTL.	P CALCD.	DEV.	PERCENT
	105.866	47.99	47.98	-0.01	-0.02
	110.423	57.72	57.72	0.00	-0.01
	114.388	67.48	67.48	-0.01	-0.01
	117.997	77.51	77.51	0.00	0.01
	121.355	87.94	87.94	0.00	0.00
	125.902	103.87	103.88	0.01	0.01
	131.081	124.86	124.86	0.00	0.00
	136.350	149.65	149.66	0.01	0.01
	141.240	176.14	176.16	0.02	0.01
	147.780	217.45	217.45	0.00	0.00
*	153.780	262.05	261.95	-0.10	-0.04
	161.031	325.28	325.27	-0.01	0.00
	168.501	402.87	402.84	-0.03	-0.01
	176.462	501.13	501.14	0.01	0.00
	185.091	628.38	628.33	-0.05	-0.01
	191.179	732.55	732.53	-0.02	0.00
	191.832	744.43	744.46	0.03	0.00
	192.441	755.69	755.73	0.05	0.01
	193.130	768.66	768.65	-0.01	0.00
	193.742	780.26	780.27	0.01	0.00

STANDARD DEVIATION = 0.02

METHYL CAPRATE $C_{11}H_{22}O_2$

ROSE A.,SUPINA W.R.: J.CHEM.ENG.DATA6,173(1961).

A = 7.18998 B = 1783.846 C = 181.605

B.P.(100) = 162.105

T	P EXPTL.	P CALCD.	DEV.	PERCENT
* 107.300	10.80	10.36	-0.44	-4.05
114.100	14.50	14.37	-0.13	-0.90
118.200	17.40	17.38	-0.02	-0.14
122.800	21.50	21.37	-0.13	-0.59
126.500	25.10	25.13	0.03	0.13
128.800	27.80	27.74	-0.06	-0.21
130.600	29.90	29.94	0.04	0.14
137.300	39.30	39.47	0.17	0.44
142.000	47.30	47.59	0.29	0.62
147.800	59.70	59.51	-0.19	-0.31
148.600	61.40	61.34	-0.06	-0.10
151.100	67.40	67.35	-0.05	-0.08
155.000	77.90	77.71	-0.19	-0.25
156.200	81.10	81.15	0.05	0.06
158.800	89.00	89.04	0.04	0.05
162.200	100.10	100.33	0.23	0.23
164.200	107.60	107.51	-0.09	-0.09
164.300	108.00	107.88	-0.12	-0.11
167.100	118.40	118.67	0.27	0.23
169.000	126.30	126.49	0.19	0.15
174.300	150.90	150.60	-0.30	-0.20
177.800	168.70	168.52	-0.18	-0.11
181.400	188.70	188.74	0.04	0.02
184.700	209.10	208.99	-0.11	-0.05
188.200	232.20	232.39	0.19	0.08

STANDARD DEVIATION = 0.17

PROPYL CAPRYLATE

C11H22O2

BONHORST C.W.,ALTHOUSE P.M.,TRIEBOLD H.O.:
IND.ENG.CHEM. 40,2379(1948).

A = 8.51672 B = 2599.488 C = 246.211

B.P.(100) = 152.684

T	P EXPTL.	P CALCD.	DEV.	PERCENT
70.500	2.00	2.04	0.04	1.84
82.300	4.00	4.02	0.02	0.40
86.000	5.00	4.92	-0.08	-1.61
90.000	6.00	6.10	0.10	1.59
94.100	8.00	7.55	-0.45	-5.58
100.000	10.00	10.19	0.19	1.94
114.100	20.00	20.05	0.05	0.26
129.800	40.00	40.12	0.12	0.31
135.000	50.00	49.85	-0.15	-0.29
139.500	60.00	59.88	-0.13	-0.21
147.000	80.00	80.50	0.50	0.62
152.600	100.00	99.68	-0.32	-0.32

STANDARD DEVIATION = 0.27

ISOPROPYL CAPRYLATE

C11H22O2

BONHORST C.W.,ALTHOUSE P.M.,TRIEBOLD H.O.:
IND.ENG.CHEM. 40,2379(1948).

A = 8.03218 B = 2213.645 C = 220.941

B.P.(100) = 146.031

T	P EXPTL.	P CALCD.	DEV.	PERCENT
65.000	2.00	1.95	-0.05	-2.38
76.700	4.00	3.93	-0.07	-1.64
80.400	5.00	4.86	-0.14	-2.90
84.100	6.00	5.96	-0.04	-0.65
89.300	8.00	7.89	-0.11	-1.41
93.800	10.00	9.98	-0.02	-0.24
108.200	20.00	20.26	0.26	1.31
123.400	40.00	40.14	0.14	0.34
128.600	50.00	50.02	0.02	0.05
133.000	60.00	59.97	-0.03	-0.06
140.100	80.00	79.60	-0.40	-0.50
146.100	100.00	100.26	0.26	0.26

STANDARD DEVIATION = 0.20

1-CHLOROUNDECANE C11H23CL

KEMME R.H.,KREPS S.I.: J.CHEM.ENG.DATA 14,98(1969).

A = 6.96762 B = 1709.398 C = 172.929

B.P.(760) = 245.343

T	P EXPTL.	P CALCD.	DEV.	PERCENT
101.400	5.50	5.45	-0.05	-0.90
114.300	10.40	10.38	-0.02	-0.18
122.300	15.10	15.05	-0.05	-0.33
128.400	19.80	19.71	-0.09	-0.44
138.500	30.10	30.11	0.01	0.04
145.600	39.80	39.91	0.11	0.28
154.500	55.60	55.84	0.24	0.43
165.100	81.00	81.41	0.41	0.50
176.800	120.50	120.18	-0.32	-0.26
193.700	203.00	201.90	-1.10	-0.54
208.700	308.00	307.90	-0.10	-0.03
227.600	499.10	500.92	1.82	0.36
227.600	499.80	500.92	1.12	0.22
245.400	763.00	760.96	-2.04	-0.27

STANDARD DEVIATION = 0.97

UNDECANE C11H24

CAMIN D.L.,ROSSINI F.D.:
J.PHYS.CHEM. 59,1173(1955).

A = 6.97712 B = 1572.752 C = 188.054

B.P.(760) = 195.890

	T	P EXPTL.	P CALCD.	DEV.	PERCENT
*	104.458	41.55	39.85	-1.70	-4.10
	110.962	52.16	52.16	0.00	0.00
	115.522	62.53	62.57	0.04	0.06
	118.963	71.54	71.52	-0.02	-0.03
	122.607	82.16	82.13	-0.03	-0.03
	127.467	98.29	98.29	0.00	0.00
	132.757	118.79	118.76	-0.03	-0.02
	138.713	145.87	145.90	0.03	0.02
	143.585	171.68	171.69	0.01	0.01
	150.437	214.14	214.17	0.03	0.01
	156.841	261.22	261.23	0.01	0.00
*	164.039	322.76	323.78	1.02	0.32
	171.724	403.38	403.33	-0.05	-0.01
	179.802	503.12	503.10	-0.02	0.00
	188.431	630.45	630.45	0.00	0.00
	194.595	736.05	736.11	0.06	0.01
	195.242	747.91	747.97	0.06	0.01
	195.794	758.17	758.20	0.03	0.00
	196.511	771.68	771.65	-0.03	0.00
	197.272	786.23	786.14	-0.09	-0.01

STANDARD DEVIATION = 0.04

CARBAZOL C12H9N

SENSEMAN C.E.,NELSON O.A.: IND.ENG.CHEM. 15,382(1923).

A = 7.08633 B = 2179.424 C = 163.514

B.P.(760) = 354.715

T	P EXPTL.	P CALCD.	DEV.	PERCENT
252.610	70.10	70.61	0.51	0.73
258.280	82.90	83.04	0.14	0.17
259.430	85.30	85.77	0.47	0.55
261.990	92.80	92.12	-0.68	-0.73
266.130	103.40	103.21	-0.19	-0.19
270.150	114.90	115.01	0.11	0.09
274.040	128.60	127.47	-1.13	-0.88
276.310	135.20	135.24	0.04	0.03
281.930	156.60	156.18	-0.42	-0.27
288.020	181.00	181.81	0.81	0.45
290.090	191.00	191.27	0.27	0.14
291.250	197.20	196.75	-0.46	-0.23
292.650	202.20	203.52	1.32	0.65
294.400	212.80	212.26	-0.54	-0.25
298.160	231.50	232.08	0.58	0.25
302.710	260.60	258.05	-2.55	-0.98
307.180	286.50	285.82	-0.68	-0.24
309.850	303.70	303.54	-0.16	-0.05
310.810	307.60	310.12	2.52	0.82
311.830	317.80	317.24	-0.56	-0.18
315.770	348.40	346.00	-2.40	-0.69
320.850	385.50	386.17	0.67	0.17
322.800	401.20	402.55	1.35	0.34
326.600	436.60	436.08	-0.52	-0.12
332.650	491.20	494.06	2.86	0.58
343.420	610.90	612.50	1.60	0.26
348.050	669.40	669.91	0.51	0.08
348.260	672.10	672.61	0.51	0.08
350.050	693.00	696.00	3.00	0.43
354.280	753.40	753.83	0.43	0.06
354.490	759.40	756.80	-2.60	-0.34
354.720	761.30	760.06	-1.24	-0.16
357.310	797.80	797.56	-0.24	-0.03
357.710	807.20	803.48	-3.72	-0.46

STANDARD DEVIATION = 1.51

ACENAPHTHENE

C12H10

MORTIMER F.S.,MURPHY R.V.:

IND.ENG.CHEM. 15,1140(1923).

A = 7.72819 B = 2534.234 C = 245.576

B.P.(760) = 277.230

T	P EXPTL.	P CALCD.	DEV.	PERCENT
147.200	19.20	18.88	-0.32	-1.65
182.400	63.60	64.08	0.48	0.76
182.400	63.60	64.08	0.48	0.76
210.200	148.00	147.20	-0.80	-0.54
210.400	149.00	148.03	-0.97	-0.65
227.200	234.60	233.27	-1.33	-0.57
233.200	272.00	272.29	0.29	0.11
246.200	374.60	375.80	1.20	0.32
246.600	377.90	379.44	1.54	0.41
247.000	382.90	383.11	0.21	0.05
252.400	434.50	435.63	1.13	0.26
252.500	434.70	436.65	1.95	0.45
264.400	573.10	573.93	0.83	0.15
264.400	573.50	573.93	0.43	0.08
275.300	733.40	729.20	-4.20	-0.57
275.400	733.40	730.78	-2.62	-0.36
275.400	733.40	730.78	-2.62	-0.36
286.800	930.30	928.85	-1.45	-0.16
287.000	932.30	932.69	0.39	0.04
287.800	943.00	948.14	5.14	0.54

STANDARD DEVIATION = 2.08

BIPHENYL C12H10

CUNNINGHAM G.B.: POWER72,374(1930).

A = 7.24541 B = 1998.725 C = 202.733

B.P.(760) = 255.208

T	P EXPTL.	P CALCD.	DEV.	PERCENT
69.200	0.78	0.79	0.01	0.75
93.300	3.10	3.12	0.02	0.54
148.700	36.25	36.14	-0.11	-0.29
160.000	54.30	54.35	0.05	0.10
171.100	79.12	79.22	0.10	0.13
182.200	112.74	112.98	0.24	0.21
193.300	158.25	157.96	-0.29	-0.19
204.400	217.20	216.84	-0.36	-0.17
215.600	293.22	293.47	0.25	0.09
226.700	389.93	390.00	0.07	0.02
237.800	510.43	510.91	0.48	0.09
248.900	661.95	660.47	-1.48	-0.22
255.300	760.21	761.53	1.32	0.17
260.000	842.95	843.36	0.41	0.05
271.100	1065.33	1064.62	-0.71	-0.07

STANDARD DEVIATION = 0.66

DIPHENYLDICHLOROSILANE C12H10CL2SI

JENKINS A.C.,CHAMBERS G.F.: IND.ENG.CHEM. 46,2367(1954).

A = 6.99903 B = 1918.201 C = 161.409

B.P.(100) = 222.306

T	P EXPTL.	P CALCD.	DEV.	PERCENT
192.100	37.40	37.40	0.00	-0.01
210.700	69.80	69.84	0.04	0.05
235.600	147.10	147.02	-0.08	-0.05
260.200	281.30	281.39	0.09	0.03
281.200	462.60	462.56	-0.04	-0.01

STANDARD DEVIATION = 0.09

DIPHENYL ETHER C12H10O

COLLERSON R.R., ET AL.:
J.CHEM.SOC. 3697(1965).

A = 7.01104 B = 1799.712 C = 177.744

B.P.(760) = 257.997

T	P EXPTL.	P CALCD.	DEV.	PERCENT
204.213	199.18	199.17	-0.01	0.00
212.102	248.07	248.07	0.00	0.00
219.226	300.19	300.21	0.02	0.01
225.105	349.61	349.61	0.00	0.00
230.561	401.15	401.13	-0.02	-0.01
235.186	449.40	449.42	0.02	0.00
239.618	499.95	499.96	0.01	0.00
243.667	550.00	550.00	0.00	0.00
247.413	599.79	599.78	-0.01	0.00
250.991	650.61	650.61	0.00	0.00
254.089	697.34	697.32	-0.02	0.00
257.458	751.14	751.09	-0.05	-0.01
260.469	801.86	801.88	0.02	0.00
263.290	851.84	851.88	0.04	0.00
265.939	901.01	901.03	0.03	0.00
268.416	949.02	948.99	-0.03	0.00
270.949	1000.08	1000.08	0.00	0.00

STANDARD DEVIATION = 0.02

2-PHENYLPHENOL C12H10O

VONTERRES E.,ET AL.:
BRENNSTOFF CHEM. 36,272(1955).

A = 6.49372 B = 1306.855 C = 83.821

B.P.(760) = 277.897

	T	P EXPTL.	P CALCD.	DEV.	PERCENT
	161.900	10.00	14.97	4.97	49.72
*	168.000	25.00	20.14	-4.86	-19.43
	188.100	50.00	48.72	-1.28	-2.56
	198.000	75.00	71.87	-3.13	-4.18
	206.800	100.00	99.30	-0.70	-0.70
	219.000	150.00	150.70	0.70	0.47
	228.100	200.00	201.38	1.38	0.69
	235.200	250.00	249.61	-0.39	-0.16
	241.600	300.00	300.49	0.49	0.16
	244.000	325.00	321.54	-3.46	-1.07
	247.400	350.00	353.31	3.31	0.94
	252.000	400.00	400.13	0.13	0.03
	257.000	450.00	456.34	6.34	1.41
	261.000	500.00	505.55	5.55	1.11
	264.500	550.00	551.89	1.89	0.34
	268.000	600.00	601.41	1.41	0.24
	270.800	650.00	643.43	-6.57	-1.01
	273.800	700.00	690.90	-9.10	-1.30
*	275.000	760.00	710.62	-49.38	-6.50

STANDARD DEVIATION = 4.35

4-PHENYLPHENOL C12H10O

VONTERRES E.,ET AL.:
BRENNSTOFF CHEM. 36,272(1955).

A = 8.65752 B = 3022.777 C = 216.060

B.P.(760) = 307.210

T	P EXPTL.	P CALCD.	DEV.	PERCENT
177.000	10.00	9.27	-0.73	-7.29
199.200	25.00	23.89	-1.11	-4.43
217.800	50.00	49.02	-0.98	-1.97
229.000	75.00	73.40	-1.60	-2.14
237.400	100.00	98.06	-1.94	-1.94
250.000	150.00	148.49	-1.51	-1.01
260.200	200.00	204.46	4.46	2.23
267.100	250.00	251.91	1.91	0.76
273.800	300.00	306.77	6.77	2.26
276.000	325.00	326.90	1.90	0.58
278.800	350.00	354.13	4.13	1.18
283.000	400.00	398.63	-1.37	-0.34
287.100	450.00	446.61	-3.39	-0.75
291.000	500.00	496.75	-3.25	-0.65
294.700	550.00	548.68	-1.32	-0.24
297.800	600.00	595.69	-4.31	-0.72
300.600	650.00	641.07	-8.93	-1.37
303.500	700.00	691.13	-8.87	-1.27
308.000	760.00	775.39	15.39	2.02

STANDARD DEVIATION = 5.81

2,2'-DIPHENOL

C12H1002

VONTERRES E.,ET AL.: BRENNSTOFF CHEM. 36,272(1955).

A = 8.19352 B = 3067.632 C = 253.146

B.P.(760) = 324.268

T	P EXPTL.	P CALCD.	DEV.	PERCENT
171.000	10.00	9.14	-0.86	-8.58
197.000	25.00	23.92	-1.08	-4.32
218.100	50.00	48.29	-1.71	-3.41
231.300	75.00	72.66	-2.34	-3.12
242.000	100.00	99.57	-0.43	-0.43
257.000	150.00	151.45	1.45	0.97
268.000	200.00	202.86	2.86	1.43
276.700	250.00	253.43	3.43	1.37
284.000	300.00	303.76	3.76	1.26
287.000	325.00	326.78	1.78	0.55
290.000	350.00	351.26	1.26	0.36
295.000	400.00	395.50	-4.50	-1.13
300.100	450.00	445.38	-4.62	-1.03
305.000	500.00	498.21	-1.79	-0.36
309.200	550.00	547.59	-2.41	-0.44
313.000	600.00	595.76	-4.24	-0.71
317.000	650.00	650.26	0.26	0.04
320.000	700.00	693.83	-6.17	-0.88
325.000	760.00	771.86	11.86	1.56

STANDARD DEVIATION = 4.32

P-DIISOPROPYLBENZENE

$C_{12}H_{18}$

MCDONALD R.A.,SHRADER S.A.,STULL D.R.:
J.CHEM.ENG.DATA 4,311(1959).

A = 6.99332 B = 1663.880 C = 194.414

B.P.(760) = 210.176

T	P EXPTL.	P CALCD.	DEV.	PERCENT
120.260	50.65	50.78	0.13	0.25
136.500	92.54	92.30	-0.25	-0.26
158.710	190.98	191.17	0.19	0.10
183.590	390.47	390.45	-0.02	-0.01
208.420	729.63	729.25	-0.38	-0.05
210.490	765.78	765.59	-0.19	-0.03
211.580	784.76	785.29	0.53	0.07

STANDARD DEVIATION = 0.38

2,4-DIISOPROPYLPHENOL C12H18O

VONTERRES E.,ET AL.: BRENNSTOFF CHEM. 36,272(1955).

A = 6.71400 B = 1506.633 C = 138.414

B.P.(760) = 254.636

T	P EXPTL.	P CALCD.	DEV.	PERCENT
122.000	10.00	8.48	-1.52	-15.19
143.000	25.00	22.92	-2.08	-8.32
161.200	50.00	48.46	-1.54	-3.07
172.900	75.00	74.89	-0.11	-0.15
180.500	100.00	97.66	-2.34	-2.34
195.000	150.00	156.74	6.74	4.50
203.000	200.00	200.02	0.02	0.01
* 213.800	250.00	273.14	23.14	9.26
217.000	300.00	298.47	-1.53	-0.51
220.500	325.00	328.28	3.28	1.01
223.200	350.00	352.84	2.84	0.81
228.000	400.00	400.09	0.09	0.02
232.200	450.00	445.41	-4.59	-1.02
237.000	500.00	502.04	2.04	0.41
240.100	550.00	541.51	-8.49	-1.54
245.000	600.00	608.80	8.80	1.47
247.000	650.00	638.07	-11.93	-1.84
251.000	700.00	699.88	-0.13	-0.02
255.000	760.00	766.23	6.23	0.82

STANDARD DEVIATION = 5.40

TERPENYL ACETATE C12H20O2

GARDNER G.S.,BREWER J.E.: IND.ENG.CHEM.29,179(1937).

A = 6.44346 B = 1377.268 C = 143.854

B.P.(10) = 109.160

T	P EXPTL.	P CALCD.	DEV.	PERCENT
37.600	0.10	0.07	-0.03	-28.67
52.500	0.20	0.27	0.07	34.34
55.000	0.30	0.33	0.03	9.73
57.400	0.30	0.40	0.10	32.71
60.800	0.50	0.52	0.02	3.45
64.200	0.80	0.67	-0.13	-16.71
74.100	1.40	1.33	-0.07	-4.88
82.800	2.30	2.33	0.03	1.21
102.300	7.00	7.05	0.05	0.74
119.800	16.60	16.58	-0.02	-0.10
134.500	31.30	31.30	0.00	0.00
150.900	59.00	59.00	0.00	0.00

STANDARD DEVIATION = 0.07

1-DODECENE C12H24

FORZIATI A.F.,CAMIN D.L.,ROSSINI F.D.:
J.RESEARCH NATL.BUR.STANDARDS 45,406(1950).

A = 6.97904 B = 1622.578 C = 182.567

B.P.(760) = 213.354

T	P EXPTL.	P CALCD.	DEV.	PERCENT
123.703	48.02	47.99	-0.03	-0.06
128.424	57.78	57.76	-0.02	-0.04
136.258	77.59	77.59	0.00	0.00
139.736	88.03	88.05	0.02	0.02
144.428	103.97	103.98	0.01	0.01
149.773	124.95	124.96	0.01	0.01
155.208	149.72	149.73	0.01	0.01
160.266	176.24	176.28	0.04	0.02
167.019	217.54	217.58	0.04	0.02
173.214	262.14	262.08	-0.06	-0.02
180.699	325.41	325.39	-0.02	-0.01
188.406	402.97	402.90	-0.07	-0.02
196.624	501.21	501.18	-0.03	-0.01
205.542	628.45	628.52	0.07	0.01
211.823	732.62	732.65	0.03	0.00
212.497	744.57	744.59	0.02	0.00
213.125	755.81	755.85	0.04	0.01
213.826	768.75	768.58	-0.17	-0.02
214.472	780.36	780.45	0.09	0.01

STANDARD DEVIATION = 0.06

TRIISOBUTYLENE C12H24

STEVENS D.R.:
IND.ENG.CHEM. 35,655(1943).

A = 7.00207 B = 1613.467 C = 212.500

B.P.(760) = 178.999

T	P EXPTL.	P CALCD.	DEV.	PERCENT
56.500	10.00	10.09	0.09	0.94
70.000	20.00	19.53	-0.47	-2.35
92.000	50.00	50.50	0.50	1.01
110.000	100.00	99.79	-0.21	-0.21
179.000	760.00	760.02	0.02	0.00

STANDARD DEVIATION = 0.52

LAURIC ACID $C_{12}H_{24}O_2$

HAMMER E.,LYDERSEN A.L.: CHEM.ENG.SCI. 7,66(1957).

A = 7.86081 B = 2159.111 C = 143.226

B.P.(10) = 171.476

T	P EXPTL.	P CALCD.	DEV.	PERCENT
106.370	0.17	0.16	0.00	-2.80
111.500	0.25	0.24	-0.01	-3.02
117.370	0.38	0.38	0.00	-0.19
122.200	0.54	0.53	-0.01	-1.76
130.950	0.96	0.97	0.01	1.26
136.760	1.43	1.41	-0.02	-1.38
142.300	1.97	1.99	0.02	1.04
149.230	3.02	3.01	-0.01	-0.43
157.830	4.86	4.89	0.03	0.55
164.050	6.85	6.83	-0.02	-0.34
171.050	9.78	9.79	0.01	0.08
175.850	12.42	12.42	0.00	-0.02

STANDARD DEVIATION = 0.02

1-CHLORODODECANE $C_{12}H_{25}CL$

KEMME R.H.,KREPS S.I.: J.CHEM.ENG.DATA 14,98(1969).

A = 6.83408 B = 1654.824 C = 155.090

B.P.(100) = 187.234

T	P EXPTL.	P CALCD.	DEV.	PERCENT
116.000	5.50	5.37	-0.13	-2.41
116.100	5.60	5.40	-0.21	-3.66
128.100	10.00	9.79	-0.21	-2.15
137.100	15.10	14.81	-0.29	-1.92
143.900	20.10	19.92	-0.18	-0.88
153.700	29.90	29.86	-0.04	-0.15
161.300	40.00	40.16	0.16	0.39
170.100	55.30	55.63	0.33	0.59
180.900	80.30	81.07	0.77	0.96
193.000	120.60	120.25	-0.35	-0.29
210.500	204.20	203.06	-1.14	-0.56
224.900	300.50	301.42	0.92	0.30
245.500	504.90	504.79	-0.11	-0.02

STANDARD DEVIATION = 0.57

DODECANE

C12H26

WILLINGHAM C.J.,TAYLOR W.J.,PIGNOCCO J.M.,ROSSINI F.D.:
J.RESEARCH NATL.BUR.STANDARDS 35,219(1945).

A = 6.98291 B = 1627.714 C = 180.521

B.P.(760) = 216.279

T	P EXPTL.	P CALCD.	DEV.	PERCENT
126.381	47.74	47.78	0.04	0.08
131.108	57.48	57.50	0.02	0.04
135.223	67.29	67.26	-0.03	-0.05
138.962	77.30	77.28	-0.02	-0.02
142.444	87.73	87.70	-0.03	-0.03
147.152	103.63	103.61	-0.02	-0.01
152.529	124.59	124.63	0.04	0.03
157.986	149.39	149.42	0.03	0.02
163.030	175.84	175.79	-0.05	-0.03
169.814	217.13	217.14	0.01	0.01
176.039	261.74	261.74	0.00	0.00
183.537	325.01	325.00	-0.01	0.00
191.255	402.44	402.44	0.00	0.00
199.488	500.67	500.68	0.01	0.00
208.417	627.81	627.90	0.09	0.01
214.709	732.02	732.01	-0.01	0.00
215.383	743.98	743.92	-0.06	-0.01
216.006	755.10	755.07	-0.03	0.00
216.712	767.95	767.86	-0.09	-0.01
217.345	779.39	779.48	0.09	0.01

STANDARD DEVIATION = 0.05

1-DODECANOL C12H26O

AMBROSE D.,SPRAKE C.H.D.:
J.CHEM.THERMODYNAMICS 2,631(1970).

A = 6.76538 B = 1503.631 C = 122.507

B.P.(760) = 264.571

T	P EXPTL.	P CALCD.	DEV.	PERCENT
152.566	19.89	19.91	0.02	0.10
155.672	22.88	22.91	0.03	0.15
160.334	28.11	28.13	0.02	0.08
164.470	33.54	33.56	0.02	0.06
168.252	39.24	39.26	0.02	0.06
172.645	46.86	46.88	0.02	0.03
177.033	55.66	55.67	0.01	0.01
181.891	66.94	66.94	0.00	0.00
186.687	79.87	79.86	-0.01	-0.02
192.069	96.74	96.72	-0.02	-0.02
197.470	116.50	116.46	-0.04	-0.03
203.454	142.10	142.06	-0.04	-0.03
207.656	162.67	162.62	-0.05	-0.03
213.435	194.81	194.77	-0.04	-0.02
220.139	238.31	238.28	-0.03	-0.01
225.798	280.80	280.79	-0.01	0.00
232.758	341.15	341.16	0.01	0.00
238.464	397.93	397.98	0.05	0.01
245.204	474.34	474.48	0.14	0.03
250.872	547.19	547.38	0.19	0.04
258.046	651.75	651.94	0.19	0.03
263.572	742.71	742.60	-0.11	-0.01
270.750	875.01	874.67	-0.34	-0.04
* 276.516	994.18	993.34	-0.84	-0.08

STANDARD DEVIATION = 0.11

BUTYL BORATE C12H27BO3

CHRISTOPHER P.M.,SHILMAN A.: J.CHEM.ENG.DATA 12,333(1967).

A = 7.40687 B = 1905.035 C = 186.134

B.P.(100) = 166.202

T	P EXPTL.	P CALCD.	DEV.	PERCENT
116.800	14.60	13.13	-1.47	-10.07
158.800	76.60	76.56	-0.04	-0.06
172.000	121.70	122.33	0.63	0.52
178.200	150.70	150.68	-0.02	-0.01
184.100	185.20	182.55	-2.65	-1.43
191.200	226.70	228.14	1.44	0.64
198.600	282.20	285.30	3.10	1.10
203.200	329.20	326.45	-2.75	-0.84
210.300	392.20	399.43	7.23	1.84
217.700	496.40	489.22	-7.18	-1.45

STANDARD DEVIATION = 4.36

ISOBUTYL BORATE C12H27BO3

CHRISTOPHER P.M.,SHILMAN A.: J.CHEM.ENG.DATA 12,333(1967).

A = 7.19710 B = 1745.774 C = 192.632

B.P.(100) = 143.281

T	P EXPTL.	P CALCD.	DEV.	PERCENT
99.300	18.50	16.48	-2.02	-10.90
106.600	24.50	23.06	-1.44	-5.87
114.600	33.50	32.72	-0.78	-2.32
126.000	50.50	52.26	1.76	3.48
142.300	95.50	96.55	1.05	1.10
155.900	153.50	154.23	0.73	0.47
166.000	210.50	213.42	2.92	1.39
173.100	266.50	265.29	-1.21	-0.45
179.500	324.50	320.50	-4.00	-1.23
185.500	380.50	380.42	-0.08	-0.02
190.200	436.50	433.45	-3.05	-0.70
195.000	492.50	493.64	1.14	0.23
199.200	548.50	551.67	3.17	0.58

STANDARD DEVIATION = 2.41

HEXAETHYLCYCLOTRISILOXANE C12H30O3SI3

JENKINS A.C.,CHAMBERS G.F.: IND.ENG.CHEM. 46,2367(1954).

A = 6.29021 B = 1212.594 C = 105.340

B.P.(100) = 177.302

T	P EXPTL.	P CALCD.	DEV.	PERCENT
161.500	55.60	55.71	0.11	0.20
186.900	138.60	138.33	-0.27	-0.20
213.000	302.30	302.76	0.46	0.15
223.900	404.90	404.76	-0.14	-0.03
232.300	500.00	499.84	-0.16	-0.03
242.800	641.40	641.42	0.02	0.00

STANDARD DEVIATION = 0.34

FLUORENE C13H10

MORTIMER F.S.,MURPHY R.V.: IND.ENG.CHEM. 15,1140(1923).

A = 7.76176 B = 2637.095 C = 243.190

B.P.(760) = 297.093

T	P EXPTL.	P CALCD.	DEV.	PERCENT
161.000	18.00	17.27	-0.73	-4.04
202.500	69.50	69.96	0.46	0.67
203.000	70.50	71.04	0.54	0.77
240.400	205.50	203.51	-1.99	-0.97
241.400	208.40	208.86	0.46	0.22
276.600	487.50	487.95	0.45	0.09
277.100	490.70	493.46	2.76	0.56
295.600	738.80	736.68	-2.12	-0.29
295.700	738.80	738.23	-0.57	-0.08
295.700	738.80	738.23	-0.57	-0.08
299.800	804.30	803.78	-0.52	-0.06
300.400	812.30	813.77	1.47	0.18

STANDARD DEVIATION = 1.51

BENZOPHENONE

C13H10O

DREISBACH R.R.,SHRADER S.A.:

IND.ENG.CHEM. 41,2879(1949).

A = 7.16294 B = 2051.855 C = 173.074

B.P.(760) = 306.093

T	P EXPTL.	P CALCD.	DEV.	PERCENT
200.500	47.16	46.82	-0.34	-0.72
206.420	57.04	57.03	-0.01	-0.01
211.160	66.39	66.50	0.11	0.17
215.460	75.86	76.19	0.33	0.44
231.570	123.76	123.64	-0.12	-0.09
266.870	315.52	315.53	0.01	0.00
287.230	507.50	507.37	-0.13	-0.02
306.100	760.00	760.10	0.10	0.01

STANDARD DEVIATION = 0.24

DIPHENYLMETHANE

C13H12

CRAFTS J.M.:

BER. 13,105(1915).

A = 6.29100 B = 1260.512 C = 105.269

B.P.(760) = 264.361

T	P EXPTL.	P CALCD.	DEV.	PERCENT
217.540	242.49	243.32	0.83	0.34
231.820	360.75	356.13	-4.62	-1.28
260.150	696.53	694.24	-2.29	-0.33
261.820	722.03	719.78	-2.25	-0.31
262.830	726.29	735.57	9.28	1.28
265.090	766.86	771.82	4.96	0.65
267.040	797.33	804.16	6.83	0.86
271.700	888.80	885.52	-3.28	-0.37
282.240	1101.79	1091.81	-9.98	-0.91

STANDARD DEVIATION = 7.07

4,4'-DIAMINODIPHENYLMETHANE C13H14N2

ZALIKIN A.A.,STREPIKHEEV YU.A.: ZH.PRIKL.KHIM. 39,2607(1966).

A = 3.17231 B = 210.490 C = -137.414

B.P.(10) = 234.311

T	P EXPTL.	P CALCD.	DEV.	PERCENT
* 198.000	2.00	0.50	-1.50	-75.05
213.000	3.50	2.44	-1.06	-30.26
218.000	5.00	3.63	-1.37	-27.33
224.000	6.00	5.51	-0.49	-8.13
231.000	8.00	8.38	0.38	4.73
234.000	10.00	9.84	-0.16	-1.60
239.000	14.00	12.60	-1.40	-10.02
246.000	17.00	17.13	0.13	0.79
247.500	18.00	18.21	0.21	1.16
249.000	19.50	19.32	-0.18	-0.93
252.000	20.50	21.64	1.14	5.58
255.000	22.50	24.11	1.61	7.16
254.000	23.00	23.27	0.27	1.19
258.500	26.00	27.16	1.16	4.47
259.000	28.00	27.61	-0.39	-1.38
262.000	29.00	30.39	1.39	4.81
265.000	33.00	33.31	0.31	0.93
266.000	34.00	34.30	0.30	0.90
268.500	37.00	36.86	-0.14	-0.38
272.000	43.00	40.58	-2.42	-5.62

STANDARD DEVIATION = 1.09

1-TRIDECENE C13H26

CAMIN D.L.,ROSSINI F.D.: J.PHYS.CHEM. 59,1173(1955).

A = 6.98167 B = 1671.743 C = 174.876

B.P.(760) = 232.780

T	P EXPTL.	P CALCD.	DEV.	PERCENT
142.603	51.91	52.00	0.09	0.17
147.442	62.44	62.38	-0.06	-0.09
160.170	98.22	98.19	-0.03	-0.03
184.559	214.16	214.12	-0.04	-0.02
198.998	323.78	323.79	0.01	0.00
207.152	403.30	403.37	0.07	0.02
224.865	630.35	630.39	0.04	0.01
231.393	735.83	735.88	0.05	0.01
232.663	758.08	757.93	-0.15	-0.02

STANDARD DEVIATION = 0.09

METHYL LAURATE

$C_{13}H_{26}O_2$

ROSE A.,SUPINA W.R.: J.CHEM.ENG.DATA6,173(1961).

A = 6.76713 B = 1589.718 C = 140.456

B.P.(100) = 193.019

T	P EXPTL.	P CALCD.	DEV.	PERCENT
157.800	27.40	27.36	-0.04	-0.15
167.400	40.10	40.11	0.01	0.03
174.800	53.10	53.03	-0.07	-0.14
177.800	59.10	59.16	0.06	0.10
181.100	66.40	66.57	0.17	0.26
184.000	74.20	73.70	-0.50	-0.67
187.400	82.70	82.85	0.15	0.18
191.000	93.40	93.53	0.13	0.14
194.200	103.70	103.95	0.25	0.24
198.100	117.90	117.91	0.01	0.01
204.500	144.40	144.10	-0.30	-0.21
208.100	160.70	160.79	0.09	0.06
212.300	182.20	182.21	0.01	0.00

STANDARD DEVIATION = 0.22

METHYL LAURATE

$C_{13}H_{26}O_2$

ALTHOUSE P.M.,TRIEBOLD H.O.: IND.ENG.CHEM.,ANAL.ED. 16,605(1944).

A = 8.43575 B = 2786.743 C = 241.739

B.P.(10) = 133.037

T	P EXPTL.	P CALCD.	DEV.	PERCENT
100.000	2.00	1.91	-0.09	-4.47
113.000	4.00	3.80	-0.20	-4.95
121.000	6.00	5.67	-0.33	-5.57
128.000	8.00	7.92	-0.08	-1.00
134.000	10.00	10.45	0.45	4.49
149.000	20.00	20.13	0.13	0.64
166.000	40.00	39.91	-0.09	-0.21

STANDARD DEVIATION = 0.31

PROPYL CAPRATE C13H26O2

BONHORST C.W.,ALTHOUSE P.M.,TRIEBOLD H.O.:
IND.ENG.CHEM. 40,2379(1948).

A = 8.70122 B = 2943.990 C = 253.654

B.P.(100) = 185.687

T	P EXPTL.	P CALCD.	DEV.	PERCENT
96.800	2.00	2.00	0.00	-0.18
109.700	4.00	3.97	-0.03	-0.81
114.200	5.00	4.98	-0.02	-0.31
117.800	6.00	5.96	-0.04	-0.67
123.700	8.00	7.93	-0.07	-0.91
128.500	10.00	9.93	-0.07	-0.66
144.300	20.00	20.09	0.09	0.46
161.200	40.00	40.22	0.22	0.55
166.600	50.00	49.62	-0.38	-0.77
171.600	60.00	59.98	-0.02	-0.03
179.600	80.00	80.51	0.51	0.64
185.600	100.00	99.69	-0.31	-0.31

STANDARD DEVIATION = 0.25

ISOPROPYL CAPRATE C13H26O2

BONHORST C.W.,ALTHOUSE P.M.,TRIEBOLD H.O.:
IND.ENG.CHEM. 40,2379(1948).

A = 9.95910 B = 4013.900 C = 326.496

B.P.(100) = 177.820

T	P EXPTL.	P CALCD.	DEV.	PERCENT
90.000	2.00	2.10	0.10	4.90
102.500	4.00	4.01	0.01	0.13
106.800	5.00	4.96	-0.04	-0.81
110.500	6.00	5.94	-0.06	-0.98
116.500	8.00	7.91	-0.09	-1.10
121.100	10.00	9.80	-0.20	-1.96
137.000	20.00	19.91	-0.09	-0.46
153.600	40.00	39.67	-0.33	-0.82
159.400	50.00	49.92	-0.08	-0.16
164.400	60.00	60.59	0.59	0.99
172.000	80.00	80.74	0.74	0.92
177.600	100.00	99.20	-0.80	-0.80

STANDARD DEVIATION = 0.44

N-TRIDECANE

C13H28

CAMIN D.L.,ROSSINI F.D.:

J.PHYS.CHEM.59,1173(1955).

A = 7.00925 B = 1693.684 C = 174.815

B.P.(760) = 235.434

T	P EXPTL.	P CALCD.	DEV.	PERCENT
139.300	41.48	41.43	-0.05	-0.12
145.160	51.99	52.01	0.02	0.03
150.011	62.45	62.39	-0.06	-0.10
157.603	82.01	82.07	0.06	0.08
162.749	98.10	98.15	0.05	0.05
174.699	145.63	145.68	0.05	0.04
187.176	214.10	214.01	-0.09	-0.04
201.634	323.69	323.69	0.00	0.00
209.788	403.29	403.20	-0.09	-0.02
218.367	503.00	503.05	0.05	0.01
227.524	630.35	630.44	0.09	0.02
234.052	735.86	735.97	0.11	0.01
235.316	757.97	757.92	-0.05	-0.01
236.065	771.28	771.17	-0.11	-0.01

STANDARD DEVIATION = 0.08

ANTRACENE C14H10

MORTIMER F.S.,MURPHY R.V.:
IND.ENG.CHEM. 15,1140(1923).

A = 7.37609 B = 2518.815 C = 218.509

B.P.(760) = 341.815

T	P EXPTL.	P CALCD.	DEV.	PERCENT
223.200	48.00	47.17	-0.83	-1.73
228.000	55.00	54.32	-0.68	-1.24
244.400	85.90	86.06	0.16	0.19
244.600	86.50	86.53	0.03	0.03
259.400	127.30	127.53	0.23	0.18
259.800	128.30	128.83	0.53	0.41
260.300	130.20	130.47	0.27	0.21
282.000	219.60	220.59	0.99	0.45
282.100	219.80	221.10	1.30	0.59
299.900	330.40	329.12	-1.28	-0.39
300.000	331.30	329.83	-1.47	-0.44
300.600	334.20	334.12	-0.08	-0.02
312.800	432.90	431.83	-1.07	-0.25
313.200	436.20	435.40	-0.80	-0.18
313.400	437.00	437.18	0.18	0.04
327.400	577.40	578.24	0.84	0.15
327.900	583.40	583.89	0.49	0.08
328.000	584.00	585.02	1.02	0.18
340.600	742.20	743.08	0.88	0.12
340.500	742.20	741.71	-0.49	-0.07
340.500	742.20	741.71	-0.49	-0.07

STANDARD DEVIATION = 0.85

PHENANTHRENE C14H10

MORTIMER F.S.,MURPHY R.V.:
IND.ENG.CHEM. 15,1140(1923).

A = 7.47774 B = 2581.568 C = 223.199

B.P.(760) = 338.386

T	P EXPTL.	P CALCD.	DEV.	PERCENT
203.600	27.20	26.86	-0.34	-1.26
233.800	67.50	67.42	-0.08	-0.12
246.000	94.30	94.55	0.25	0.27
246.100	94.60	94.81	0.21	0.22
271.500	182.90	181.67	-1.23	-0.67
271.500	183.00	181.67	-1.33	-0.73
293.100	299.70	300.33	0.63	0.21
293.200	299.80	301.00	1.20	0.40
293.200	299.90	301.00	1.10	0.37
306.400	399.20	401.02	1.82	0.46
306.500	399.70	401.87	2.17	0.54
324.500	584.70	581.11	-3.59	-0.61
324.900	588.70	585.73	-2.97	-0.50
325.400	592.60	591.55	-1.05	-0.18
337.100	741.00	741.74	0.74	0.10
337.100	741.00	741.74	0.74	0.10
345.100	864.40	861.22	-3.18	-0.37
345.700	870.90	870.77	-0.13	-0.01
346.800	883.90	888.51	4.61	0.52

STANDARD DEVIATION = 2.09

C14H14O

O-PHENYLETHYLPHENOL

GOLDBLUM K.B.,MARTIN R.W.,YOUNG R.B.:
IND.ENG.CHEM. 39,1474(1947).

A = 4.50602 B = 516.763 C = -32.058

B.P.(100) = 238.267

T	P EXPTL.	P CALCD.	DEV.	PERCENT
169.200	6.10	5.47	-0.63	-10.34
177.200	9.30	8.82	-0.48	-5.13
179.100	10.40	9.81	-0.59	-5.68
183.200	12.90	12.22	-0.68	-5.29
187.600	14.90	15.26	0.36	2.45
193.100	20.10	19.82	-0.28	-1.38
195.600	23.70	22.19	-1.51	-6.36
196.800	22.80	23.40	0.60	2.63
202.400	31.50	29.67	-1.83	-5.80
211.600	42.70	42.44	-0.26	-0.60
212.200	43.50	43.39	-0.11	-0.26
217.400	51.70	52.22	0.52	1.01
217.600	52.30	52.59	0.29	0.55
223.800	63.80	64.70	0.90	1.42
224.300	65.60	65.76	0.16	0.24
225.000	65.00	67.25	2.25	3.46
227.600	74.80	73.00	-1.80	-2.41
228.200	76.20	74.37	-1.83	-2.40
234.600	87.70	90.08	2.38	2.72
234.800	91.60	90.60	-1.00	-1.09
235.400	92.60	92.19	-0.41	-0.45
235.600	88.90	92.72	3.82	4.30
239.600	103.00	103.78	0.78	0.75
239.700	102.40	104.06	1.66	1.62
241.900	109.80	110.51	0.71	0.64
242.100	111.60	111.11	-0.49	-0.44
247.200	128.10	127.07	-1.03	-0.80
247.600	128.20	128.39	0.19	0.14
250.300	140.90	137.46	-3.44	-2.44

STANDARD DEVIATION = 1.51

P-PHENYLETHYLPHENOL

C14H14O

GOLDBLUM K.B.,MARTIN R.W.,YOUNG R.B.:
IND.ENG.CHEM. 39,1474(1947).

A = 4.30413 B = 459.321 C = -52.425

B.P.(10) = 191.439

T	P EXPTL.	P CALCD.	DEV.	PERCENT
174.400	4.20	3.46	-0.75	-17.74
176.000	4.80	3.87	-0.93	-19.47
194.000	12.00	11.48	-0.52	-4.37
199.800	16.40	15.40	-1.00	-6.11
200.300	17.00	15.78	-1.22	-7.20
201.600	17.10	16.79	-0.31	-1.81
215.400	31.60	30.60	-1.00	-3.15
216.000	31.50	31.34	-0.16	-0.50
216.500	32.10	31.97	-0.13	-0.42
217.000	32.50	32.60	0.10	0.30
223.800	42.40	42.07	-0.33	-0.79
228.900	50.20	50.28	0.08	0.16
229.300	50.60	50.96	0.36	0.72
238.200	63.70	67.87	4.17	6.55
238.400	64.10	68.29	4.19	6.53
241.800	74.70	75.63	0.93	1.24
242.000	75.70	76.07	0.37	0.49
243.400	79.70	79.25	-0.45	-0.56
243.700	80.10	79.94	-0.16	-0.20
246.200	86.70	85.85	-0.85	-0.98
246.500	87.10	86.58	-0.52	-0.60
250.400	98.10	96.39	-1.71	-1.74
250.500	98.90	96.65	-2.25	-2.27

STANDARD DEVIATION = 1.59

2-(2-BIPHENYLYLOXY)-ETHANOL

C14H14O2

MCDONALD R.A.,SHRADER S.A.,STULL D.R.:
J.CHEM.ENG.DATA 4,311(1959).

A = 8.00587 B = 2776.761 C = 206.914

B.P.(100) = 255.427

T	P EXPTL.	P CALCD.	DEV.	PERCENT
240.600	63.86	63.24	-0.62	-0.97
245.600	73.84	74.06	0.22	0.30
249.650	83.61	83.95	0.34	0.40
253.220	93.34	93.58	0.24	0.26
267.170	141.07	140.85	-0.22	-0.15
299.540	333.52	333.52	0.00	0.00

STANDARD DEVIATION = 0.46

P-(1,1,3,3-TETRAMETHYLBUTYL)-PHENOL C14H22O

MCDONALD R.A.,SHRADER S.A.,STULL D.R.:
J.CHEM.ENG.DATA 4,311(1959).

A = 6.94940 B = 1785.731 C = 148.497

B.P.(760) = 290.410

T	P EXPTL.	P CALCD.	DEV.	PERCENT
199.320	65.31	65.36	0.05	0.07
203.510	75.44	75.23	-0.21	-0.28
207.500	85.71	85.75	0.04	0.05
210.810	95.27	95.38	0.11	0.12
256.340	345.41	345.46	0.05	0.02
274.890	539.01	539.09	0.08	0.02
288.570	731.94	730.59	-1.35	-0.18
290.490	760.79	761.28	0.49	0.06
291.620	779.11	779.81	0.70	0.09

STANDARD DEVIATION = 0.66

1-TETRADECENE C14H28

CAMIN D.L.,ROSSINI F.D.:
J.PHYS.CHEM. 59,1173(1955).

A = 7.02170 B = 1746.331 C = 170.627

B.P.(760) = 251.102

T	P EXPTL.	P CALCD.	DEV.	PERCENT
158.464	51.91	51.90	-0.01	-0.02
163.516	62.44	62.43	-0.01	-0.01
188.869	145.81	145.87	0.06	0.04
201.642	214.15	214.12	-0.03	-0.02
216.455	323.78	323.72	-0.06	-0.02
224.822	403.31	403.30	-0.01	0.00
242.996	630.35	630.47	0.12	0.02
249.689	735.82	736.02	0.20	0.03
250.984	758.09	757.98	-0.11	-0.02
251.750	771.36	771.20	-0.16	-0.02

STANDARD DEVIATION = 0.12

MYRISTIC ACID

$C_{14}H_{28}O_2$

HAMMER E.,LYDERSEN A.L.:

CHEM.ENG.SCI. 7,66(1957).

A = 7.98072 B = 2333.688 C = 142.494

B.P.(10) = 191.811

T	P EXPTL.	P CALCD.	DEV.	PERCENT
* 111.880	0.11	0.06	-0.04	-39.00
124.750	0.17	0.18	0.00	2.39
129.300	0.25	0.25	0.00	-0.79
133.680	0.34	0.34	0.00	1.30
139.900	0.52	0.52	0.00	0.18
145.400	0.75	0.75	0.00	-0.09
149.580	0.97	0.98	0.01	0.90
149.600	1.00	0.98	-0.02	-2.00
156.250	1.47	1.48	0.01	0.40
161.750	2.03	2.04	0.01	0.64
162.500	2.15	2.13	-0.02	-0.76
169.800	3.22	3.22	0.00	0.03
178.350	5.10	5.09	-0.01	-0.10
185.100	7.20	7.19	-0.01	-0.08
185.500	7.32	7.34	0.02	0.27
* 185.820	7.26	7.46	0.20	2.73
191.900	10.05	10.04	-0.01	-0.07

STANDARD DEVIATION = 0.01

1-CHLOROTETRADECANE C14H29CL

KEMME R.H.,KREPS S.I.: J.CHEM.ENG.DATA 14,98(1969).

A = 7.20070 B = 2018.879 C = 170.591

B.P.(760) = 296.755

T	P EXPTL.	P CALCD.	DEV.	PERCENT
141.700	5.20	5.44	0.24	4.70
155.100	9.90	10.04	0.14	1.46
164.600	14.90	15.05	0.15	1.03
172.200	20.40	20.47	0.07	0.35
181.800	29.50	29.62	0.12	0.41
190.700	41.00	41.00	0.00	-0.01
199.100	55.20	54.92	-0.28	-0.51
210.500	80.00	79.99	-0.01	-0.01
223.800	122.40	120.70	-1.70	-1.39
242.600	203.80	206.36	2.56	1.26
258.200	311.20	310.73	-0.47	-0.15
277.700	498.40	497.96	-0.44	-0.09
297.000	764.20	763.97	-0.23	-0.03

STANDARD DEVIATION = 1.01

TETRADECANE C14H30

CAMIN D.L.,ROSSINI F.D.: J.PHYS.CHEM. 59,1173(1955).

A = 7.02216 B = 1747.452 C = 168.437

B.P.(760) = 253.516

T	P EXPTL.	P CALCD.	DEV.	PERCENT
154.860	41.49	41.41	-0.08	-0.20
165.911	62.44	62.47	0.03	0.06
173.637	81.97	81.99	0.02	0.02
191.234	145.64	145.77	0.13	0.09
204.019	214.11	214.02	-0.09	-0.04
218.840	323.71	323.59	-0.12	-0.04
236.013	503.00	503.02	0.02	0.00
245.408	630.36	630.48	0.12	0.02
252.104	735.86	736.04	0.18	0.03
253.401	758.02	758.02	0.00	0.00
254.165	771.44	771.20	-0.24	-0.03

STANDARD DEVIATION = 0.14

1-TETRADECANOL $C_{14}H_{30}O$

KEMME R.H., KREPS S.I.:

J.CHEM.ENG.DATA 14,98(1969).

A = 6.74820 B = 1588.831 C = 114.603

B.P.(760) = 296.225

T	P EXPTL.	P CALCD.	DEV.	PERCENT
151.600	5.20	6.02	0.82	15.79
163.000	10.50	10.59	0.09	0.83
171.000	15.30	15.31	0.01	0.10
177.000	19.90	19.93	0.03	0.17
188.400	32.10	31.96	-0.14	-0.44
199.100	48.40	48.24	-0.16	-0.33
213.300	80.10	79.94	-0.16	-0.20
225.700	120.00	120.04	0.04	0.03
243.100	202.80	202.50	-0.30	-0.15
257.800	302.60	303.22	0.62	0.21
277.900	501.20	501.47	0.27	0.05
295.900	755.20	754.64	-0.56	-0.07

STANDARD DEVIATION = 0.42

BISPENTAMETHYLDISILANOXYDISILANE $C_{14}H_{42}O_2Si_6$

CRAIG A.D., ET AL.:

J.CHEM.SOC. 1962,548.

A = 8.55664 B = 3051.316 C = 258.850

B.P.(10) = 144.942

T	P EXPTL.	P CALCD.	DEV.	PERCENT
168.900	26.50	26.50	0.00	0.00
176.800	35.70	35.69	-0.01	-0.02
179.000	38.70	38.71	0.01	0.02
186.500	50.70	50.72	0.02	0.03
191.000	59.40	59.39	-0.01	-0.02
201.000	83.40	83.40	0.00	0.00

STANDARD DEVIATION = 0.01

DIPHENYLMETHANE-4,4'-DIISOCYANATE C15H10N2O2

ZALIKIN A.A.,STREPIKHEEV YU.A.: ZH.PRIKL.KHIM. 39,2607(1966).

A = 3.56416 B = 265.914 C = -105.591

B.P.(10) = 209.295

T	P EXPTL.	P CALCD.	DEV.	PERCENT
* 169.000	2.00	0.23	-1.77	-88.26
197.000	6.00	4.52	-1.48	-24.67
200.000	7.00	5.59	-1.41	-20.12
208.000	10.00	9.28	-0.72	-7.20
211.000	11.00	11.00	0.00	0.02
216.000	14.00	14.31	0.31	2.23
217.000	15.00	15.04	0.04	0.29
219.000	17.00	16.57	-0.43	-2.51
222.000	19.00	19.05	0.05	0.25
223.000	21.00	19.92	-1.08	-5.14
227.000	23.00	23.66	0.66	2.85
229.000	26.00	25.67	-0.33	-1.27
233.000	30.00	30.00	0.00	-0.01
241.000	38.00	39.84	1.84	4.85
247.000	45.00	48.27	3.27	7.27
249.000	49.00	51.28	2.28	4.65
256.000	64.00	62.55	-1.45	-2.27
257.000	67.00	64.25	-2.75	-4.10

STANDARD DEVIATION = 1.60

DECYLCYCLOPENTANE $C_{15}H_{30}$

CAMIN D.L.,FORZIATI A.F.,ROSSINI F.D.:
J.PHYS.CHEM. 58,440(1954).

A = 6.99912 B = 1822.046 C = 163.051

B.P.(760) = 279.375

T	P EXPTL.	P CALCD.	DEV.	PERCENT
181.867	52.01	52.07	0.06	0.11
191.085	71.48	71.46	-0.02	-0.03
200.861	98.25	98.24	-0.01	-0.01
213.782	145.93	145.87	-0.06	-0.04
219.377	171.73	171.67	-0.06	-0.03
227.251	214.10	214.20	0.10	0.05
242.859	323.76	323.83	0.07	0.02
251.676	403.47	403.41	-0.06	-0.01
260.943	503.23	503.24	0.01	0.00
277.907	736.33	736.37	0.04	0.01
279.283	758.55	758.49	-0.06	-0.01

STANDARD DEVIATION = 0.07

1-PENTADECENE $C_{15}H_{30}$

CAMIN D.L.,ROSSINI F.D.:
J.PHYS.CHEM. 59,1173(1955).

A = 7.01697 B = 1783.268 C = 162.746

B.P.(760) = 268.395

T	P EXPTL.	P CALCD.	DEV.	PERCENT
173.613	51.91	51.91	0.00	0.01
232.940	323.78	323.74	-0.04	-0.01
241.504	403.34	403.34	0.00	0.00
250.494	503.04	503.09	0.05	0.01
260.094	630.35	630.40	0.05	0.01
266.940	735.79	735.88	0.09	0.01
268.273	758.10	757.95	-0.15	-0.02

STANDARD DEVIATION = 0.09

METHYL MYRISTATE $C_{15}H_{30}O_2$

ROSE A., SUPINA W.R.: J.CHEM.ENG.DATA6,173(1961).

A = 7.62230   B = 2283.952   C = 184.813

B.P.(100) = 221.414

	T	P EXPTL.	P CALCD.	DEV.	PERCENT
	166.000	13.00	12.94	-0.06	-0.47
	171.100	15.90	16.04	0.14	0.88
	179.200	22.20	22.28	0.08	0.38
	179.600	22.70	22.64	-0.06	-0.27
	185.000	27.90	27.95	0.05	0.18
	192.300	36.90	36.81	-0.09	-0.25
	197.100	44.00	43.86	-0.14	-0.32
	202.900	53.90	53.89	-0.01	-0.02
	207.300	62.80	62.75	-0.05	-0.08
	210.500	70.00	69.95	-0.05	-0.08
	214.700	80.30	80.45	0.15	0.18
*	218.800	98.70	91.96	-6.74	-6.83
	221.600	100.40	100.59	0.19	0.19
	225.600	114.10	114.11	0.01	0.01
	228.000	123.10	122.94	-0.16	-0.13
	230.600	133.00	133.14	0.14	0.11
	234.700	150.80	150.68	-0.12	-0.08
	237.800	165.20	165.19	-0.01	-0.01

STANDARD DEVIATION = 0.12

METHYL MYRISTATE $C_{15}H_{30}O_2$

BONHORST C.W., ALTHOUSE P.M., TRIEBOLD H.O.: IND.ENG.CHEM. 40,2379(1948).

A = 8.43942   B = 2953.604   C = 235.254

B.P.(10) = 161.766

T	P EXPTL.	P CALCD.	DEV.	PERCENT
127.100	2.00	1.94	-0.06	-2.90
141.300	4.00	3.94	-0.06	-1.47
146.300	5.00	4.99	-0.01	-0.12
150.300	6.00	6.01	0.01	0.14
156.600	8.00	7.98	-0.02	-0.27
161.800	10.00	10.01	0.01	0.14
178.600	20.00	20.07	0.07	0.36
196.800	40.00	40.11	0.11	0.27
202.800	50.00	49.76	-0.24	-0.48
208.200	60.00	60.11	0.11	0.19

STANDARD DEVIATION = 0.12

PROPYL LAURATE C15H30O2

BONHORST C.W.,ALTHOUSE P.M.,TRIEBOLD H.O.: IND.ENG.CHEM. 40,2379(1948).

A = 8.06888 B = 2692.400 C = 222.494

B.P.(10) = 158.387

T	P EXPTL.	P CALCD.	DEV.	PERCENT
123.700	2.00	1.96	-0.04	-2.12
138.100	4.00	4.00	0.00	0.06
143.000	5.00	5.04	0.04	0.80
147.000	6.00	6.06	0.06	0.93
153.300	8.00	8.02	0.02	0.28
156.800	10.00	9.34	-0.66	-6.58
175.800	20.00	20.37	0.37	1.86
194.000	40.00	40.22	0.22	0.55
200.200	50.00	50.03	0.03	0.07
205.400	60.00	59.80	-0.20	-0.34

STANDARD DEVIATION = 0.31

ISOPROPYL LAURATE C15H30O2

BONHORST C.W.,ALTHOUSE P.M.,TRIEBOLD H.O.: IND.ENG.CHEM. 40,2379(1948).

A = 8.53260 B = 2950.615 C = 240.669

B.P.(10) = 151.044

T	P EXPTL.	P CALCD.	DEV.	PERCENT
117.400	2.00	1.96	-0.04	-2.00
131.300	4.00	3.98	-0.02	-0.43
135.800	5.00	4.95	-0.05	-0.91
139.800	6.00	5.99	-0.01	-0.17
145.800	8.00	7.90	-0.10	-1.21
151.000	10.00	9.98	-0.02	-0.19
167.400	20.00	20.04	0.04	0.20
185.000	40.00	39.89	-0.11	-0.27
191.400	50.00	50.53	0.53	1.06
196.000	60.00	59.63	-0.37	-0.61

STANDARD DEVIATION = 0.25

PENTADECANE C15H32

CAMIN D.L.,ROSSINI F.D.: J.PHYS.CHEM. 59,1173(1955).

A = 7.03121 B = 1795.065 C = 161.891

B.P.(760) = 270.613

T	P EXPTL.	P CALCD.	DEV.	PERCENT
169.686	41.49	41.45	-0.04	-0.10
180.919	62.44	62.36	-0.08	-0.13
188.905	81.96	82.05	0.09	0.11
206.886	145.64	145.74	0.10	0.07
219.982	214.11	214.06	-0.05	-0.03
235.150	323.72	323.67	-0.05	-0.02
252.703	503.00	502.94	-0.06	-0.01
262.310	630.36	630.33	-0.03	0.00
269.164	735.86	735.96	0.10	0.01
270.499	758.08	758.07	-0.01	0.00

STANDARD DEVIATION = 0.08

FLUORANTHENE C16H10

TSYPKINA O.YA.: ZH.PRIKL.KHIM. 28,185(1955).

A = 6.37310 B = 1756.355 C = 118.428

B.P.(760) = 384.496

T	P EXPTL.	P CALCD.	DEV.	PERCENT
197.000	4.90	6.38	1.48	30.24
209.000	7.90	10.21	2.31	29.23
228.500	19.40	20.44	1.04	5.37
238.100	28.40	27.98	-0.42	-1.48
247.700	40.40	37.67	-2.73	-6.76
255.000	49.40	46.75	-2.65	-5.37
261.300	59.40	55.95	-3.45	-5.81
270.900	77.40	72.75	-4.65	-6.00
281.500	100.40	95.81	-4.59	-4.57
305.000	152.40	167.94	15.54	10.20
314.500	205.40	207.10	1.70	0.83
382.900	743.00	740.78	-2.22	-0.30
384.200	760.00	756.40	-3.60	-0.47

STANDARD DEVIATION = 5.83

PYRENE C16H10

TSYPKINA O.YA.:

ZH.PRIKL.KHIM. 28,185(1955).

A = 5.61838 B = 1122.025 C = 15.158

B.P.(760) = 394.704

T	P EXPTL.	P CALCD.	DEV.	PERCENT
200.400	2.60	2.59	-0.01	-0.41
220.800	6.90	7.30	0.40	5.77
242.700	18.30	18.49	0.20	1.07
256.400	32.80	30.66	-2.14	-6.52
270.000	48.70	48.27	-0.44	-0.89
277.000	60.00	59.97	-0.03	-0.06
288.700	81.50	84.29	2.79	3.43
293.000	94.50	94.91	0.41	0.43
306.000	134.50	133.26	-1.24	-0.92
316.000	170.00	169.90	-0.10	-0.06
394.700	760.00	759.95	-0.05	-0.01

STANDARD DEVIATION = 1.35

DIBUTYL PHTHALATE C16H22O4

HAMMER E.,LYDERSEN A.L.:
CHEM.ENG.SCI. 7,66(1957).

A = 6.63980 B = 1744.197 C = 113.685

B.P.(10) = 195.581

T	P EXPTL.	P CALCD.	DEV.	PERCENT
125.700	0.24	0.23	-0.01	-5.93
131.550	0.37	0.34	-0.03	-8.46
139.210	0.55	0.55	0.00	-0.14
147.020	0.90	0.89	-0.01	-1.09
156.700	1.54	1.55	0.01	0.34
158.390	1.71	1.69	-0.02	-0.90
162.120	2.08	2.07	-0.01	-0.53
172.760	3.60	3.55	-0.05	-1.29
180.180	5.03	5.06	0.03	0.66
180.360	5.00	5.11	0.11	2.12
186.870	6.91	6.86	-0.05	-0.67
195.880	10.15	10.13	-0.02	-0.23
202.050	13.04	13.05	0.01	0.06

STANDARD DEVIATION = 0.04

DECYLBENZENE C16H26

CAMIN D.L.,FORZIATI A.F.,ROSSINI F.D.:
J.PHYS.CHEM. 58,440(1954).

A = 7.03596 B = 1903.983 C = 160.330

B.P.(760) = 297.893

T	P EXPTL.	P CALCD.	DEV.	PERCENT
202.987	62.47	62.43	-0.04	-0.06
211.392	82.04	82.01	-0.03	-0.03
217.156	98.17	98.20	0.03	0.03
230.476	145.83	145.89	0.06	0.04
244.331	214.18	214.20	0.02	0.01
260.372	323.82	323.76	-0.06	-0.02
278.950	503.12	503.07	-0.05	-0.01
296.370	736.16	736.13	-0.03	0.00
297.799	758.41	758.50	0.09	0.01

STANDARD DEVIATION = 0.06

DECYLCYCLOHEXANE C16H32

CAMIN D.L.,FORZIATI A.F.,ROSSINI F.D.:
J.PHYS.CHEM. 58,440(1954).

A = 7.01937 B = 1899.329 C = 161.349

B.P.(760) = 297.586

T	P EXPTL.	P CALCD.	DEV.	PERCENT
196.812	52.08	52.04	-0.04	-0.07
206.361	71.50	71.46	-0.04	-0.05
216.465	98.22	98.22	0.00	0.00
229.848	145.88	145.95	0.07	0.04
235.626	171.71	171.74	0.03	0.01
243.758	214.15	214.24	0.09	0.04
259.864	323.79	323.72	-0.07	-0.02
268.976	403.47	403.32	-0.15	-0.04
296.058	736.32	736.18	-0.14	-0.02
297.507	758.50	758.74	0.24	0.03

STANDARD DEVIATION = 0.13

1-HEXADECENE C16H32

CAMIN D.L., FORZIATI A.F., ROSSINI F.D.: J.PHYS.CHEM. 58,440(1954).

A = 7.04354 B = 1842.888 C = 157.839

B.P.(760) = 284.873

T	P EXPTL.	P CALCD.	DEV.	PERCENT
188.152	52.06	52.13	0.07	0.14
197.257	71.56	71.40	-0.16	-0.22
206.981	98.06	98.18	0.12	0.12
219.806	145.81	145.74	-0.07	-0.05
225.397	171.68	171.70	0.02	0.01
233.203	214.17	214.18	0.01	0.00
248.690	323.84	323.82	-0.02	-0.01
257.440	403.38	403.48	0.10	0.02
266.590	503.01	502.91	-0.10	-0.02
283.402	736.10	736.09	-0.01	0.00
284.768	758.24	758.26	0.02	0.00

STANDARD DEVIATION = 0.09

TETRAISOBUTYLENE C16H32

STEVENS D.R.: IND.ENG.CHEM. 35,655(1943).

A = 8.89527 B = 3216.812 C = 299.023

B.P.(100) = 167.501

T	P EXPTL.	P CALCD.	DEV.	PERCENT
108.500	10.00	10.04	0.04	0.39
124.500	20.00	19.95	-0.05	-0.26
148.000	50.00	50.03	0.03	0.05
167.500	100.00	99.99	-0.01	-0.01

STANDARD DEVIATION = 0.07

1-CHLOROHEXADECANE C16H33CL

KEMME R.H.,KREPS S.I.:
J.CHEM.ENG.DATA 14,98(1969).

A = 7.28203 B = 2152.607 C = 162.725

B.P.(760) = 326.368

T	P EXPTL.	P CALCD.	DEV.	PERCENT
165.600	5.10	5.32	0.22	4.26
180.200	10.00	10.11	0.11	1.12
190.300	15.20	15.29	0.09	0.60
197.200	20.00	20.01	0.01	0.07
208.300	30.20	30.22	0.02	0.06
216.100	39.80	39.78	-0.02	-0.04
225.600	54.80	54.79	-0.01	-0.01
237.500	79.90	80.08	0.18	0.23
251.300	122.20	121.01	-1.19	-0.98
270.200	204.60	204.08	-0.52	-0.26
285.100	295.50	298.69	3.19	1.08
306.900	501.50	499.30	-2.20	-0.44
326.800	766.60	766.82	0.22	0.03

STANDARD DEVIATION = 1.30

HEXADECANE C16H34

CAMIN D.L.,FORZIATI A.F.,ROSSINI F.D.:
J.PHYS.CHEM. 58,440(1954).

A = 7.03519 B = 1835.240 C = 154.968

B.P.(760) = 286.792

T	P EXPTL.	P CALCD.	DEV.	PERCENT
190.054	52.09	52.00	-0.09	-0.18
195.301	62.48	62.47	-0.01	-0.02
199.273	71.49	71.52	0.03	0.04
203.437	82.14	82.15	0.01	0.01
208.962	98.24	98.25	0.01	0.01
215.000	118.69	118.75	0.06	0.05
221.780	145.87	145.86	-0.01	-0.01
227.336	171.66	171.68	0.02	0.01
235.145	214.20	214.19	-0.01	0.00
242.432	261.15	261.26	0.11	0.04
250.605	323.81	323.69	-0.12	-0.04
259.336	403.36	403.18	-0.18	-0.05
268.540	503.11	503.22	0.11	0.02
278.333	630.54	630.52	-0.02	0.00
285.337	736.32	736.34	0.02	0.00
286.704	758.50	758.54	0.04	0.00

STANDARD DEVIATION = 0.08

1-HEXADECANOL C16H34O

AMBROSE D.,SPRAKE C.H.S.:
J.CHEM.THERMODYNAMICS 2,631(1970).

A = 7.01338 B = 1863.150 C = 126.508

B.P.(100) = 245.123

T	P EXPTL.	P CALCD.	DEV.	PERCENT
225.309	52.15	52.20	0.05	0.09
231.209	63.81	63.83	0.02	0.02
235.702	74.07	74.06	-0.01	-0.01
243.601	95.38	95.36	-0.02	-0.02
250.105	116.55	116.50	-0.05	-0.05
257.000	143.01	142.97	-0.04	-0.03
261.020	160.59	160.56	-0.03	-0.02
265.535	182.42	182.40	-0.02	-0.01
274.191	230.94	231.04	0.10	0.04
281.076	276.70	276.84	0.14	0.05
289.126	339.53	339.44	-0.09	-0.03
295.904	400.62	400.58	-0.04	-0.01

STANDARD DEVIATION = 0.07

TITANIUM TETRA-TERT-BUTOXIDE C16H3604TI

BRADLEY D.C.,SWANWICK J.D.:

J.CHEM.SOC. 1958,3207.

A = 6.53979 B = 1526.601 C = 166.539

B.P.(10) = 109.031

T	P EXPTL.	P CALCD.	DEV.	PERCENT
49.150	0.30	0.29	-0.01	-3.10
49.650	0.30	0.30	0.00	0.29
51.750	0.36	0.35	-0.01	-2.82
54.550	0.45	0.43	-0.02	-3.92
58.250	0.58	0.56	-0.02	-2.70
58.950	0.59	0.59	0.00	-0.62
61.750	0.71	0.71	0.01	0.74
61.950	0.74	0.72	-0.02	-2.05
65.350	0.91	0.90	0.00	-0.27
72.450	1.41	1.42	0.01	1.01
74.150	1.58	1.57	-0.01	-0.53
76.350	1.79	1.80	0.01	0.52
79.050	2.11	2.11	0.00	-0.13
81.750	2.46	2.46	0.01	0.25
82.450	2.55	2.56	0.02	0.63
84.750	2.92	2.92	0.00	-0.09
85.750	3.07	3.08	0.01	0.35
87.850	3.50	3.46	-0.04	-1.08
88.450	3.60	3.57	-0.02	-0.68
91.950	4.27	4.30	0.03	0.78
95.550	5.16	5.19	0.03	0.53
95.750	5.23	5.24	0.02	0.30
98.550	6.04	6.04	0.00	0.06
109.150	10.11	10.06	-0.06	-0.55
114.550	12.82	12.85	0.03	0.20

STANDARD DEVIATION = 0.02

ZIRCONIUM TETRA-TERT-BUTOXIDE C16H36O4ZR

BRADLEY D.C.,SWANWICK J.D.: J.CHEM.SOC. 1959,748.

A = 6.81225 B = 1583.239 C = 169.009

B.P.(100) = 159.992

T	P EXPTL.	P CALCD.	DEV.	PERCENT
101.650	9.46	9.18	-0.28	-3.00
110.550	14.42	14.09	-0.33	-2.29
119.450	21.67	21.07	-0.60	-2.77
129.150	32.49	31.78	-0.71	-2.17
144.150	58.10	57.09	-1.01	-1.74
145.550	61.10	60.12	-0.98	-1.60
147.050	62.80	63.52	0.72	1.15
150.250	71.00	71.31	0.31	0.44
158.650	94.90	95.56	0.66	0.70
166.750	124.10	124.98	0.88	0.71
174.950	159.70	161.91	2.21	1.38
180.850	192.20	193.59	1.39	0.72
183.850	216.30	211.53	-4.77	-2.21
184.750	215.30	217.16	1.86	0.86
204.450	372.40	373.99	1.59	0.43
207.750	412.30	407.37	-4.93	-1.20
213.150	464.60	467.06	2.46	0.53

STANDARD DEVIATION = 2.25

ZIRCONIUM TETRA-TERT-BUTOXIDE C16H36O4ZR

BRADLEY D.C.,SWANWICK J.D.:

J.CHEM.SOC. 1959,748.

A = 6.87666 B = 1625.200 C = 173.978

B.P.(10) = 102.574

T	P EXPTL.	P CALCD.	DEV.	PERCENT
45.950	0.31	0.31	0.00	-1.01
49.150	0.40	0.39	-0.01	-2.08
51.450	0.46	0.46	0.00	1.04
53.950	0.58	0.56	-0.02	-3.86
54.950	0.60	0.60	0.00	-0.16
56.150	0.63	0.65	0.02	3.55
57.450	0.74	0.71	-0.03	-3.41
58.250	0.79	0.76	-0.03	-4.34
58.750	0.81	0.78	-0.03	-3.42
61.050	0.90	0.92	0.02	1.74
62.350	1.03	1.00	-0.03	-2.96
63.650	1.11	1.09	-0.02	-1.81
64.950	1.16	1.19	0.03	2.36
66.250	1.32	1.29	-0.03	-2.09
67.950	1.42	1.44	0.02	1.55
70.350	1.71	1.68	-0.03	-1.84
70.450	1.66	1.69	0.03	1.75
74.650	2.24	2.19	-0.05	-2.34
75.050	2.20	2.24	0.04	1.87
76.350	2.40	2.42	0.02	0.96
76.850	2.50	2.50	0.00	-0.14
80.450	3.07	3.08	0.01	0.43
80.750	3.12	3.14	0.02	0.54
81.250	3.17	3.23	0.06	1.85
84.450	3.85	3.87	0.02	0.55
87.750	4.78	4.65	-0.13	-2.79
89.850	5.18	5.21	0.03	0.52
92.950	6.17	6.14	-0.03	-0.50
95.550	6.93	7.03	0.10	1.42
97.050	7.65	7.59	-0.06	-0.79
102.250	9.79	9.84	0.05	0.54
106.450	12.10	12.06	-0.04	-0.36
107.950	12.95	12.94	-0.01	-0.05
111.250	15.14	15.09	-0.05	-0.32
114.750	17.60	17.69	0.09	0.53
115.950	18.72	18.67	-0.05	-0.27
121.750	24.05	24.05	0.00	-0.01
122.050	24.28	24.36	0.08	0.32
129.250	32.93	32.88	-0.05	-0.14

STANDARD DEVIATION = 0.05

TETRABUTYLTIN C16H36SN

UMILIN V.A.,TSINOVOI YU.N.,DEVYATYKH G.G.:
ZH.FIZ.KHIM. 42,2320(1968).

A = 6.54534 B = 1648.837 C = 147.704

B.P.(760) = 302.241

T	P EXPTL.	P CALCD.	DEV.	PERCENT
100.700	1.80	0.81	-0.99	-55.09
108.200	2.30	1.27	-1.03	-44.99
119.000	4.00	2.31	-1.69	-42.32
135.500	9.00	5.29	-3.71	-41.25
142.200	10.70	7.21	-3.49	-32.64
145.200	12.50	8.24	-4.26	-34.06
152.000	15.50	11.06	-4.44	-28.64
158.400	18.50	14.41	-4.09	-22.08
166.800	24.00	20.08	-3.92	-16.35
171.400	28.00	23.89	-4.11	-14.67
178.000	33.00	30.41	-2.59	-7.86
183.600	39.30	37.03	-2.27	-5.78
189.300	46.00	44.95	-1.05	-2.29
194.200	53.00	52.82	-0.18	-0.33
199.500	63.00	62.58	-0.42	-0.67
204.000	70.50	71.98	1.48	2.10
208.000	78.00	81.27	3.27	4.19
215.100	95.50	100.15	4.65	4.86
220.500	110.50	116.76	6.26	5.66
229.300	147.00	148.53	1.53	1.04
235.000	166.00	172.56	6.56	3.95
239.600	191.00	194.14	3.14	1.65
243.000	199.00	211.43	12.43	6.25
249.500	250.00	247.87	-2.13	-0.85
256.300	305.00	291.14	-13.87	-4.55
262.500	345.30	335.57	-9.73	-2.82
268.500	381.00	383.47	2.47	0.65
273.000	418.00	422.77	4.77	1.14
275.700	450.00	447.81	-2.19	-0.49
278.500	479.00	474.98	-4.02	-0.84
285.300	545.10	546.30	1.20	0.22
290.600	613.70	607.40	-6.30	-1.03
297.400	695.50	693.34	-2.16	-0.31
300.600	730.00	736.87	6.87	0.94

STANDARD DEVIATION = 5.23

METHYL PALMITATE $C_{17}H_{34}O_2$

BONHORST C.W., ALTHOUSE P.M., TRIEBOLD H.O.:
IND.ENG.CHEM. 40,2379(1948).

A = 9.59441 B = 4146.426 C = 297.761

B.P.(10) = 184.695

T	P EXPTL.	P CALCD.	DEV.	PERCENT
148.500	2.00	2.01	0.01	0.44
163.200	4.00	3.97	-0.03	-0.65
168.400	5.00	5.01	0.01	0.14
172.600	6.00	6.01	0.01	0.19
179.200	8.00	7.96	-0.04	-0.49
184.800	10.00	10.04	0.04	0.43
202.200	20.00	19.99	-0.01	-0.03

STANDARD DEVIATION = 0.03

PROPYL MYRISTATE C17H34O2

BONHORST C.W.,ALTHOUSE P.M.,TRIEBOLD H.O.:
IND.ENG.CHEM. 40,2379(1948).

A = 9.21685 B = 3744.675 C = 272.865

B.P.(10) = 182.866

T	P EXPTL.	P CALCD.	DEV.	PERCENT
147.000	2.00	1.99	-0.01	-0.67
161.700	4.00	3.98	-0.02	-0.52
166.800	5.00	5.01	0.01	0.18
171.000	6.00	6.03	0.03	0.50
177.600	8.00	8.02	0.02	0.20
182.800	10.00	9.97	-0.03	-0.27
200.200	20.00	20.00	0.00	0.01

STANDARD DEVIATION = 0.03

ISOPROPYL MYRISTATE C17H34O2

BONHORST C.W.,ALTHOUSE P.M.,TRIEBOLD H.O.:
IND.ENG.CHEM. 40,2379(1948).

A = 10.41799 B = 4866.480 C = 341.172

B.P.(10) = 175.549

T	P EXPTL.	P CALCD.	DEV.	PERCENT
140.200	2.00	2.03	0.03	1.71
154.400	4.00	3.96	-0.04	-0.91
159.400	5.00	4.97	-0.03	-0.65
163.600	6.00	5.98	-0.02	-0.25
170.400	8.00	8.04	0.04	0.49
175.600	10.00	10.02	0.02	0.21
192.600	20.00	19.99	-0.01	-0.04

STANDARD DEVIATION = 0.04

HEPTADECANE C17H36

KRAFFT F.:

BER. 15,1687(1882).

A = 6.97509 B = 1851.699 C = 149.263

B.P.(760) = 303.002

T	P EXPTL.	P CALCD.	DEV.	PERCENT
163.500	11.00	11.34	0.34	3.10
170.000	15.00	14.97	-0.03	-0.21
187.500	30.00	29.96	-0.04	-0.13
201.500	50.00	49.66	-0.34	-0.67
223.000	100.00	100.21	0.21	0.21
303.000	760.00	759.96	-0.04	0.00

STANDARD DEVIATION = 0.30

DIBUTYL SEBACATE C18H34O4

HAMMER E.,LYDERSEN A.L.:

CHEM.ENG.SCI. 7,66(1957).

A = 7.58766 B = 2364.890 C = 147.541

B.P.(10) = 211.447

T	P EXPTL.	P CALCD.	DEV.	PERCENT
128.450	0.10	0.10	0.00	2.41
134.200	0.15	0.16	0.01	3.48
141.210	0.25	0.25	0.00	-0.08
146.410	0.35	0.35	0.00	-0.37
155.410	0.60	0.60	0.00	0.27
160.040	0.79	0.79	0.00	0.31
164.670	1.03	1.03	0.01	0.53
175.360	1.85	1.84	-0.01	-0.78
186.020	3.16	3.15	-0.01	-0.43
194.980	4.80	4.82	0.02	0.47
208.140	8.69	8.68	-0.01	-0.06

STANDARD DEVIATION = 0.01

OCTADECANE C18H38

KRAFFT F.: BER. 15,1687(1882).

A = 7.14067 B = 2012.745 C = 155.492

B.P.(760) = 316.999

T	P EXPTL.	P CALCD.	DEV.	PERCENT
174.500	11.00	11.00	0.00	-0.02
181.500	15.00	14.72	-0.28	-1.85
200.000	30.00	30.12	0.12	0.39
214.500	50.00	50.20	0.20	0.40
236.000	100.00	99.87	-0.13	-0.13
317.000	760.00	760.01	0.01	0.00

STANDARD DEVIATION = 0.22

1-OCTADECANOL C18H38O

AMBROSE D.,SPRAKE C.H.S.: J.CHEM.THERMODYNAMICS 2,631(1970).

A = 7.10796 B = 2003.695 C = 123.643

B.P.(100) = 268.626

T	P EXPTL.	P CALCD.	DEV.	PERCENT
221.940	20.39	20.41	0.02	0.12
225.430	23.33	23.33	0.00	0.00
229.234	26.89	26.90	0.01	0.04
235.065	33.27	33.27	0.00	0.01
240.038	39.67	39.67	0.00	0.00
245.487	47.90	47.84	-0.06	-0.12
249.846	55.38	55.35	-0.03	-0.05
255.741	67.08	67.07	-0.01	-0.02
261.453	80.28	80.32	0.04	0.06
268.037	98.25	98.24	-0.01	-0.01
273.588	115.79	115.82	0.03	0.03
281.285	144.29	144.44	0.15	0.10
285.362	161.98	161.81	-0.17	-0.11
292.467	196.19	196.18	-0.01	-0.01
301.155	246.09	246.11	0.02	0.01

STANDARD DEVIATION = 0.07

METHYL LINOLATE

$C_{19}H_{34}O_2$

ALTHOUSE P.M.,TRIEBOLD H.O.:
IND.ENG.CHEM.,ANAL.ED. 16,605(1944).

A = 6.11114 B = 1660.097 C = 118.750

B.P.(10) = 206.050

T	P EXPTL.	P CALCD.	DEV.	PERCENT
166.500	2.00	1.96	-0.04	-2.21
182.400	4.00	3.97	-0.03	-0.79
193.000	6.00	6.11	0.11	1.83
199.900	8.00	7.97	-0.03	-0.40
206.000	10.00	9.98	-0.02	-0.18

STANDARD DEVIATION = 0.09

METHYL OLEATE

$C_{19}H_{36}O_2$

ALTHOUSE P.M.,TRIEBOLD H.O.:
IND.ENG.CHEM.,ANAL.ED. 16,605(1944).

A = 7.54412 B = 2656.887 C = 200.659

B.P.(10) = 205.337

T	P EXPTL.	P CALCD.	DEV.	PERCENT
166.200	2.00	2.00	0.00	0.19
182.000	4.00	3.99	-0.01	-0.27
192.000	6.00	5.99	-0.01	-0.10
199.500	8.00	8.03	0.03	0.33
205.300	10.00	9.99	-0.01	-0.14

STANDARD DEVIATION = 0.02

METHYL STEARATE $C_{19}H_{38}O_2$

ROSE A., SUPINA W.R.: J.CHEM.ENG.DATA6,173(1961).

A = 2.35703 B = 68.916 C = -156.556

B.P.(10) = 207.320

T	P EXPTL.	P CALCD.	DEV.	PERCENT
204.100	9.50	8.09	-1.41	-14.81
205.800	10.20	9.08	-1.12	-10.97
208.400	11.10	10.67	-0.43	-3.85
212.100	12.90	13.08	0.18	1.43
216.600	15.30	16.21	0.91	5.92
222.300	19.00	20.38	1.38	7.24
227.300	23.10	24.16	1.06	4.60
229.400	25.10	25.78	0.68	2.69
230.700	26.00	26.78	0.78	3.00
237.900	31.70	32.36	0.66	2.09
239.700	36.20	33.76	-2.44	-6.75

STANDARD DEVIATION = 1.36

PROPYL PALMITATE $C_{19}H_{38}O_2$

BONHORST C.W., ALTHOUSE P.M., TRIEBOLD H.O.: IND.ENG.CHEM. 40,2379(1948).

A = 14.12916 B = 9759.180 C = 539.689

B.P.(10) = 203.631

T	P EXPTL.	P CALCD.	DEV.	PERCENT
166.000	2.00	1.99	-0.01	-0.26
181.600	4.00	3.97	-0.03	-0.71
187.000	5.00	5.01	0.01	0.13
191.400	6.00	6.03	0.03	0.51
198.200	8.00	8.01	0.01	0.06
203.600	10.00	9.99	-0.01	-0.13

STANDARD DEVIATION = 0.03

ISOPROPYL PALMITATE C19H38O2

BONHORST C.W.,ALTHOUSE P.M.,TRIEBOLD H.O.:
IND.ENG.CHEM. 40,2379(1948).

A = 10.91642 B = 5571.977 C = 364.809

B.P.(10) = 197.085

T	P EXPTL.	P CALCD.	DEV.	PERCENT
160.000	2.00	1.99	-0.01	-0.41
175.400	4.00	4.00	0.00	-0.03
180.600	5.00	5.02	0.02	0.30
184.800	6.00	6.00	0.00	0.04
191.600	8.00	7.98	-0.02	-0.20
197.100	10.00	10.01	0.01	0.06

STANDARD DEVIATION = 0.01

TITANIUM TETRA-TERT-AMYLOXIDE C20H44O4TI

BRADLEY D.C., SWANWICK J.D.: J.CHEM.SOC. 1958, 3207.

A = 6.80527 B = 1870.498 C = 164.144

B.P.(10) = 158.063

T	P EXPTL.	P CALCD.	DEV.	PERCENT
88.050	0.25	0.24	-0.01	-2.57
93.050	0.35	0.34	-0.01	-2.61
97.950	0.48	0.47	-0.01	-2.68
104.550	0.69	0.70	0.01	1.15
105.650	0.75	0.75	0.00	-0.52
112.150	1.08	1.08	0.00	0.34
115.350	1.29	1.30	0.01	0.44
115.750	1.32	1.33	0.00	0.34
118.250	1.54	1.52	-0.02	-1.43
118.450	1.53	1.54	0.00	0.29
119.350	1.61	1.61	0.00	0.04
119.350	1.62	1.61	-0.01	-0.58
121.650	1.80	1.82	0.02	1.12
124.850	2.16	2.15	-0.01	-0.42
125.950	2.26	2.28	0.02	0.67
126.550	2.35	2.35	0.00	0.00
128.150	2.54	2.55	0.00	0.17
128.250	2.55	2.56	0.01	0.28
131.550	3.05	3.02	-0.04	-1.20
131.750	3.05	3.05	-0.01	-0.22
135.750	3.68	3.70	0.02	0.43
137.050	3.94	3.94	-0.01	-0.19
138.050	4.10	4.13	0.02	0.56
142.250	5.01	5.02	0.00	0.05
143.650	5.34	5.35	0.00	0.07
147.950	6.49	6.48	0.00	0.00
149.850	7.07	7.05	-0.02	-0.22
150.050	7.11	7.11	0.01	0.09

STANDARD DEVIATION = 0.01

ZIRCONIUM TETRA-TERT-AMYLOXIDE

C20H44O4ZR

BRADLEY D.C.,SWANWICK J.D.:

J.CHEM.SOC. 1959,748.

A = 2.46158 B = 86.288 C = -87.446

B.P.(10) = 146.483

T	P EXPTL.	P CALCD.	DEV.	PERCENT
121.550	2.28	0.85	-1.43	-62.54
123.250	2.45	1.13	-1.32	-54.04
126.850	3.02	1.87	-1.15	-38.09
127.950	3.16	2.14	-1.02	-32.15
136.250	4.85	4.94	0.09	1.81
136.350	4.86	4.98	0.12	2.45
137.150	5.02	5.32	0.30	5.89
140.250	7.67	6.72	-0.95	-12.37
146.250	11.47	9.87	-1.60	-13.97
155.450	12.96	15.59	2.63	20.26
158.350	13.35	17.56	4.21	31.56
158.850	14.68	17.91	3.23	22.01
161.150	18.14	19.54	1.40	7.70
166.350	25.02	23.33	-1.69	-6.74
188.150	41.70	40.25	-1.45	-3.48

STANDARD DEVIATION = 2.08

ISOPROPYL STEARATE

C21H42O2

BONHORST C.W.,ALTHOUSE P.M.,TRIEBOLD H.O.:

IND.ENG.CHEM. 40,2379(1948).

A = 0.07931 B = 10.408 C = -220.592

B.P.(10) = 209.287

T	P EXPTL.	P CALCD.	DEV.	PERCENT
181.600	2.00	2.22	0.22	10.97
197.500	4.00	3.39	-0.61	-15.28
203.000	5.00	4.69	-0.31	-6.24
207.000	6.00	7.00	1.00	16.67

STANDARD DEVIATION = 1.23

CHLORINE PENTAFLUORIDE CLF5

PILIPOVICH D.,ET AL.: INORG.CHEM. 6,1918(1967).

A = 6.51806 B = 748.974 C = 219.819

B.P.(760) = -13.902

T	P EXPTL.	P CALCD.	DEV.	PERCENT
-79.200	16.50	15.55	-0.95	-5.74
-63.200	55.00	54.44	-0.56	-1.02
-50.000	128.00	128.13	0.13	0.10
-45.300	166.00	168.43	2.43	1.46
-35.000	290.00	292.14	2.14	0.74
-26.200	453.50	446.46	-7.04	-1.55
-23.700	507.00	500.14	-6.86	-1.35
-17.900	634.50	643.86	9.36	1.47
-14.500	734.50	741.67	7.17	0.98
-0.200	1293.00	1281.54	-11.46	-0.89
24.800	2855.00	2859.35	4.35	0.15

STANDARD DEVIATION = 7.05

TETRAFLUORODICHLORODISILOXANE CL2F4OSI2

BOOTH H.S.,OSTEN R.A.:

J.AM.CHEM.SOC. 67,1092(1945).

A = 7.21811 B = 1121.976 C = 241.455

B.P.(760) = 17.226

T	P EXPTL.	P CALCD.	DEV.	PERCENT
-38.300	61.00	49.59	-11.41	-18.71
-35.600	64.90	58.59	-6.31	-9.73
-30.500	78.00	79.35	1.35	1.73
-25.000	102.60	108.32	5.72	5.57
-20.300	135.80	139.59	3.79	2.79
-14.600	178.80	187.21	8.41	4.70
-9.800	229.50	237.03	7.53	3.28
-4.900	292.40	298.63	6.23	2.13
0.200	371.40	376.04	4.64	1.25
4.900	459.70	461.11	1.41	0.31
9.600	561.90	561.14	-0.76	-0.14
12.100	625.20	621.06	-4.14	-0.66
15.100	705.90	699.64	-6.26	-0.89
16.800	756.60	747.59	-9.01	-1.19
17.300	776.70	762.18	-14.52	-1.87

STANDARD DEVIATION = 7.95

TRICHLOROTRIFLUORODISILOXANE CL3F3OSI2

BOOTH H.S.,OSTEN R.A.:

J.AM.CHEM.SOC. 67,1092(1945).

A = 8.36052 B = 1863.672 C = 297.262

B.P.(760) = 42.842

T	P EXPTL.	P CALCD.	DEV.	PERCENT
-38.000	16.60	14.86	-1.74	-10.46
-34.000	19.90	19.11	-0.79	-3.95
-23.200	35.30	36.34	1.04	2.93
-20.400	43.30	42.57	-0.73	-1.69
-14.400	57.90	59.14	1.24	2.14
0.300	124.30	125.13	0.83	0.67
6.600	168.30	168.74	0.44	0.26
10.800	204.10	204.57	0.47	0.23
14.700	244.20	243.49	-0.71	-0.29
22.800	346.50	344.88	-1.62	-0.47
27.300	414.80	415.34	0.54	0.13
28.500	437.40	436.07	-1.33	-0.30
30.300	469.60	468.80	-0.80	-0.17
34.800	560.60	559.88	-0.72	-0.13
35.100	568.10	566.45	-1.65	-0.29
39.200	660.20	662.96	2.76	0.42
42.400	745.80	747.61	1.81	0.24
43.200	770.20	770.13	-0.07	-0.01

STANDARD DEVIATION = 1.36

TETRACHLOROSILANE CL4SI

JENKINS A.C.,CHAMBERS G.F.:
IND.ENG.CHEM. 46,2367(1954).

A = 6.94467 B = 1185.999 C = 234.602

B.P.(760) = 57.238

T	P EXPTL.	P CALCD.	DEV.	PERCENT
2.200	86.30	86.35	0.05	0.06
25.100	239.00	238.73	-0.27	-0.11
34.200	340.70	340.81	0.11	0.03
40.600	431.70	431.64	-0.06	-0.01
46.500	531.00	531.58	0.58	0.11
56.900	752.20	751.79	-0.41	-0.05

STANDARD DEVIATION = 0.45

DISILANYL CHLORIDE CLH5SI2

CRAIG A.D.,URENOVITCH J.V.,MACDIARMID A.G.: J.CHEM.SOC. (1962)548.

A = 7.10484 B = 1211.778 C = 245.196

B.P.(100) = -7.817

T	P EXPTL.	P CALCD.	DEV.	PERCENT
-46.200	10.30	10.36	0.06	0.58
-42.700	13.90	13.20	-0.70	-5.03
-37.400	18.90	18.76	-0.14	-0.74
-32.200	25.90	26.04	0.14	0.54
-28.200	32.90	33.15	0.25	0.76
-25.200	39.60	39.50	-0.10	-0.24
-21.000	50.00	50.10	0.10	0.20
-17.000	61.80	62.31	0.51	0.83
-10.400	87.70	87.87	0.17	0.20
-6.000	109.50	109.34	-0.16	-0.15
0.000	146.40	145.46	-0.94	-0.64
4.200	175.90	176.19	0.29	0.16
8.000	208.30	208.40	0.10	0.05
18.000	316.60	316.77	0.17	0.05

STANDARD DEVIATION = 0.42

TETRACHLOROSILANE CL4SI

CAPKOVA A.,FRIED V.: COLLECTION 29,336(1964).

A = 6.85726 B = 1138.922 C = 228.877

B.P.(100) = 5.602

T	P EXPTL.	P CALCD.	DEV.	PERCENT
0.550	78.20	78.17	-0.03	-0.04
5.280	98.70	98.48	-0.23	-0.23
12.370	136.50	136.86	0.36	0.26
19.820	189.50	189.54	0.04	0.02
29.530	281.60	281.69	0.09	0.03
34.410	340.20	339.99	-0.21	-0.06
40.320	423.40	423.10	-0.30	-0.07
52.530	645.40	645.67	0.27	0.04

STANDARD DEVIATION = 0.28

SULPHUR TETRAFLUORIDE F4S

BROWN F.,ROBINSON P.L.:

J.CHEM.SOC. 1955,3147.

A = 6.11320 B = 529.225 C = 199.958

B.P.(100) = -71.293

T	P EXPTL.	P CALCD.	DEV.	PERCENT
-112.300	1.56	1.19	-0.37	-23.67
-102.700	5.14	4.70	-0.44	-8.63
-92.000	16.56	16.26	-0.30	-1.82
-83.600	36.78	36.72	-0.06	-0.15
-75.000	74.36	75.51	1.15	1.54
-66.300	142.46	142.45	-0.01	-0.01
-59.800	217.10	217.41	0.31	0.14
-54.700	299.58	295.02	-4.56	-1.52
-49.200	397.20	400.65	3.45	0.87

STANDARD DEVIATION = 2.40

HEXAFLUORODISILOXANE F6OSI2

BOOTH H.S.,OSTEN R.A.:

J.AM.CHEM.SOC. 67,1092(1945).

A = 7.47122 B = 1169.320 C = 278.058

B.P.(760) = -23.327

T	P EXPTL.	P CALCD.	DEV.	PERCENT
-38.600	385.60	387.28	1.68	0.44
-37.400	405.80	409.61	3.81	0.94
-35.900	438.30	439.01	0.71	0.16
-34.300	472.20	472.25	0.05	0.01
-33.200	497.10	496.27	-0.83	-0.17
-32.200	520.00	518.97	-1.03	-0.20
-31.200	545.10	542.51	-2.59	-0.48
-30.300	570.00	564.44	-5.56	-0.98
-29.200	596.70	592.21	-4.49	-0.75
-27.600	634.90	634.59	-0.31	-0.05
-26.400	666.20	667.97	1.77	0.27
-24.900	711.30	711.68	0.38	0.05
-23.400	756.90	757.69	0.79	0.11
-22.600	777.50	783.20	5.70	0.73

STANDARD DEVIATION = 3.22

TRICHLOROSILANE HCL3SI

JENKINS A.C.,CHAMBERS G.F.:
IND.ENG.CHEM. 46,2367(1954).

A = 6.77385 B = 1009.048 C = 227.248

B.P.(760) = 31.945

T	P EXPTL.	P CALCD.	DEV.	PERCENT
2.100	236.80	236.71	-0.09	-0.04
10.100	333.40	333.04	-0.36	-0.11
18.700	466.90	468.98	2.08	0.45
22.500	542.90	541.48	-1.42	-0.26
27.000	638.90	638.40	-0.50	-0.08
31.600	750.70	750.96	0.26	0.03

STANDARD DEVIATION = 1.51

HYDRAZINE H4N2

SCOTT D.W.,ET AL.:
J.AM.CHEM.SOC. 71,2293(1949).

A = 7.81292 B = 1684.041 C = 228.005

B.P.(100) = 61.701

T	P EXPTL.	P CALCD.	DEV.	PERCENT
15.000	7.65	7.64	-0.01	-0.18
20.000	10.55	10.53	-0.02	-0.16
25.000	14.38	14.35	-0.03	-0.23
30.000	19.29	19.31	0.02	0.10
35.000	25.67	25.70	0.03	0.10
40.000	33.82	33.83	0.01	0.03
45.000	44.08	44.09	0.01	0.03
50.000	56.91	56.93	0.02	0.03
55.000	72.85	72.84	-0.01	-0.02
60.000	92.43	92.40	-0.03	-0.04
65.000	116.30	116.26	-0.04	-0.03
70.000	145.12	145.17	0.05	0.03

STANDARD DEVIATION = 0.03

OXYGEN O_2

BROWER G.T.,THODOS G.:
J.CHEM.ENG.DATA13,262(1968).

A = 6.68748 B = 318.692 C = 266.683

B.P.(760) = -182.963

T	P EXPTL.	P CALCD.	DEV.	PERCENT
-218.790	1.13	1.08	-0.05	-4.47
-216.360	2.28	2.26	-0.02	-0.78
-213.080	5.53	5.52	-0.01	-0.15
-208.150	17.60	17.49	-0.11	-0.62
-205.190	31.90	31.98	0.08	0.25
-201.290	65.20	65.16	-0.04	-0.06
-198.120	109.20	109.47	0.27	0.25
-193.140	225.60	225.98	0.38	0.17
-187.690	450.00	449.82	-0.18	-0.04
-182.990	759.92	757.88	-2.04	-0.27
-182.970	760.00	759.47	-0.53	-0.07
-178.680	1164.10	1164.38	0.28	0.02
-175.550	1549.10	1550.51	1.41	0.09
-172.990	1931.40	1932.07	0.67	0.03

STANDARD DEVIATION = 0.81

XENON XE

MICHELS A.,WASSENNAR T.: PHYSICA 16,253(1950).

A = 38.26364 B = 54624.500 C = 1651.838

B.P.(760) = -108.024

T	P EXPTL.	P CALCD.	DEV.	PERCENT
-111.452	624.94	633.98	9.04	1.45
-110.822	649.08	655.50	6.42	0.99
-110.819	649.12	655.61	6.49	1.00
-109.621	696.49	698.51	2.02	0.29
-109.619	696.94	698.59	1.65	0.24
-109.157	715.87	715.88	0.01	0.00
-108.724	733.98	732.42	-1.56	-0.21
-108.723	733.87	732.47	-1.40	-0.19
-108.445	746.07	743.31	-2.76	-0.37
-108.197	756.98	753.11	-3.87	-0.51
-107.933	768.44	763.66	-4.78	-0.62
-107.098	805.74	798.03	-7.71	-0.96
-97.506	1346.63	1319.02	-27.61	-2.05
-92.587	1711.64	1702.64	-9.00	-0.53
-88.450	2072.27	2107.75	35.48	1.71

STANDARD DEVIATION =14.09

INDEX

A

B

1-Chlorohexadecane	$C_{16}H_{33}Cl$	566
1-Chlorohexane	$C_6H_{13}Cl$	288
Chlorohexylisocyanate	$C_7H_{12}Cl\ NO$	336
Chloromethoxytrichlorosilane	$CH_2Cl_4O\ Si$	27
2-Chloro-2-methylpropane	C_4H_9Cl	171
1-Chlorononane	$C_9H_{19}Cl$	462
1-Chlorooctane	$C_8H_{17}Cl$	408
Chloropentafluorobenzene	$C_6Cl\ F_5$	235
p-Chlorophenetole	$C_8H_9Cl\ O$	373
beta-Chloro-beta-phenyl ethyl alcohol	$C_8H_9Cl\ O$	373
1-Chlorophenylisocyanate	$C_7H_4Cl\ N\ O$	316
m-Chlorophenylisocyanate	$C_7H_4Cl\ N\ O$	316
Chloroprene	C_4H_5Cl	151
1-Chloropropane	C_3H_7Cl	126
2-Chloropropane	C_3H_7Cl	126
gama-Chloropropyltrichlorosilane	$C_3H_6Cl_4Si$	120
1-Chlorotetradecane	$C_{14}H_{29}Cl$	555
Chloro-2,4,6-trinitrobenzene	$C_6H_2Cl\ N_3O_6$	243
1-Chloroundecane	$C_{11}H_{23}Cl$	527
o-Chlorovinylbenzene	C_8H_7Cl	365
p-Chlorovinylbenzene	C_8H_7Cl	366
beta-Chlorovinyldichloroarsine	$C_2H_2AsCl_3$	66
cis-2-Chlorovinyldichloroarsine	$C_2H_2AsCl_3$	66
trans-Chlorovinyldichloroarsine	$C_2H_2AsCl_3$	67

Dichloromethane	CH_2Cl_2	26,27
2-(2,4)Dichlorophenoxy)ethanol	$C_8H_8Cl_2O_2$	369
3,4-Dichlorophenylisocyanate	$C_7H_3Cl_2NO$	314
1,2-Dichloropropane	$C_3H_6Cl_2$	119
Diethanolamine	$C_4H_{11}NO_2$	185
1,1-Diethoxyethane	$C_6H_{14}O_2$	300
Diethoxymethane	$C_5H_{12}O_2$	228
Diethylalumininium chloride	$C_4H_{10}Al\ Cl$,	175
Diethylamine	$C_4H_{11}N$	184
N,N-Diethylaniline	$C_{10}H_{15}N$	499
1,2-Diethylbenzene	$C_{10}H_{14}$	491
1,3-Diethylbenzene	$C_{10}H_{14}$	492
1,4-Diethylbenzene	$C_{10}H_{14}$	493
Diethyldichlorosilane	$C_4H_{10}Cl_2Si$	175
Diethyl ether	$C_4H_{10}O$	177
Diethylene glycol	$C_4H_{10}O_2$	180
Diethyl ethylphosphonate	$C_6H_{15}O_3P$	308
N,N-Diethylformamide	$C_5H_{11}NO$	221
Diethyl ketone	$C_5H_{10}O$	210
3,3-Diethylpentane	C_9H_{20}	463
3,5-Diethylphenol	$C_{10}H_{14}O$	497
Diethyl propylphosphonate	$C_7H_{17}O_3P$	360
Difluoromethane	CH_2F_2	28
cis-1,2-Diiodoethylene	$C_2H_2J_2$	74
trans-1,2-Diiodoethylene	$C_2H_2J_2$	74

Diisoamyl sulfide	$C_{10}H_{22}S$	517
p-Diisopropylbenzene	$C_{12}H_8$	536
Diisopropyl ether	$C_6H_{14}O$	294
2,4-Diisopropylphenol	$C_{12}H_{18}O$	537
Diisopropyl methylphosphonate	$C_7H_{17}O_3P$	361
1,1-Dimethoxyethane	$C_4H_{10}O_2$	181
N,N-Dimethylacetamide	C_4H_9NO	172
Dimethylamine	C_2H_7N	101
N-Dimethylaminodiborane	$C_2H_{11}B_2N$	108
Dimethylaminomethylborane cyclic dimer	$C_6H_{20}B_2N_2$	311
Dimethylaminotrifluorosilane	$C_2H_6F_3NSi$	95
Dimethylaniline	$C_8H_{11}N$	390
Dimethyl berilium	C_2H_6Be	93
1,4-Dimethyl-bicyclo(2,2,1)heptane	C_9H_{16}	453
trans-2,3-Dimethyl-bicyclo(2,2,1) heptane	C_9H_{16}	453
2,3-Dimethyl-1,3-butadiene	C_6H_{10}	273
2,2-Dimethylbutane	C_6H_{14}	290
2,3-Dimethylbutane	C_6H_{14}	291
2,3-Dimethyl-2-butanethiol	$C_6H_{14}S$	301
2,3-Dimethyl-2-butene	C_6H_{12}	280
3,3-Dimethyl-1-butene	C_6H_{12}	279
Dimethyl cadmium	C_2H_6Cd	94
1,1-Dimethylcyclohexane	C_8H_{16}	396
cis-1,2-Dimethylcyclohexane	C_8H_{16}	397

1-Propanethiol	C_3H_8S	134
2-Propanethiol	C_3H_8S	134
1-Propanol	C_3H_8O	130
2-Propanol	C_3H_8O	131
1-Propen-3-ol	C_3H_6O	122
Propionic acid	$C_3H_6O_2$	124
Propionic anhydride	$C_6H_{10}O_3$	275
Propionitrile	C_3H_5N	116
Propiophenone	$C_9H_{10}O$	430
Propyl acetate	$C_5H_{10}O_2$	215
1-Propylamine	C_3H_9N	138
2-Propylamine	C_3H_9N	139
N-Propylamine trimethylboron	$C_6H_{18}BN$	309
Propylbenzene	C_9H_{12}	436
Propyl borate	$C_9H_{21}BO_3$	472
Propyl caprate	$C_{13}H_{26}O_2$	547
Propyl caproate	$C_9H_{18}O_2$	462
Propyl caprylate	$C_{11}H_{22}O_2$	526
Propyl cellosolve	$C_5H_{12}O_2$	228
Propylcyclohexane	C_9H_{18}	457
Propylcyclopentane	C_8H_{16}	401
Propylene	C_3H_6	119
Propyl formate	$C_4H_8O_2$	167
Propyl laurate	$C_{15}H_{30}O_2$	560
Propyl myristate	$C_{17}H_{34}O_2$	573

Perfluorocyclohexane	C_6F_{12}	238
Perfluorocyclopentane	C_5F_{10}	189
Perfluoro-2,3-dimethylbutane	C_6F_{14}	240
Perfluoro-1,2-diemthylcyclobutane	C_6F_{12}	239
Perfluorodimethylcyclohexane	C_8F_{16}	362
Perfluoroethylcyclohexane	C_8F_{16}	363
Perfluoroheptane	C_7F_{16}	313
Perfluorohexane	C_6F_{14}	239
Perfluoro-2-methylbutane	C_5F_{12}	190
Perfluoromethylcyclohexane	C_7F_{14}	312
Perfluoro-2-methylpentane	C_6F_{14}	240
Perfluoro-3-methylpentane	C_6F_{14}	241
Perfluorooctane	C_8F_{18}	363
Perfluoropentane	C_5F_{12}	190
Perfluoropiperidine	$C_5F_{11}N$	190
Perfluoropropane	C_3F_8	110
Perfluoropropene	C_3F_6	109
Phenantrene	$C_{14}H_{10}$	55^0
Phenol	C_6H_6O	258
beta-Phenyl ethyl acetate	$C_{10}H_{12}O_2$	486
alpha-Phenyl ethyl alcohol	$C_8H_{10}O$	381
beta-Phenyl ethyl alcohol	$C_8H_{10}O$	382
o-Phenylethylphenol	$C_{14}H_{14}O$	551
p-Phenylethylphenol	$C_{14}H_{14}O$	552
Phenyl isocyanate	C_7H_5NS	318